Genetics: Concepts and Applied Principles

Genetics: Concepts and Applied Principles

Editor: Callum Gray

CALLISTO REFERENCE

www.callistoreference.com

Callisto Reference,
118-35 Queens Blvd., Suite 400,
Forest Hills, NY 11375, USA

Visit us on the World Wide Web at:
www.callistoreference.com

ISBN: 978-1-64116-547-1 (Hardback)

Cataloging-in-Publication Data

Genetics : concepts and applied principles / edited by Callum Gray.
    p. cm.
Includes bibliographical references and index.
ISBN 978-1-64116-547-1
1. Genetics. 2. Genetics--Methodology. I. Gray, Callum.
QH430 .G46 2022
576.5--dc23

Table of Contents

Preface

Genetics can be defined as the study of genes, inheritance of traits and variation in living organisms. It is an essential aspect of biology and provides insights into how evolution occurs. The inheritance of traits in genetics is understood with the tools of pedigree charts, Punnett squares and twin studies. The understanding of genes requires comprehension of gene expression, extra nuclear inheritance, mutation, epigenetics and control of development. Some of the modern focus areas of genetics are alternative splicing, mobile genetic elements, genomics, genetic engineering and horizontal gene transfer. The genetics of human behavior is another area of constant research. The genetic contribution to the development of schizophrenia, autism, ADHD and drug abuse are some of the conditions that are being actively investigated. This book includes some of the vital pieces of work being conducted across the world, on various topics related to genetics. It provides significant information of this discipline to help develop a good understanding of the concepts and applied principles of genetics.

This book is the end result of constructive efforts and intensive research done by experts in this field. The aim of this book is to enlighten the readers with recent information in this area of research. The information provided in this profound book would serve as a valuable reference to students and researchers in this field.

At the end, I would like to thank all the authors for devoting their precious time and providing their valuable contribution to this book. I would also like to express my gratitude to my fellow colleagues who encouraged me throughout the process.

Editor

Formulation of Genetic Counseling Format for Adult Bangladeshi Patients with Acute Myeloid Leukemia

M. Z. Rahman, L. Nishat, Z. A. Yesmin, and L. A. Banu ⓘ

Department of Anatomy, Bangabandhu Sheikh Mujib Medical University (BSMMU), Dhaka, Bangladesh

Correspondence should be addressed to L. A. Banu; dr.lailabanu@gmail.com

Academic Editor: Francine Durocher

With the advancement of medical genetics, particular emphasis is given on the genetic counseling worldwide. In Bangladesh, genetic counseling services are not yet developed. Acute myeloid leukemia (AML) is a malignant disease of the myeloid cells of bone marrow. Like other malignant diseases, it may result from a mutation in the DNA. A genetic counseling format will educate the AML patients and provide appropriate medical and emotional support. The aim of this descriptive cross-sectional study was to develop a genetic counseling format for adult Bangladeshi patients with AML. Taking this into account, a draft format was prepared by reviewing relevant documents available online which was later analyzed by an expert panel through a group discussion and thus a proposed format was developed. To make the format effective in the perspective of Bangladeshi population, the proposed format was applied in counseling, and thus a final format was developed in the English language. This format will educate the counselors, clinicians, and patients about the utility and importance of the genetic counseling and genetic tests. Also, the patients feel comfort regarding the whole counseling process and going for postcounseling treatments and advice. Though it is written in English, it may be translated into mother tongue for better communication during counseling.

1. Introduction

Acute myeloid leukemia (AML) is a malignant disease of the myeloid cells of bone marrow in which precursors of blood cells are arrested in an early stage of development [1]. The synonyms of AML are acute myelogenous leukemia and acute nonlymphocytic leukemia [2]. According to WHO, Bangladesh is experiencing increasing cancer burden with an estimated 122,715 new cancer cases in 2012. In case of Bangladesh, AML was the most frequent hematological malignancy at the rate of 28.3% [3]. Survival with AML depends on age [4]. The incidence of therapy related (secondary) AML is 10–20% of all AML cases [5]. According to Hossain et al. [3] the median age at onset of AML in Bangladesh is 35 years which is an active age for society. Untreated AML cases are typically fatal within a short period of time [6]. The chance of an individual developing cancer depends on both genetic and nongenetic factors. AML "is a genetic disease" [7] and, like other cancers, it may results from mutations in the DNA. Familial AML is a rare type of inherited leukemia which is transmitted by a

nonsex chromosome in a dominant fashion. Several genes have been identified with their incidence, treatment, and prognostic outcomes. In spite of the rapid development of genetic techniques for diagnosis, genetic disorders remain undetected for several years [8]. So, there are some complex issues that need to be communicated with patients. "Genetic counseling is a communicative process" [9]. As genetic testing is a useful tool in the clinical management of disease, the importance of genetic counseling is increasing worldwide with the advancement in the field of genetics. In context of genetic counseling, the aim is to communicate information regarding the personal impact of disease, so that individuals can take decision from the informed choices regarding options for risk management, disease surveillance, and predictive genetic testing and also adapt to the emotional, psychological, medical, social, and economic consequences of the test results with treatment. Thus, genetic counseling enables one to assess the risk of cancer without the use of genetic tests. As a result, genetic counseling acts as an integral part for the cancer risk assessment process [10]. In this context, family history holds important information about an individual's past and future life. For this reason, need is felt to communicate the complex cancer genetic information to the AML patients using everyday language. The workshop Genetic Counseling/Consultations in Southeast Asia at the 10th Asia Pacific Conference on Human Genetics in Kuala Lumpur, Malaysia, in December 2012 aimed at addressing culture and context-specific genetic counseling/consultation practices in Southeast Asia. There are a large number of AML patients in Bangladesh and so far no organized counseling service is available for the patients or their relatives. Responding to the global trends, the human genetics research in Bangabandhu Sheikh Mujib Medical University (BSMMU) has already been modernized with the help of Higher Education Quality Enhancement Project (HEQEP) of University Grants Commission of Bangladesh. It would, therefore, be a great opportunity to modernize our medical knowledge in this field for the betterment of the patient care. The unique aspects of AML genetic information can guide a course of action to minimize distress and maximize benefit for both the patient and the family. This study aims to present formats of practice recommendations for genetic counselors conveying AML genetic counseling with surveillance and management in a manner so that all patients can be provided with the same information by using a structured, valid, and reliable genetic counseling format. So, the purpose of this study was to develop a genetic counseling intake format for AML patients. Presently in Bangladesh, there is no standard genetic counseling format for counseling patients. If a genetic counseling format is available for any disease, then any one from the counseling team can counsel the patients with minimum training and deliver the same information to the patients.

2. Methods

The present research is a cross-sectional descriptive study. It was carried out in the Department of Anatomy, BSMMU, and the counseling was done in the Hematology Department of the university. The study was approved by the Institutional Review Board (IRB) of BSMMU.

The study involves many steps which are shown in Figure 1.

The AML counseling formats available online, other cancer related counseling formats, journal articles, and books related to AML counseling were reviewed and a "draft format" for genetic counseling was developed for the adult Bangladeshi patients with AML. A group discussion for analysis of the draft format was organized. Five specialists (two hematologists, one educationist, one psychiatrist, and one counselor) were selected for the discussion. Relevant documents were sent to the participants seven days prior to the group discussion. The participants provided a written feedback form after the discussion. A "proposed genetic counseling format" was developed for AML patients by modifying the draft format according to the feedback of the group discussion participants. Both the draft and the proposed formats were in English. Genetic counseling of 24 adult Bangladeshi AML patients using the proposed format was conducted by a four-member counseling team. A "final

FIGURE 1: Steps of development of genetic counseling format for AML patients.

genetic counseling format" was developed both in English and in Bengali (the mother tongue) after counseling.

3. Ethical Issue

The study was conducted after receiving ethical approval from the Institutional Review Board (IRB) of BSMMU. All the participants of the group discussion and the patients participated in the genetic counseling voluntarily. A written informed consent form was provided to the participating patients, and counseling was conducted only after getting their consent in that form. Anonymity was strictly maintained for both the patients and the group discussion participants. Confidentiality of the information of the patients was maintained.

4. Results and Discussion

The draft format for genetic counseling was prepared by analyzing journal articles related to the AML counseling, counseling formats available online for AML, and other cancers and books. The draft format was constructed on the following parameters:

(i) Instructional note related to the way of conversation and filling up the form

(ii) General information of the patient

(iii) Present medical history of the patient

(iv) Three generations' family history of the patient and pedigree chart

(v) Findings of the laboratory investigations of the patient

(vi) Information to be provided by the counselor to the consult

(vii) Summary of the case and card for the consult.

The development of genetic counseling format aims to assist AML patients and genetic counselors in understanding the

TABLE 1: Outcome of the group discussion on the draft format.

Area where modifications are required	Modification
Overall organization of the draft format	Need to be organized with headings & subheadings
Parameters	
Instructional note	Should be omitted
	Should be divided into two broad headings:
	(A) General information
	(B) Personal information
General information	Identification (ID) number, serial number, and emergency contact number should be included in general information
	Birth history, consanguinity of marriage, and history of substance abuse should be included in personal information
	Should be divided into two broad headings:
Present medical history	(A) Medical history and
	(B) Environmental factor exposure history
Family history	Will be arranged according to the degree of relation

occurrence, probable cause, and available genetic test with related management of the AML. The format acts as a guideline for the counselor. According to Nishat [11] and Yesmin [12] analysis of available relevant documents is an important step for formulating any new format. Journal articles related to counseling, counseling formats for AML and other cancers available online, and books on genetic counseling were helpful for idea generation and construction of the draft format.

The draft format was presented to the specialists for group discussion. The participants discussed various aspects of the draft format. The opinion of the group discussion participants on the draft format is presented in Table 1.

Group discussion is an effective process in qualitative research [13]. Information related to the disease, educational and psychological aspects, and the counselor's view about the draft format was collected in the group discussion. Necessary modification of the draft format was done according to the findings of the group discussion and the "proposed format" for genetic counseling was prepared. The proposed format contains the following parameters:

(A) General information

(B) Personal information

(C) Medical history

(D) Environmental factor exposure history

(E) Family history

(F) Laboratory investigations

(G) Summary of the case

(H) Comment of the genetic counselor.

Family history includes patient's first-degree (50% shared gene), second-degree (25% shared gene), and third-degree (12.5% shared gene) relatives. In case of the pedigree chart, only three generations of the index case were included.

The researchers (the first and third authors), a geneticist, and a nurse conducted the genetic counseling of 24 adult Bengali Bangladeshi patients with AML using the proposed format. Genetic counseling was performed to validate the proposed counseling format. The counselors and the patients experienced few linguistic problems during counseling. The proposed format was in English, but the mother tongue of the patients as well as counselors is Bengali. Payne et al. [14] reported that genetic counseling is a communication process to provide information and support patients so that they could understand the medical facts. On the basis of the counseling feedback, the proposed format was finalized without any major modifications. The psychological orientation of a person is significantly affected by the mother tongue or known language [15]. But it was observed that it is quite difficult to communicate with patients in the English version of the format. So the final format was also translated into Bengali to overcome the linguistic barrier and for better communication with patients during counseling. The English version of the final format is attached herewith (available here).

5. Conclusion

The final genetic counseling format will help a variety of professionals to counsel AML patients. The face-to-face description and genetic explanation of AML to each patient make them aware about the genetic background of the disease and sharing the information with the family member and other members of the society.

Abbreviations

AML: Acute myeloid leukemia
BSMMU: Bangabandhu Sheikh Mujib Medical University
HEQEP: Higher Education Quality Enhancement Project.

Acknowledgments

The authors thank the CPSF- (Complete Proposal Supplementary Fund-) 2057 of the Higher Education Quality Enhancement Project (HEQEP) of University Grants Commission of Bangladesh for funding.

References

[1] S. Karen, "Acute Myelogenous Leukemia, 2015," View at Google Scholar.

[2] American Cancer Society (ACS), "What is acute myeloid leukemia, US 2016," View at Google.

[3] M. S. Hossain, M. S. Iqbal, M. A. Khan et al., "Diagnosed hematological malignances in Bangladesh - A retrospective analysis of over 5000 cases from 10 specialized hospitals," *BMC Cancer*, vol. 14, no. 1, article no. 438, 2014.

[4] B. Deschler and M. Lübbert, "Acute myeloid leukemia: Epidemiology and etiology," *Cancer*, vol. 107, no. 9, pp. 2099–2107, 2006.

[5] G. Leone, L. Pagano, D. Ben-Yehuda, and M. T. Voso, "Therapy-related leukemia and myelodysplasia: Susceptibility and incidence," *Haematologica*, vol. 92, no. 10, pp. 1389–1398, 2007.

[6] H. Eleftheria, G. Georgios, B. Leonidas, and B. Evangelos, "Gene mutations and molecularly targeted therapies in acute myeloid leukemia," *American Journal of Blood Research*, vol. 3, no. 1, pp. 29–51, 2013.

[7] D. Grimwade, A. Ivey, and B. J. P. Huntly, "Molecular landscape of acute myeloid leukemia in younger adults and its clinical relevance," *Blood*, vol. 127, no. 1, pp. 29–41, 2016.

[8] National Center for Biotechnology Information (NCBI), "Understanding genetics: a district of Columbia guide for patients and health professionals, 2010," View at Pubmed.

[9] S. R. Phadke, A. Pandey, R. D. Puri, and S. J. Patil, "Genetic counseling: The impact in Indian milieu," *The Indian Journal of Pediatrics*, vol. 71, no. 12, pp. 1079–1082, 2004.

[10] A. Trepanier, M. Ahrens, W. McKinnon et al., "Genetic cancer risk assessment and counseling: recommendations of the national society of genetic counselors," *Journal of Genetic Counseling*, vol. 13, no. 2, pp. 83–114, 2004.

[11] L. Nishat, *Needs Analyses and Preparation of Documents for Developing a Clinically-Oriented Neuroanatomy Course for Four Postgraduate Programs, master thesis*, BSMMU, Dhaka, Bangladesh, 2013.

[12] Z. A. Yesmin, *Developing curriculum for Advanced Genetics as an elective subject in the MS Anatomy residency programmes of BSMMU, master thesis*, BSMMU, Dhaka, Bangladesh, 2015.

[13] S. Powell and G. Easton, "Student perceptions of GP teachers' role in community-based undergraduate surgical education: a qualitative study," *JRSM Short Reports*, vol. 3, no. 8, pp. 1–8, 2012.

[14] K. Payne, S. Nicholls, M. McAllister, R. MacLeod, D. Donnai, and L. M. Davies, "Outcome measurement in clinical genetics services: A systematic review of validated measures," *Value in Health*, vol. 11, no. 3, pp. 497–508, 2008.

[15] S. P. Corder, *Language transfer in language learning*, John Benjamins Publishing Company, 1992, https://ebooks.google.com.

Leveraging Comparative Genomics to Identify and Functionally Characterize Genes Associated with Sperm Phenotypes in *Python bivittatus* (Burmese Python)

Kristopher J. L. Irizarry[1] and Josep Rutllant[2]

[1]*The Applied Genomics Center, Graduate College of Biomedical Sciences, College of Veterinary Medicine, Western University of Health Sciences, 309 East Second Street, Pomona, CA 91766, USA*
[2]*Molecular Reproduction Laboratory, College of Veterinary Medicine, Western University of Health Sciences, 309 East Second Street, Pomona, CA 91766, USA*

Correspondence should be addressed to Kristopher J. L. Irizarry; kirizarry@westernu.edu

Academic Editor: Jerzy Kulski

Comparative genomics approaches provide a means of leveraging functional genomics information from a highly annotated model organism's genome (such as the mouse genome) in order to make physiological inferences about the role of genes and proteins in a less characterized organism's genome (such as the Burmese python). We employed a comparative genomics approach to produce the functional annotation of *Python bivittatus* genes encoding proteins associated with sperm phenotypes. We identify 129 gene-phenotype relationships in the python which are implicated in 10 specific sperm phenotypes. Results obtained through our systematic analysis identified subsets of python genes exhibiting associations with gene ontology annotation terms. Functional annotation data was represented in a semantic scatter plot. Together, these newly annotated *Python bivittatus* genome resources provide a high resolution framework from which the biology relating to reptile spermatogenesis, fertility, and reproduction can be further investigated. Applications of our research include (1) production of genetic diagnostics for assessing fertility in domestic and wild reptiles; (2) enhanced assisted reproduction technology for endangered and captive reptiles; and (3) novel molecular targets for biotechnology-based approaches aimed at reducing fertility and reproduction of invasive reptiles. Additional enhancements to reptile genomic resources will further enhance their value.

1. Introduction

Reptiles represent a diverse and biologically distinct group of vertebrates for which most species have yet to be systematically studied. Over the last few decades, reptiles in general, and snakes in particular, have grown in popularity among owners and breeders. It is worth noting that reptiles are of ecological interest as both endangered and invasive species. For example, in 2015, California Department of Fish and Wildlife listed ten distinct reptile species as either endangered or threatened, including four species of snakes: *Charina bottae* (southern rubber boa), *Thamnophis gigas* (giant garter snake), *Thamnophis sirtalis tetrataenia* (San Francisco garter snake), and *Masticophis lateralis euryxanthus* (alameda whipsnake). At the same time, California Department of Fish and

Wildlife classifies other reptiles as invasive species, such as *Nerodia fasciata* (southern watersnake). The identification of genetic and genomic resources for use in reptiles can accelerate research and ultimately enhance knowledge of their unique biology. Deciphering reptile reproductive biology can provide avenues for facilitating successful breeding in endangered species and, at the same time, may offer insights into reducing the reproduction of invasive species.

Two species of the genus *Python* have recently become the focus of genomics level investigations. Castoe et al. sequenced the genome of the Burmese python as a reptilian model organism with the expectation that the sequence data would provide insight into unique aspects of reptilian physiology and evolution [1]. Similarly, whole transcriptome studies in ball python have led to the production of genomics resources which can be used to further study this species [2]. Recent investigations into python physiology have identified rapid gene expression changes associated with basic physiological processes such as eating [3]. For example, the Burmese python exhibits unique organ and metabolic adaptations which have been characterized recently at the molecular and genetic levels [4].

Python regius (ball python or royal python) is a relatively small member of the Python family which has become an extremely popular pet over the last 15 to 20 years due to the tremendous expansion of color variations that have been produced. *P. regius* is a relatively small species in length (rarely over 2 meters) and adapts easily to being raised in captivity. The name "ball python" is derived from the fact that this species curls up in a ball whenever it is approached or handled, making it particularly easy to manage in captivity.

Python molurus bivittatus (Burmese python) has been a popular pet in the United States since the 1990s due to their attractive color patterns, docile nature, and large size [5]. Size-wise, the Burmese python is one of the largest snakes in the world and can reach over 23 feet in length and weights of over 200 pounds. Some of these giant snakes have been illegally released by their owners into the wild due to the difficulty to handle them and the lack of alternate housing or sheltering. Nowadays, although native to Southeast Asia, these snakes are exotic (nonnative) species in areas like South Florida (e.g., Everglades National Park) and they are also considered invasive species [6] since they are not constrained by natural factors. Consequently, due to their potential to harm invaded environments (wildlife and ecosystem), efforts are underway to reduce their numbers in these sensitive environments. Because of their large size, Burmese pythons have few predators, and subsequently their predation upon native species is decreasing the native populations to the level of being threatened or endangered [7, 8]. The impact of the invasive Burmese pythons on the normal wild life is serious concern. Dorcas et al. describe the severe reduction in mammals in the Florida Everglades due to the pythons growing population size [9]. Novel genetic strategies are being employed to monitor the Burmese python. Recently, PCR-based detection methods have been employed to detect Burmese python DNA in environmental water samples, such as marshes, streams, swamps, and lakes [10].

In contrast to invasive reptiles, endangered reptiles are exhibiting decreased numbers in their native ecosystems. Populations of endangered reptiles may suffer from inbreeding due to reduced populations or even reproduction in a captive setting such as a zoo [11]. Hussain et al. developed methods to quantify damage to semen at specific steps in the preservation process [12]. The work was carried out in mammals; however, the approach is viable for reptiles as well. Ruiz-Lopez et al. describe the relationship between homozygosity, heterozygosity, and inbreeding depression [13]. These undesirable genetic issues arise when populations are endangered and can contribute to decreased reproductive fitness, including poor sperm function, reduced motility, and decreased sperm numbers in the endangered population. Birds have been used as models for developing reproductive technology such as artificial insemination and extenders

capable of improving the longevity and value of cryopreserved semen from endangered species. For example, a 2009 study evaluated post-thaw semen quality in wild-caught Griffon vultures and determined that cryopreservation of semen is a useful tool in the conservation of endangered species genetic resources [14].

Assisted reproductive technology helped immensely to improve genetic pools in farm and domestic animals for several decades; however, although recognized as an important strategy to enhance diversity and increase captive populations of endangered animals, these techniques are rarely applied to reptiles [15] and much less to the specific field of snakes [16]. Reproduction in snakes differs from mammalian reproduction in many distinct ways, but the most significant difference is that spermatozoa can be stored in the female genital tract for months, if not years, before fertilization [17]. Studies related to the development of assisted reproductive techniques in snakes have been ignored and only few reports on semen collection [18, 19], sperm preservation [20], and artificial insemination [15] have been the focus of large research efforts.

Comparative genomics has been successfully employed in previous studies to identify physiologically important genes in one organism based on the annotation provided by a model organism genome. The comparative genomics approach was developed and heavily leveraged in the 1990s to facilitate functional annotation of large-scale human EST data produced during the effort to sequence the human genome [21]. Functional information about gene-phenotype relationships in model organisms were shown to be extremely valuable in deciphering the consequence of identified mutations in human genes [22]. Comparative genomics approaches became more widespread following the sequencing of multiple genomes and the emerging need to characterize unknown genes [23]. For example, genes sequenced from hamster testis were used to identify and characterize genes previously uncharacterized in human, mouse, rat, and pufferfish genomes [24]. Beyond simply facilitating the identification of novel genes, comparative genomics approaches have been effectively used to characterize individual proteins involved in sperm phenotypes, such as the sperm mitochondrial cysteine-rich protein, SMCP [25]. Such approaches have also been effective in identifying economically important reproductive traits in agriculturally important species, such as the pig [26]. Similarly, the characterization of 1227 genes in the domestic cat was achieved using a comparative genomics approach in which cat genes having phenotypically characterized orthologs in the mouse were annotated with developmental, clinical, and nutritional phenotypes [27].

A number of bioinformatics resources have been developed to facilitate the use of comparative and functional genomics. Ontologies, which are controlled vocabularies organized around specific parent-sibling relationships and maintained in a graph structure, provide standardized nomenclature and relationships among biologically relevant terminology for applications in bioinformatics and functional genomics [28]. One of the most widely used ontologies in genomics is the gene ontology [29]. Gene set enrichment,

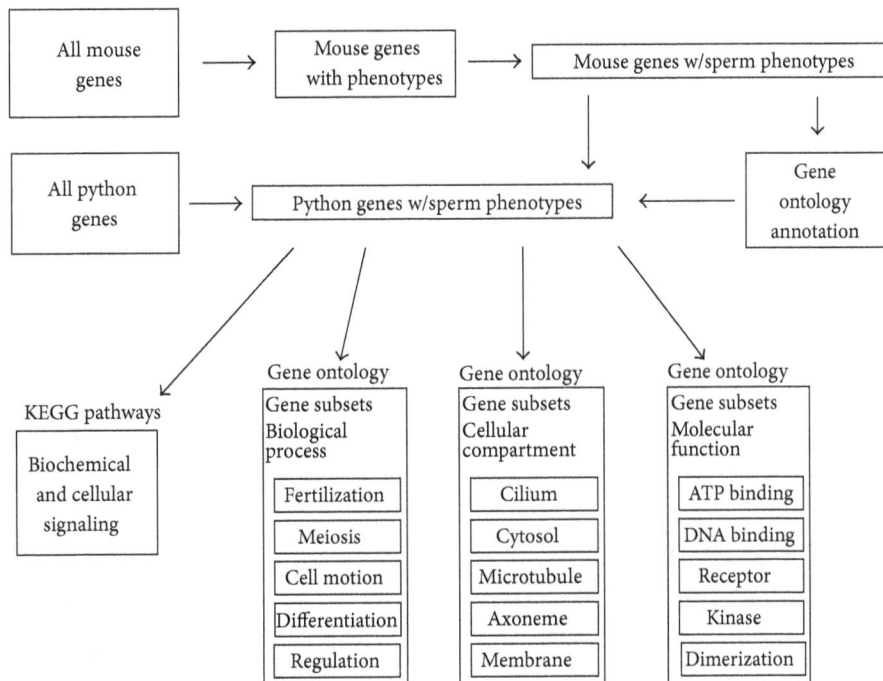

FIGURE 1: Overview of the comparative genomics approach to identify and characterize python genes associated with sperm phenotypes. The set of mouse protein coding genes was used to select the subset of mouse genes for which phenotype annotation information was available. Starting with all mouse genes having phenotype annotation, we identified the subset corresponding to protein coding genes associated with only sperm phenotypes. This set of mouse protein sequences was subsequently used to identify the corresponding protein coding sequences in *Python bivittatus* (i.e., orthologous genes). For each protein coding sequence shared between mouse and python, a pairwise protein sequence alignment was generated and measures of sequence identity and significance were calculated. Gene ontology (GO) annotation provides gene level information about biological processes, cellular locations, and molecular functions of gene products. The existing GO annotation for each mouse gene was "added" to each python orthologous gene. Then set of python genes was analyzed for statistically significant enrichment of genes associated with particular GO annotation terms across the three GO categories (biological process, cellular component, and molecular function). This resulting set of annotated python genes provides additional biological, physiological, cellular, and molecular information about the roles of these genes in sperm production and function. Moreover, the annotation also offers an independent set of annotation information to help validate the python genes as truly being associated with sperm biology. Additionally, cellular pathways which are associated with the sperm associated gene set were identified along with human disorders caused by human orthologs of these genes.

such as the identification of shared biological processes among a set of differentially regulated genes in a gene expression experiment, represents one of the most widely used applications of gene ontology [30]. Gene set enrichment has been previously applied to the identification and analysis of genes associated with phenotypes [14, 31]. Other ontologies have been developed that are useful for functional annotation of genes, including the mammalian phenotype ontology [32], the human phenotype ontology [33], and the mouse-human anatomy ontology [34].

In addition to ontological resources, pathway mapping resources provide additional means of functionally annotating genes based on biologically relevant information. One of the most widely utilized pathway resources for gene annotation is the KEGG pathway database [35] which provides knowledge-based representations of biochemical pathways and protein interaction networks. Tools have been developed that facilitate the identification of KEGG pathway members from a set of genes [36, 37]. One of the most widely used tools, based on more than 21,000 citations, is the Database for Annotation, Visualization, and Integrated Discovery [38] that

provides an interface for functional gene annotation including gene ontology and KEGG pathway analysis.

The relatively recent production of reptile genomic resources has opened the door to leveraging genome-scale information to address the challenges associated with endangered and invasive reptiles. In an attempt to expand the set of genetic resources available for investigating reptile biology, we have chosen to utilize a comparative genomics strategy to identify python genes likely to play a significant role in sperm development and function (Figure 1). We describe the selection of a set of mouse genes that have been previously demonstrated to modulate sperm phenotypes and use that mouse set to identify and functionally characterize a corresponding set of python genes. Although reptiles and mammals share many aspects of biology with each other, they each also have evolved unique adaptations that are found only in their respective lineages. By leveraging mammalian reproductive genetics to initiate the construction of reptile reproductive genomics tools, the long and productive history of mouse biology and genetics can be brought to bear on the emerging field of reptile reproductive genetics.

2. Materials and Methods

2.1. Comparative Genomics Approach to Identify Python Genes Associated with Sperm Phenotypes.

The comparative genomics approach (Figure 1) leveraged the set of mouse genes for which phenotype information was available. Mouse genes annotated with sperm specific phenotypes were identified and python orthologs were identified as described below. Gene ontology enrichment was performed to identify gene ontology biological process, cell component, and molecular function terms associated with each gene set identified by a sperm associated phenotype. The resulting set of python genes mapped to sperm phenotypes is available as a supplemental data file, as is a second supplemental file containing the mapping of the mouse genes to the corresponding python orthologs, all of which are mapped to sperm phenotypes.

2.2. Identification of P. regius and P. bivittatus Sequences from NCBI.

Publicly available DNA, mRNA, and protein *Python bivittatus* sequences were downloaded from the NCBI nucleotide and protein sequence databases (http://www.ncbi.nlm.nih.gov/) by searching NCBI for "Python bivittatus." A total of 25,944 protein sequences were identified. *P. regius* sequences were also downloaded from NCBI using a similar query for which "Python regius" was used as search term. Although considerable sequence data exists for *P. bivittatus* (20,392 genes, 25,944 protein sequences, and 105,311 nucleotide sequences), relatively few sequence resources are available for *P. regius* (21 genes, 141 protein sequences, and 123 nucleotide sequences). Interestingly, some of the identified *P. regius* sequences in the database correspond to viral sequences, such as the ball python nidovirus (8 genes, 3 nucleotide sequences, and 17 protein sequences). Even so, we identified a set of *P. regius* sequences for which orthologous *P. bivittatus* sequences were available. Because we are specifically interested in python molecular reproductive genetics, we selected (from among the set of 21 genes) those for which published papers had previously implicated the gene as (1) involved in sperm related biology or (2) being associated with variation in sperm parameters, or (3) being associated with variation in fertility in another species (Table 1).

2.3. Selection of Mouse Genes Associated with Sperm Phenotypes.

Mouse gene and protein identifiers along with protein sequences were downloaded from the Ensembl database (http://www.ensembl.org/). Data was loaded into a table in a relational database (MySQL) containing phenotype annotation from the Mammalian Phenotype Browser (http://www.informatics.jax.org/searches/MP_form.shtml). Database queries in SQL provided a mechanism for selecting all mouse gene identifiers linked to specific phenotypes. The output from the database contained individual gene-phenotype relationships. Since a single gene could be associated with more than one phenotype, the results contained unique gene-phenotype relationships even though the same genes were listed under multiple phenotypes. The list of 129 sperm related phenotypes mapped to each mouse gene (listed as mouse ensemble gene identifier) is contained in Supplemental File 1.

2.4. Mapping Mouse Genes Associated with Sperm Phenotypes to P. bivittatus Protein Identifiers.

Each mouse gene ID was used to query the database to retrieve the gene description field which contains a variety of header information including the complete title of the gene. The mouse gene title information was used to manually search through NCBI's protein database (http://www.ncbi.nlm.nih.gov/protein/?term=Python%20bivittatus) for orthologous python protein sequences. Because the goal of the project was to insure the collection of genes having sperm phenotypes, any genes which were not the obvious ortholog of the mouse gene were excluded from the final dataset. For example, in some cases, a specific family member of a multigene family was sought, but the exact family member could not be found even though a number of paralogous *P. bivittatus* genes from the same family were located.

Every successfully identified *P. bivittatus* gene was downloaded and stored in FASTA format for later use. Once the complete set of mouse gene identifiers was mapped to all available python orthologs, a file containing the complete set of snake protein sequences was used to develop a blast database using the *formatdb* tool with the −p option set to true to indicate amino-acid sequences versus nucleotide sequences.

Finally, the blastall program was used to perform blastp analysis of the protein sequences associated with sperm phenotypes. For each orthologous mouse-python gene pair, blastp calculated the bit score, generated an alignment, and calculated the percent identity. A tab-delimited text file containing each sperm associated phenotype, the Ensembl mouse gene identifier, official gene symbol, the Burmese python NCBI gene identifier, the Burmese python NCBI refId, and the description of each gene is contained in Supplemental File 1.

2.5. Pairwise Alignments, Multiple Sequence Alignments, and Phylogenetic Trees.

Epididymal protein E1 sequence was obtained for the following species: *Gallus gallus* (chicken), *Alligator mississippiensis* (alligator), *Chrysemys picta bellii* (painted turtle), *Pantherophis guttatus* (corn snake), *Echis coloratus* (palestine saw-scaled viper), *Python bivittatus* (burmese python), *Python regius* (ball python), *Anolis carolinensis* (Carolina anole), *Equus caballus* (domestic horse), *Canis familiaris*, *Sus scrofa*, *Homo sapiens*, *Pan troglodytes*, *Felis catus*, *Mus musculus* (mouse), *Bos taurus* (cow), *Oncorhynchus mykiss* (trout), *Bombus impatiens* (eastern bumble bee), *Thelohanellus kitauei* (fish parasite). The mitochondrial cytochrome b protein sequence was obtained for multiple sequence alignment across 17 species (Figure 2(c)). The equine sequence was taken from *Equus przewalskii*, the feral horse, instead of the domestic horse. Alignments were generated using CLC Sequence Viewer 6 (http://www.clcbio.com/products/clc-sequence-viewer/). Phylogenetic trees were produced using the tree construction algorithm in CLC Viewer 6.

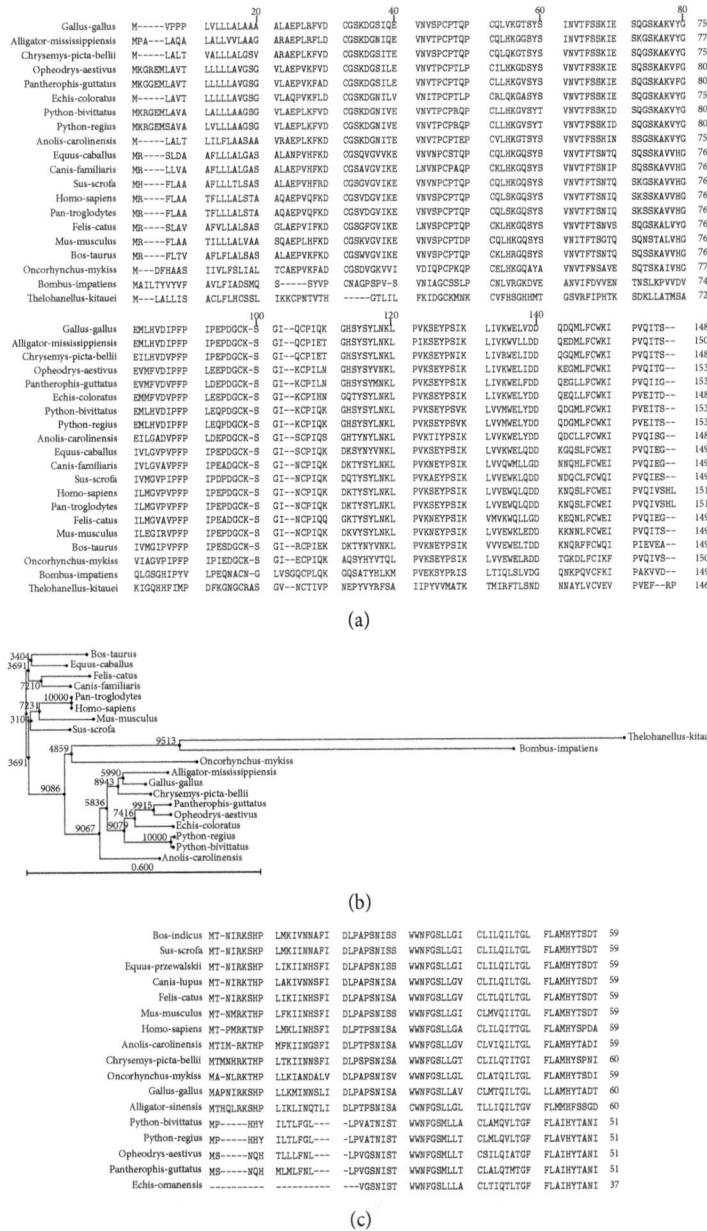

FIGURE 2: (a) Multiple sequence alignment of epididymal protein E1. Twenty species representing fish, parasite, insect, snake, turtle, lizard, alligator, chicken, mouse, cow, horse, dog, cat, chimpanzee, and human conserved regions of the alignment are indicated in blue while nonconserved regions are shown in red. Sequences from the snake species *Opheodrys aestivus*, *Pantherophis guttatus*, *Echis coloratus*, *Python regius*, and *Python bivittatus* all contain a 5-amino-acid insertion (KRGEM in python species) within the first ten amino acids of the alignment in comparison to other reptiles and mammalian species. Similarly, the primate lineage, represented by *Homo sapiens* and *Pan troglodytes*, exhibits a 2-amino-acid insertion (HL) at the very end of the alignment. *Thelohanellus kitauei* (an aquatic invertebrate parasite) and *Bombus impatiens* (bee) provide evidence for an ancient role of this gene in the common ancestor of insects, parasites, and vertebrates. (b) Phylogenetic tree of epididymal protein E1. The multiple sequence alignment shown in Figure 3(a) was used to generate a phylogenetic tree using a bootstrapping approach. The number near each root or interior node of the tree indicates how many times the same subtree, as shown in the image, was obtained when the input sequences were sampled during the bootstrapping. In this tree, *Python regius* and *Python bivittatus* are grouped together in all 10,000 trees produced during the phylogenetic tree construction process. Similarly, the *Homo sapiens* and *Pan troglodytes* subtree also exhibits a count of 10000. The scale bar below the tree provides an estimate for sequence evolution rate between the taxa. (c) Multiple sequence alignment of mitochondrial cytochrome b protein. The cytochrome b protein was aligned across multiple species to visualize the sequence relationship between *Python regius* and *Python bivittatus* in the context of other species. Conserved regions of the alignment are indicated in blue while less conserved regions are shown in red. The species representing snakes exhibit an absence of the 5 amino acids within the first ten amino acids of the alignment. This sequence feature is not observed in other reptiles, *Gallus gallus* (chicken) and *Oncorhynchus mykiss* (trout), or in mammals. Thus, there is the possibility that certain aspects of biology and/or physiology may be uniquely shared among snakes.

Number of genes in each phenotype

(a)

Average percent identity in phenotype

(b)

StdDev (average percent identity) in phenotype

(c)

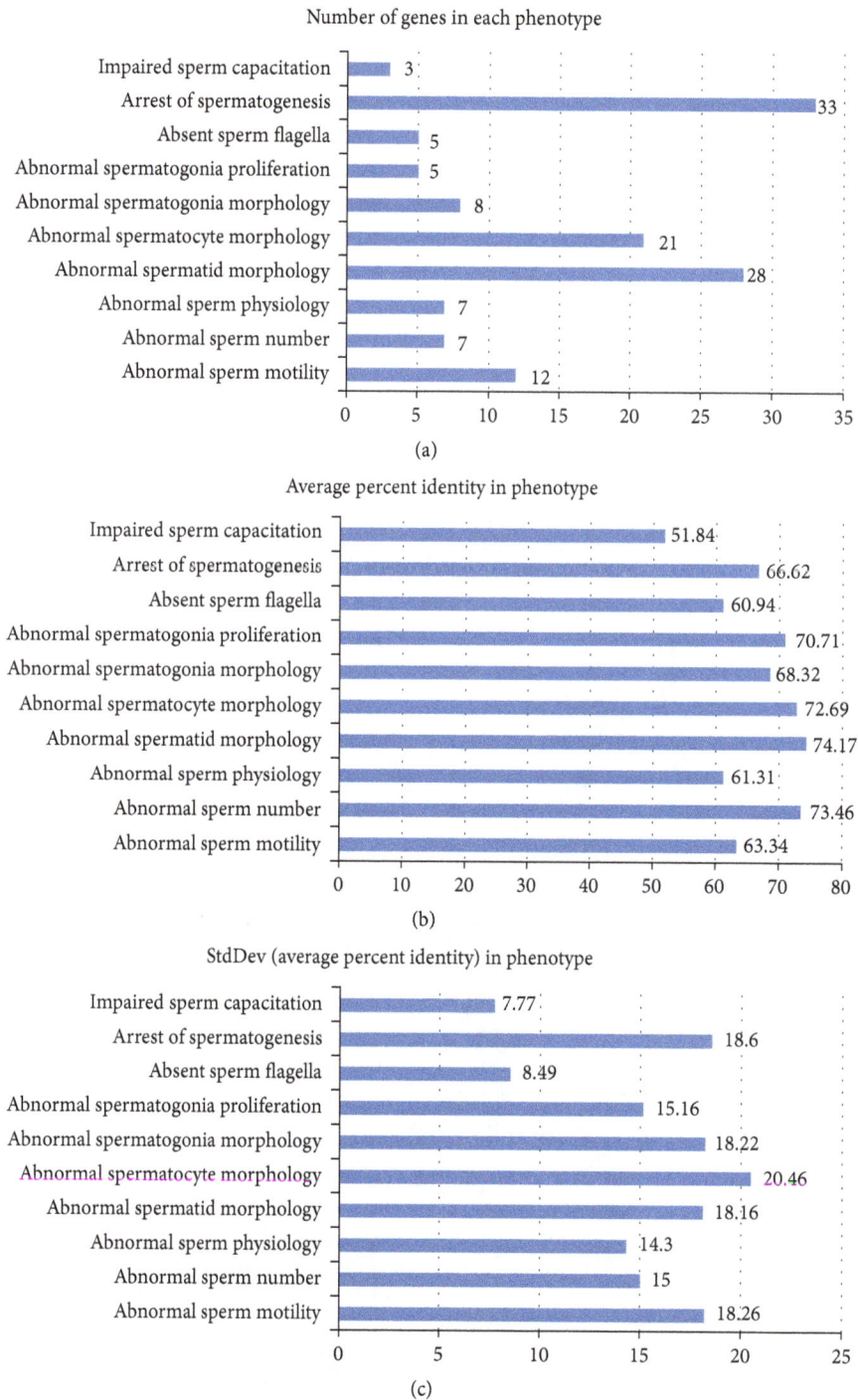

FIGURE 3: (a) Number of genes in each phenotype. The number of genes in each phenotype is shown in histogram. Within each phenotype, no gene is duplicated. However, a gene may appear in more than one phenotype. The sum of the counts is 129 corresponding to 129 gene-phenotype relationships. The number of distinct genes is 98. (b) Average percent identity for the set of genes within each phenotype. The average percent identity was calculated for each phenotype. The percent identity for each *P. bivittatus* protein sequence was determined by using BLASTP to align each *P. bivittatus* protein sequence against its corresponding mouse protein ortholog. Only the single top scoring blast hit was used to determine identity. Average identity was calculated for each phenotype by summing the individual identities within each phenotype and dividing by the total number of genes within each phenotype. (c) Standard deviation of the "average percent identity" in each phenotype. The population standard deviation of percent identity was calculated for each phenotype. The standard deviations range from a low of 7.7 to a high of 20.46. The two phenotypes with the lowest standard deviations are "impaired sperm capacitation" and "absent sperm flagella."

TABLE 1: Python orthologs of genes previously associated with sperm function and/or traits along with literature references.

Gene name	*P. regius* (Genbank IDs)	Identity & *e*-value	*P. bivittatus* (Genbank IDs)	Reference
Epididymal secretory protein type E1	gi|698375281 gb|JAC94921.1	99% $8e - 114$	gi|602671876 ref|XP_007441448.1	[39]
Complement C3	gi|698375324 gb|JAC94937.1	95% 0.0	gi|602634163 ref|XP_007423955.1	[40]
Phospholipase B	gi|698375293 gb|JAC94925.1	96% 0.0	gi|602641944 ref|XP_007427768.1	[41]
L-Amino acid oxidase	gi|698375301 gb|JAC94928.1	N/A	N/A	[42]
Cysteine-rich secretory protein A	gi|698375279 gb|JAC94920.1	95% $1e - 166$	gi|602662716 ref|XP_007436999.1	[43]
Kallikrein	gi|698375306 gb|JAC94930.1	97% 0.0	gi|602667543 ref|XP_007439332.1	[44]
Factor X	gi|669634856 gb|JAC88993.1	94% 0.0	gi|602632281 ref|XP_007423033.1	[45]
Alpha enolase	gi|5305427 gb|AAD41646.1	88% 0.0	gi|602651461 ref|XP_007432114.1	[46]
Cystatin F	gi|698375284 gb|JAC94922.1	97% $7e - 115$	gi|602642301 ref|XP_007427941.1	[47]
Cathepsin D	gi|698375257 gb|JAC94912.1	98% 0.0	gi|602659329 ref|XP_007435358.1	[48]
NADH dehydrogenase subunit 4 (mitochondrion)	gi|74310582 ref|YP_313703.1	89% 0.0	gi|511768956 ref|YP_008083608.1	[49]
Cytochrome B (mitochondrion)	gi|74310585 ref|YP_313706.1	94% $2e - 63$	gi|224815091 gb|ACN65710.1	[50]
ATP synthase F0 subunit 6 (mitochondrion)	gi|74310578 ref|YP_313699.1	90% $5e - 126$	gi|511768952 ref|YP_008083604.1	[51]

Python orthologs of genes previously associated with sperm function and/or traits along with literature references. Among a set of 21 *P. regius* gene sequences available within the NCBI database, 13 were implicated in sperm function and/or traits based on previous publications in other species. The table contains the *P. regius* NCBI identifiers and corresponding orthologous *P. bivittatus* gene identifiers (for genes in which the ortholog was identified). For each orthologous pair, the protein percent identity and *e*-value obtained from the BLASTP analysis of the sequences are provided. Additionally, each *P. regius* gene identified has a reference implicating the gene in sperm related biology in another species. Together these genes represent evidence that *P. bivittatus* and *P. regius* exhibit high sequence identity among these genes. Specifically, 4 out of the 12 identified ortholog pairs have identity of 97% or greater (epididymal secretory protein E1, Kallikrein, Cystatin F, and cathepsin D), while just two ortholog pairs have percent identity below 90% (alpha enolase having 88% and mitochondrial NADH dehydrogenase subunit 4 with 89%).

2.6. Functional Gene Ontology Enrichment Analysis of Genes Associated with Sperm Phenotypes. Gene set analysis was performed on the set of 98 genes associated with phenotypes using the online bioinformatics package DAVID (https://david.ncifcrf.gov/). The DAVID bioinformatics resource provides genome level functional annotation of genes and data sets through a web-based interface [52]. It is an ideal resource that facilitates the identification of biologically relevant signals in large-scale genomics data sets [53].

Mapping of mouse gene identifiers to python protein identifiers provided a means to accomplish analysis of the python genes using the DAVID mouse gene identifiers since the DAVID database relies upon established gene ontology associations with gene identifiers. At the time of writing, the DAVID database does not handle analyses using python gene or protein identifiers. Running the functional genomic analysis using the mouse gene identifiers as surrogates for the orthologous python genes enabled the gene ontology analysis to be accomplished with the DAVID software.

Gene ontology enrichment terms were identified using a *p* value <0.05 and/or Benjamini probability <0.05. Because the purpose of the DAVID software suite is to identify biological annotations that are enriched in gene sets from gene expression studies, it is of value to use stringent statistical measures of enrichment. However, unlike a gene expression experiment, the genes identified in this study each are selected due to their individual phenotype relating to sperm development, function, and morphology. Therefore, even in cases where the *p* value is >0.05, the functional annotation associated with a specific gene is still the true biology underlying that gene's involvement in reproductive biology.

2.7. Characterization of Phenotype-Specific Functional Annotation in Python Sperm Genes. Additional rounds of gene set functional analysis were employed to phenotype-specific biological annotations. Specifically, gene ontology enrichment terms were identified using a *p* value <0.05 and/or Benjamini probability < 0.05 for each of the ten sperm related phenotypes. The analysis was carried out using the DAVID bioinformatics software tool. Gene sets from each of the ten sperm phenotypes were analyzed and the results were compared to the functional annotation identified in the 98-gene set. As was considered in the analysis of the 98-gene set, annotation terms relating to reproduction and fertility were included in the functional annotation of genes even in cases where the p value was >0.05.

2.8. Visualization of Phenotype-Specific Functional Annotation in Python Sperm Genes. Visualization of gene ontology annotation was accomplished using Revigo (Supek 2011) which creates two-dimensional scatterplots and tree maps that organize the annotations via their semantic relationships with one another. The gene ontology *biological process*, *cellular compartment*, and *molecular function* terms identified via the characterization of phenotype-specific annotation were used as input for Revigo. The p values were converted to $-\log(p$ value) and included as scores for each GO annotation. Parameters were set to reflect that $-\log(p$ value) entered as the score with the GO annotation was considered higher scoring if the associated $-\log(p$ value) was higher (this is in contrast to p values in which lower values are better scores). Additionally, the analysis was performed using the GO database from *Homo sapiens* because it contained the greatest number of genes from the set of sperm associated phenotypes (97 out of 98) compared to *Mus musculus* and *Gallus gallus*. Since enrichment metrics utilize the hypergeometric distribution for calculating statistical significance, *H. sapiens* was considered more appropriate based on inclusion of sperm associated genes represented in the Revigo resource. The SimRel measure of semantic similarity was used for the analysis.

2.9. KEGG Pathway Enrichment Analysis of Genes Associated with Sperm Phenotypes. KEGG database provides a repository for genes associated with cellular and signaling pathways [54] which can be used to decipher gene functions. Pathway enrichment analysis was performed on the set of 98 genes associated with sperm phenotypes using the online bioinformatics package DAVID (https://david.ncifcrf.gov/) [55, 56].

2.10. Public Release of Data and Functional Annotation Associated with This Study. The authors of this project believe that the benefit of genomics and genetic resources is best accomplished when such resources are freely made available to the research community. Subsequently, the set of mouse and python gene identifiers, along with their orthologous mappings and functional associations with specific phenotypes, have been made freely available to the research community through the supplemental data associated with this publication (Supplemental File 1 and Supplemental File 2). Specifically these supplemental resources include tab-delimited files with gene ontology (biological process, cellular compartment, and molecular function) enrichment results for the sperm associated genes (Supplemental File 3, Supplemental File 4, and Supplemental File 5, resp.) as well as a FASTA file containing the protein sequences for *Python bivittatus* along

with the corresponding sperm associated phenotype for each sequence (Supplemental File 6).

3. Results

3.1. Analysis of Protein Sequence Identity between Python regius and Python bivittatus. To explore the possibility that *P. regius* and *P. bivittatus* exhibit sufficient genetic similarity to justify using one species as a genetic model for the other species, we assessed the level of sequence identity among a set of protein coding sequences, mitochondrial DNA sequences, and protein sequences. We aligned the mRNA for epididymal protein E1 and assessed the extent of identity between the two species. The mRNAs were aligned using pairwise nucleotide BLASTN resulting in an alignment length of 1213 nucleotides with a percent identity of 98% corresponding to 1189 identities with 6 gaps. This alignment produced a bit score of 2102 and an *e*-value of 0. Pairwise BLASTP was used to assess the identity for the pairwise protein alignment for epididymal protein E1 between the two python species. The length of protein sequence was 153 amino acids long in both species. The aligned sequences covered the full length of each protein with an identity of 98% corresponding to 151 identical amino acids aligned with 0 gaps. A set of 12 proteins implicated in sperm related functions were selected for pairwise alignment between *P. regius* and *P. bivittatus* orthologs in order to assess the extent of protein sequence identity between the two species (Table 1). Together these orthologous alignments provide insight into the genetic relationship between these two python species. Four out of the twelve genes identified exhibited identity of 97% or greater (epididymal secretory protein E1, Kallikrein, Cystatin F, and cathepsin D), while just two ortholog pairs have percent identity below 90% (alpha enolase having 88% and mitochondrial NADH dehydrogenase subunit 4 with 89%).

3.2. Multispecies Sequence Alignments and Construction of Phylogenetic Trees across Taxa. To gain a better appreciation for the relationship that exists among taxa, with regard to the proteins implicated in sperm function, multiple sequence analysis was performed for two protein coding sequences: epididymal protein E1 (Figures 2(a) and 2(b)) and mitochondrial cytochrome b (Figure 2(c)). Sequences from the snake species (*Opheodrys aestivus*, *Pantherophis guttatus*, *Echis coloratus*, *Python regius*, and *Python bivittatus*) all contain a 5-amino-acid insertion (KRGEM in python species) within the first ten amino acids of the epididymal protein E1 alignment (Figure 2(a)) in comparison to other reptiles and mammalian species. Similarly, the primate lineage, represented by *Homo sapiens* and *Pan troglodytes*, exhibits a 2-amino-acid insertion (HL) at the C-terminal end of the alignment (Figure 2(a)). *Thelohanellus kitauei* (an aquatic invertebrate parasite) and *Bombus impatiens* (bee) provide evidence for an ancient role of this protein coding gene in the common ancestor of insects, parasites, and vertebrates. The multiple sequence alignment shown in Figure 2(a) was used to generate a phylogenetic tree using a bootstrapping approach (Figure 2(b)). The number near each root or interior node of the tree indicates how many times the same subtree, as shown in the image, was obtained when the input sequences were sampled during the bootstrapping. Larger numbers indicate a greater percent of bootstrapped trees contained in the same tree organization as depicted in the figure. *Python regius* and *Python bivittatus* are grouped together in all 10,000 trees produced during the phylogenetic bootstrapping tree construction process. The only other species, for which 10,000 iterations of tree construction resulted in the two species being paired together 100% of the time, are *Homo sapiens* and *Pan troglodytes*. The common evolutionary relationship between the snake species exhibiting the 5-amino-acid insertion at the beginning of the alignment is also characterized by the magnitude of the number at the subnode of the tree which contains these species.

The mitochondrial cytochrome b protein sequence was obtained for multiple sequence alignment across 17 species (Figure 2(c)). The equine sequence was taken from *Equus przewalskii*, the feral horse, instead of the domestic horse. Upon inspecting the alignment, it was apparent that the snake sequences diverged from the nonsnake species. However, in the case of mitochondrial cytochrome b, the snakes (*P. bivittatus*, *P. regius*, *P. guttatus*, and *E. omanensis*) exhibit two short deletions within the first 30 amino acids of the alignment, in contrast to the insertion identified in epididymal protein E1. Although sequence divergence within these regions of the alignment is evident among the other species, neither the nonsnake reptiles nor the mammals exhibit the gapped alignment pattern observed in the snakes.

3.3. Identification of P. bivittatus Protein Sequences Associated with Sperm Phenotypes. Through the comparative genomics approach employed, 129 gene-phenotype relationships were identified in *P. bivittatus* genes (Supplemental File 1). Initially we sought to identify 152 gene-phenotype relationships based on the phenotype annotation in the mouse. However, while attempting to identify orthologous genes in the python, 13 orthologs could not be adequately identified due to ambiguity in resolving whether some python genes were truly the orthologs, or whether what was identified in the database was a paralogous sequence.

In some cases, the identified python gene contained the annotation term "partial" in the fasta header line in NCBI databases. These sequences were still included in the final gene set (even though the sequence may not be complete). Python genes lacking the term "partial" were considered to be full length; however during our analysis it became apparent that some genes lacking the annotation "partial" did not represent full length sequences.

The final set of *P. bivittatus* gene sequences associated with sperm phenotypes included 98 distinct genes (Supplemental File 2) mapping to ten classes of phenotype (Supplemental File 6). In order to carefully maintain the relationship between phenotype and gene, our approach treated each gene-phenotype relationship as a unique data point. Subsequently the 129 gene-phenotype relationships collapsed down to 98 distinct genes once duplicate genes were excluded. The number of *P. bivittatus* genes associated with each phenotype is shown in Figure 3(a). The average percent identity for each phenotype is shown in Figure 3(b) and the standard

deviation for the average percent identity within each phenotype is shown in Figure 3(c).

Genes in each of the four mature sperm phenotypes gene sets (sperm number [7 genes], sperm motility [12 genes], sperm physiology [7 genes], and capacitation [3 genes]) were analyzed for overlap across the phenotypes (Figure 4(a)). A total of 23 unique genes were distributed among the phenotypes with 7 genes being exclusive to sperm motility, an additional 7 genes were unique to the sperm number phenotype, 3 genes were specific to sperm physiology, and a single gene was associated with capacitation. Three genes were common between sperm motility and sperm physiology while just a single gene was found to be associated with both motility and capacitation. Interestingly, one gene was associated with the motility, physiology, and capacitation phenotypes. The majority of genes were unique to specific phenotypes.

Genes in each of the three abnormal morphological phenotypes associated with spermatogonia [8 genes], spermatocytes [21 genes], and spermatids [28 genes] were analyzed for overlap across the distinct phenotypes (Figure 4(b)). A total of 45 genes were distributed among the phenotypes with 22 unique to spermatids, 11 unique to spermatocytes, and just 2 genes unique to spermatogonia. Four genes were common among spermatids and spermatocytes while another four genes were common spermatocytes and spermatogonia. Only two genes were associated with all three phenotypes.

3.4. Functional Analysis of Sperm Phenotype-Specific Gene Sets Using Gene Ontology Annotation. Gene ontology (GO) enrichment was performed to assess the biological role of the sperm associated python genes (Table 2). Among biological process annotation, highly significant terms were identified relating to reproduction including "gamete generation", "spermatogenesis", "germ cell development", "spermatid differentiation", and "meiosis". Many of these terms were associated with p values as low as $7.20E - 40$ and $7.70E - 37$. Among the enriched GO terms representing cellular component information were cilium, cell projection, acrosomal vesicle, and microtubule cytoskeleton. Within the molecular function GO terms enriched themes of transcriptional factor regulation and DNA binding were identified as well as ATP binding and kinase activity. The complete set of GO annotation data, including a list of genes enriched for each identified GO annotation term, is available in Supplemental File 3 (biological process), Supplemental File 4 (cellular compartment), and Supplemental File 5 (molecular function). A two-dimensional semantic scatter plot was generated from the gene ontology biological process annotation (Figure 5) in order to facilitate visualization of the GO enrichment data. Semantic relationships within the gene ontology annotation terms provide evidence of common themes relating to spermatogenesis and sperm motility and function in the context of reproduction.

3.5. KEGG Pathways Enriched for P. bivittatus Genes Associated with Sperm Phenotypes. KEGG pathways enriched for genes within the set of 98 *Python bivittatus* genes associated with sperm phenotypes were identified. Six pathways were

Table 2: Gene ontology enrichment.

Category	GO identifier	Annotation term	Count	Enrichment	p value
Biological process	GO:0019953	Sexual reproduction	45	13.99	$7.20E - 40$
Biological process	GO:0007276	Gamete generation	41	14.78	$7.70E - 37$
Biological process	GO:0048232	Male gamete generation	37	17.11	$3.46E - 35$
Biological process	GO:0007283	Spermatogenesis	37	17.11	$3.46E - 35$
Biological process	GO:0032504	Multicellular organism reproduction	42	12.28	$1.54E - 34$
Biological process	GO:0048609	Reproductive process in a multicellular organism	42	12.28	$1.54E - 34$
Biological process	GO:0003006	Reproductive developmental process	27	14.67	$2.70E - 23$
Biological process	GO:0048610	Reproductive cellular process	22	19.34	$2.55E - 21$
Biological process	GO:0007281	Germ cell development	18	25.38	$2.20E - 19$
Biological process	GO:0048515	Spermatid differentiation	14	34.98	$6.97E - 17$
Biological process	GO:0007286	Spermatid development	13	34.28	$1.62E - 15$
Biological process	GO:0007548	Sex differentiation	15	14.15	$2.05E - 12$
Biological process	GO:0045137	Development of primary sexual characteristics	12	13.46	$1.15E - 09$
Biological process	GO:0008406	Gonad development	11	13.99	$5.09E - 09$
Biological process	GO:0048608	Reproductive structure development	11	12.43	$1.60E - 08$
Biological process	GO:0046661	Male sex differentiation	9	17.56	$3.73E - 08$
Biological process	GO:0051327	M phase of meiotic cell cycle	9	13.08	$3.79E - 07$
Biological process	GO:0007126	Meiosis	9	13.08	$3.79E - 07$
Biological process	GO:0051321	Meiotic cell cycle	9	12.82	$4.43E - 07$
Biological process	GO:0046546	Development of primary male sexual characteristics	7	15.34	$5.69E - 06$
Biological process	GO:0009566	Fertilization	7	12.62	$1.77E - 05$
Biological process	GO:0008584	Male gonad development	6	16.43	$2.92E - 05$
Cellular component	GO:0005929	Cilium	7	8.89	$1.24E - 04$
Cellular component	GO:0031514	Motile secondary cilium	3	122.90	$2.13E - 04$
Cellular component	GO:0042995	Cell projection	14	3.29	$2.44E - 04$
Cellular component	GO:0001669	Acrosomal vesicle	4	17.72	$1.41E - 03$
Cellular component	GO:0019861	Flagellum	4	14.25	$2.65E - 03$
Cellular component	GO:0030141	Secretory granule	6	5.46	$4.57E - 03$
Cellular component	GO:0016023	Cytoplasmic membrane-bounded vesicle	10	2.98	$5.65E - 03$
Cellular component	GO:0031988	Membrane-bounded vesicle	10	2.89	$6.93E - 03$
Cellular component	GO:0005625	Soluble fraction	7	3.66	$1.14E - 02$
Cellular component	GO:0031410	Cytoplasmic vesicle	10	2.55	$1.47E - 02$
Cellular component	GO:0015630	Microtubule cytoskeleton	9	2.69	$1.72E - 02$
Cellular component	GO:0031982	Vesicle	10	2.45	$1.89E - 02$
Cellular component	GO:0000267	Cell fraction	13	1.97	$2.79E - 02$
Cellular component	GO:0033391	Chromatoid body	2	54.62	$3.56E - 02$
Cellular component	GO:0060293	Germ plasm	2	46.82	$4.14E - 02$
Cellular component	GO:0045495	Pole plasm	2	46.82	$4.14E - 02$
Cellular component	GO:0043186	P granule	2	46.82	$4.14E - 02$
Cellular component	GO:0034464	BBSome	2	46.82	$4.14E - 02$
Cellular component	GO:0060170	Cilium membrane	2	40.97	$4.72E - 02$
Molecular function	GO:0046983	Protein dimerization activity	13	3.71	$1.60E - 04$
Molecular function	GO:0042802	Identical protein binding	14	3.38	$1.96E - 04$
Molecular function	GO:0043565	Sequence-specific DNA binding	12	3.06	$1.61E - 03$
Molecular function	GO:0003707	Steroid hormone receptor activity	4	12.62	$3.76E - 03$
Molecular function	GO:0015631	Tubulin binding	5	7.73	$3.82E - 03$
Molecular function	GO:0030554	Adenyl nucleotide binding	20	1.96	$4.37E - 03$
Molecular function	GO:0001883	Purine nucleoside binding	20	1.93	$5.16E - 03$
Molecular function	GO:0042803	Protein homodimerization activity	8	3.70	$5.46E - 03$
Molecular function	GO:0001882	Nucleoside binding	20	1.92	$5.55E - 03$

TABLE 2: Continued.

Category	GO identifier	Annotation term	Count	Enrichment	p value
Molecular function	GO:0004879	Ligand-dependent nuclear receptor activity	4	10.66	$6.04E-03$
Molecular function	GO:0030528	Transcription regulator activity	19	1.94	$6.36E-03$
Molecular function	GO:0003700	Transcription factor activity	14	2.22	$8.63E-03$
Molecular function	GO:0046982	Protein heterodimerization activity	6	4.46	$1.06E-02$
Molecular function	GO:0005524	ATP binding	18	1.88	$1.12E-02$
Molecular function	GO:0047485	Protein N-terminus binding	4	8.35	$1.18E-02$
Molecular function	GO:0032559	Adenyl ribonucleotide binding	18	1.86	$1.27E-02$
Molecular function	GO:0004672	Protein kinase activity	10	2.55	$1.50E-02$
Molecular function	GO:0016563	Transcription activator activity	8	3.02	$1.58E-02$
Molecular function	GO:0004674	Protein serine/threonine kinase activity	8	2.88	$2.00E-02$

Gene Ontology Enrichment. The set of *P. bivittatus* identified as having sperm associated phenotypes were further explored to identify enrichment of gene ontology (GO) terms. A subset of the results are shown above, and the complete set of data is available in Supplemental File 3, Supplemental File 4, and Supplemental 5. Enrichment within the three categories of gene ontology (biological process, cell component, and molecular function) was identified. The displayed GO annotation terms correspond to the most significant *p* values in each of the three categories. The number of genes associated with each annotation term is included, as are the *p* value and the fold enrichment. The enrichment analysis was performed using the DAVID resource. Bonferroni corrected *p* values and false discovery rate values are included in the tab-delimited supplemental files as well as the list of genes associated with each GO annotation term.

TABLE 3: Identification of KEGG pathways enriched for genes within *Python bivittatus* genes associated with sperm phenotypes.

Pathway-ID	Pathway name	Count	Genes	Enrichment	p value
hsa05200	Pathways in cancer	10	LAMA2, FOS, AR, PDGFA, RXRB, BAX, NOS2, KIT, FAS, CCNA1	3.45	$1.64E-03$
hsa04115	p53 signaling pathway	4	BAX, APAF1, FAS, ATM	6.65	$2.05E-02$
hsa05222	Small cell lung cancer	4	LAMA2, RXRB, APAF1, NOS2	5.38	$3.54E-02$
hsa04210	Apoptosis	4	BAX, APAF1, FAS, ATM	5.20	$3.87E-02$
hsa04020	Calcium signaling pathway	5	ATP2B4, CAMK4, PLCD4, NOS2, BDKRB2	3.21	$6.44E-02$
hsa04060	Cytokine-cytokine interaction	6	LEP, AMHR2, AMH, PDGFA, KIT, FAS	2.59	$7.38E-02$

Identification of KEGG pathways enriched for genes within *Python bivittatus* genes associated with sperm phenotypes. KEGG pathways were identified using the bioinformatics resource DAVID. The KEGG pathway identifier and pathway name are provided along with the number of genes identified in the pathway, the gene symbol for each identified gene, the fold enrichment, and the associated *p* value.

identified (Table 3). Among the 33 gene-pathway relationships identified were ten genes associated with pathways in cancer (p value = $1.64E-03$), four genes implicated in p53 signaling (p value = $2.05E-02$), and six genes enriched for cytokine-cytokine receptor interactions (p value = $7.38E-02$). Although some of the results were associated with p values slightly larger than 0.05, they were still included in the table because the fold enrichment was greater than 2.5 for each enriched pathway, which provides supplemental support for their inclusion as they offer insight into the cellular and molecular processes underlying sperm differentiation, activation, and function in the python.

4. Discussion

The discovery of python genes associated with sperm phenotypes provides a tremendously important genetic resource for future use in studying reproduction and fertility in endangered and invasive reptile species. The results reported here highlight the value of comparative genomics and its application in species for which genomic resources are available. Through the mapping of python genes to specific phenotypes, it is now possible to develop more focused genetic research projects aimed at identifying genes associated with poor male fertility and reproductive success in endangered populations.

Our analysis of nucleotide and protein sequence similarity between *P. regius* and *P. bivittatus* provides evidence of similarity ranging from 86% to 98% at the nucleotide level and even higher when considering similarity at the protein level. Subsequently each of these two species can serve as a model organism for the other. For example, the genomic resources available for *P. bivittatus* can be used as a model for *P. regius*, such as for applications like PCR primer design. Housley et al. assessed PCR success among cross-species PCR primers and identified the relatedness of the target species and index species as one of the most important factors underlying PCR success [57]. Similarly, genetic dissection of *P. regius* phenotypes can be leveraged for applications in *P. bivittatus*. Unlike *P. bivittatus*, *P. regius* is a much smaller and

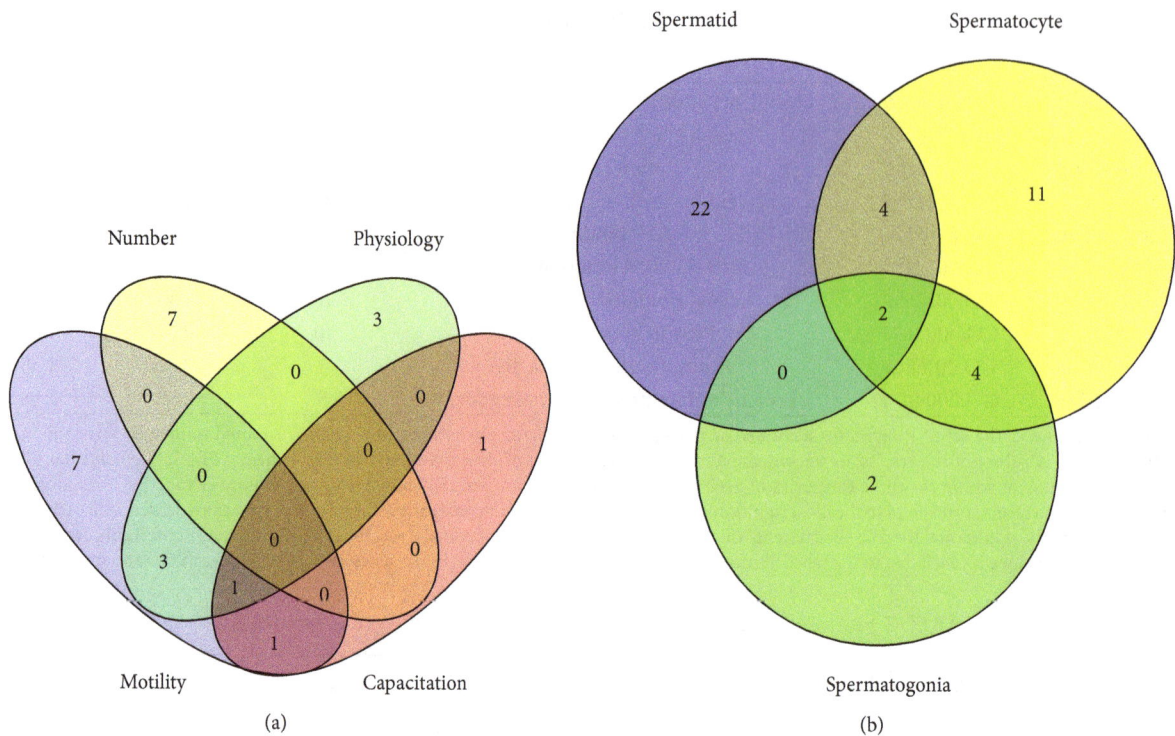

FIGURE 4: (a) Venn diagram illustrating relationship of genes among phenotypes associated with mature sperm. Genes in each of the four mature sperm phenotypes gene sets (sperm number [7 genes], sperm motility [12 genes], sperm physiology [7 genes], and capacitation [3 genes]) were analyzed for overlap across the phenotypes. A total of 23 unique genes were distributed among the phenotypes with 7 genes being exclusive to sperm motility, an additional 7 genes were unique to the sperm number phenotype, 3 genes were specific to sperm physiology, and just a single gene was only associated with capacitation. Three genes were common between sperm motility and sperm physiology while just a single gene was found to be associated with both motility and capacitation. Interestingly, one gene was associated with the motility, physiology, and capacitation phenotypes. The majority of genes were unique to specific phenotypes. (b) Venn diagram illustrating relationship of genes among phenotypes associated with morphological phenotypes in sperm precursors. Genes in each of the three abnormal morphological phenotypes associated with spermatogonia [8 genes], spermatocytes [21 genes], and spermatids [28 genes] were analyzed for overlap across the distinct phenotypes. A total of 45 genes were distributed among the phenotypes with 22 unique to spermatids, 11 unique to spermatocytes, and just 2 genes unique to spermatogonia. Four genes were common among spermatids and spermatocytes while another four genes were common spermatocytes and spermatogonia. Only two genes were associated with all three phenotypes.

more docile species which is amenable to reproductive studies as they are easily maintained in captivity and are known for being easy to breed in captivity.

The protein sequences selected to create alignments and phylogenetic trees were based on the limiting number of *P. regius* sequences available in NCBI (fewer than 30 protein sequences). Nonetheless, those selected were all implicated in sperm physiology or related to sperm biology. For example, the mitochondrial protein cytochrome b has been associated with decreased sperm mobility in association with specific haplotypes [58] and mutations in cytochrome b have been linked with asthenospermia [59]. Interestingly, Chen et al. demonstrated that cytochrome b is differentially expressed in X-chromosome versus Y-chromosome containing sperm [60].

The gene encoding epididymal protein E1 participates in sperm physiology and is associated with important sperm parameters. Giacomini et al. identified a threefold decrease in the expression of epididymal protein E1 in oligoastheno-zoospermia compared to normozoospermia [61]. Epididymal

protein E1 has been identified as a seminal plasma protein across species, such as boars [39] and bulls [62]. Moreover, a study in rams identified this protein as a factor that when added to frozen/thawed semen increased motility through repair of sperm damage that occurred during the cryopreservation process [63]. This protein exhibited marked conservation across taxa in our data. Specifically, it was conserved across mammals, birds, fish, reptiles, and even insects.

The protein cathepsin D is also involved in sperm biology and appears to play a role in its maturation. In mice, cathepsin D expression has been observed in the testis and cathepsin D has been detected on the surface of mouse sperm during epididymal maturation [48]. Similar results have been observed in humans. For example, cathepsin D expression has been observed in human Sertoli cells and Leydig cells and was shown to be anchored to the sperm surface in the postacrosomal region [64]. Evidence that cathepsin D may exhibit an evolutionarily conserved role in sperm physiology comes from a recent study in which

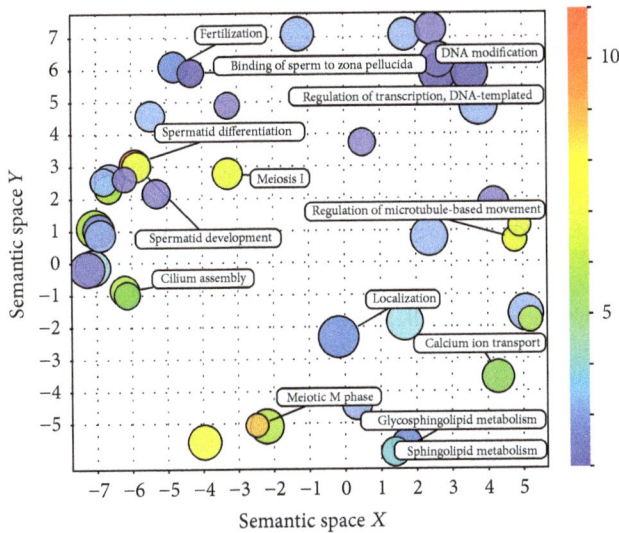

FIGURE 5: Semantic space scatter plot of gene ontology biological process annotation (node labels minimized). Gene ontology (GO) biological process annotations associated with subsets of genes within each sperm associated phenotype were visualized using a two-dimensional semantic space scatterplot. The spatial organization of annotations is based on semantic similarity. The number of node labels is minimized to allow visualization of the node colors on the scatterplot. The score = − log(p value) for each GO annotation term node. Blue nodes indicate less significant p values and red nodes indicate more significant p values.

cathepsin D was identified as a seminal plasma protein in carp [65].

Based on our multiple sequence alignments across taxa, proteins we investigated each exhibited unique patterns of conservation and divergence across the species. On the one hand, the conservation suggests that these factors have been conserved for millions of years to maintain their specific role in male reproduction. On the other hand, the distinct patterns of divergence between snakes and other reptiles, and even between reptiles and mammals, suggest that the biology may vary between different groups of taxa. Subsequently it is of value to expand our knowledge of the molecular basis of reproduction in reptiles and in particular snakes that can serve as research models for both endangered and invasive species.

In an attempt to expand molecular and genetic knowledge in reptile reproductive biology, we have leveraged a comparative genomics approach to identify P. bivittatus genes likely to be associated with sperm phenotypes. Although there are hundreds of genes implicated in reproductive biology, we specifically chose to limit our effort to just those for which each single gene we identified has been previously shown to cause a sperm phenotype in the mouse.

Our rationale for such stringency in our approach was that we prefer to identify a small set of genes, for which the proportion of true positives is very high, compared to a large number of predictions for which the false positive rate is high. When making bioinformatics predictions, one must always consider the trade-off that exists between false negatives and

false positives [66, 67]. In our particular case, we wanted to provide a public resource that can facilitate efforts aimed at elucidating reptile reproduction. Since the time and cost of validating bioinformatics predictions in the laboratory are proportional to the number of predictions made, we chose to maximize the number of true positives and subsequently minimize the laboratory cost per true positive identified. The complete set of publicly released data is available in the form of six supplemental files associated with this study.

Although this work leverages bioinformatics and comparative genomics approaches, the genes identified are merely predicted to have sperm related phenotypes in the python species. As evidenced by the phylogenetic analysis performed as part of this study, protein sequences can diverge greatly across taxa. Using mouse genomic annotation to characterize reptile reproductive biology is a challenging process. Nonetheless, the identification of these genes is exciting and provides new avenues of subsequent investigation. However, one must proceed cautiously as it is likely some of the gene-phenotype associations we report may not be present in reptiles. Since it is well known that the reproductive biology of reptiles differs from mammals, the true value of this gene set and the accuracy of the functional predictions will require further study. Even so, the results obtained provide an important first step in expanding reproductive genomics to pythons. Future investigations into these ecologically important species will undoubtedly elucidate a variety of conserved and divergent reproductive properties between reptiles and mammals. Perhaps, some of these discoveries may offer novel pharmacological targets for developing novel reproductive technologies, not only in reptiles, but also in mammals.

5. Conclusion

To explore the possibility that P. regius and P. bivittatus exhibit sufficient genetic similarity to justify using one species as a genetic model for the other species, we assessed the level of sequence identity among a set of protein coding sequences, mitochondrial DNA sequences, and protein sequences between Python bivittatus (Burmese python) and Python regius (ball python) to assess the extent of sequence identity between the two species in genes implicated in sperm maturation and function. Alignments of epididymal protein E1 demonstrated the molecular similarity between these species with 98% identity when comparing the mRNA and 98% identity when comparing the protein sequences. In multiple sequence alignments and phylogenetic analysis we identify snake specific patterns of sequence similarity in epididymal protein E1 and cytochrome b which support the use of P. regius as a model for P. bivittatus and other snakes. Most importantly, we employed a comparative genomics strategy to identify python genes enriched for association with sperm related phenotypes. Our approach identified 129 gene-phenotype relationships corresponding to 98 unique genes representing ten specific sperm associated phenotypes. We characterized these genes using gene ontology enrichment to annotate the biological processes, cellular compartments, and molecular functions associated with these reproductively important python genes. Our analysis also identified KEGG pathways enriched for these genes. Specifically, these genes are involved in the regulation of cell cycle, apoptosis, cancer, cytokine-cytokine signaling, and calcium regulation. To our knowledge, our results provide the first comprehensive view of genes associated with sperm development, sperm morphology, and sperm function in pythons. By making our data sets and findings publicly available, in the form of six supplemental files, we hope to facilitate the elucidation of reptile reproduction and promote effective conservation and management of reptiles worldwide.

Competing Interests

The authors declare that there are no competing interests regarding the publication of this paper.

Acknowledgments

This work was supported by Western University of Health Sciences IMR Fund 12349V, Developing Resources for Reptilian Genomics. The authors wish to thank Dr. Steven Henriksen for his support of this project and the resources he has contributed to facilitate its success. The authors would also like to acknowledge the success of this project in spite of the obstacles. Most importantly, the authors wish to acknowledge Castoe TA, de Koning AP, Hall KT, Card DC, Schield DR, Fujita MK, Ruggiero RP, Degner JF, Daza JM, Gu W, Reyes-Velasco J, Shaney KJ, Castoe JM, Fox SE, Poole AW, Polanco D, Dobry J, Vandewege MW, Li Q, Schott RK, Kapusta A, Minx P, Feschotte C, Uetz P, Ray DA, Hoffmann FG, Bogden R, Smith EN, Chang BS, Vonk FJ, Casewell NR, Henkel CV, Richardson MK, Mackessy SP, Bronikowski AM, Yandell M, Warren WC, Secor SM, and Pollock DD for sequencing the *Python bivittatus* genome and making the sequences publicly available to the research community at large. Without their hard work and contribution to reptile genomics, their project would not be possible.

References

[1] T. A. Castoe, A. J. de Koning, K. T. Hall et al., "Sequencing the genome of the Burmese python (*Python molurus bivittatus*) as a model for studying extreme adaptations in snakes," *Genome Biology*, vol. 12, no. 7, article no. 406, 2011.

[2] C. E. Wall, S. Cozza, C. A. Riquelme et al., "Whole transcriptome analysis of the fasting and fed Burmese python heart: insights into extreme physiological cardiac adaptation," *Physiological Genomics*, vol. 43, no. 2, pp. 69–76, 2011.

[3] A. L. Andrew, D. C. Card, R. P. Ruggiero et al., "Rapid changes in gene expression direct rapid shifts in intestinal form and function in the Burmese python after feeding," *Physiological Genomics*, vol. 47, no. 5, pp. 147–157, 2015.

[4] T. A. Castoe, A. P. J. De Koning, K. T. Hall et al., "The Burmese python genome reveals the molecular basis for extreme adaptation in snakes," *Proceedings of the National Academy of Sciences of the United States of America*, vol. 110, no. 51, pp. 20645–20650, 2013.

[5] R. G. Harvey, M. L. Brien, Michael et al., "Burmese pythons in south Florida: scientific support for invasive species management," Tech. Rep. WEC242, University of Florida, IFAS Extension, Gainesville, Fla, USA, 2008.

[6] R. N. Reed, J. D. Willson, G. H. Rodda, and M. E. Dorcas, "Ecological correlates of invasion impact for Burmese pythons in Florida," *Integrative Zoology*, vol. 7, no. 3, pp. 254–270, 2012.

[7] R. W. Snow, M. L. Brien, M. S. Cherkiss, L. Wilkins, and F. J. Mazzotti, "Dietary habits of the *Burmese python, Python molurus* bivittatus, in Everglades National Park, Florida," *Herpetological Bulletin*, vol. 101, pp. 5–7, 2007.

[8] R. W. Snow, K. L. Krysko, K. M. Enge, L. Oberhofer, A. Warren-Bradley, and L. Wilkins, "Introduced populations of Boa constrictor (Boidae) and *Python molurus* bivittatus (Pythonidae) in southern Florida," in *The Biology of Boas and Pythons*, R. W. Henderson and R. Powell, Eds., pp. 416–438, Eagle Mountain Publishing, 2007.

[9] M. E. Dorcas, J. D. Willson, R. N. Reed et al., "Severe mammal declines coincide with proliferation of invasive Burmese pythons in Everglades National Park," *Proceedings of the National Academy of Sciences of the United States of America*, vol. 109, no. 7, pp. 2418–2422, 2012.

[10] A. J. Piaggio, R. M. Engeman, M. W. Hopken et al., "Detecting an elusive invasive species: a diagnostic PCR to detect Burmese python in Florida waters and an assessment of persistence of environmental DNA," *Molecular Ecology Resources*, vol. 14, no. 2, pp. 374–380, 2014.

[11] D. G. Chapple, A. Birkett, K. A. Miller, C. H. Daugherty, and D. M. Gleeson, "Phylogeography of the endangered Otago skink, Oligosoma otagense: population structure, hybridisation and genetic diversity in captive populations," *PLoS ONE*, vol. 7, no. 4, Article ID e34599, 2012.

[12] S. A. Hussain, C. Lessard, and M. Anzar, "Quantification of damage at different stages of cryopreservation of endangered North American bison (Bison bison) semen and the effects of extender and freeze rate on post-thaw sperm quality," *Animal Reproduction Science*, vol. 129, no. 3-4, pp. 171–179, 2011.

[13] M. J. Ruiz-Lopez, D. P. Evenson, G. Espeso, M. Gomendio, and E. R. S. Roldan, "High levels of DNA fragmentation in spermatozoa are associated with inbreeding and poor sperm quality in endangered ungulates," *Biology of Reproduction*, vol. 83, no. 3, pp. 332–338, 2010.

[14] M. Madeddu, F. Berlinguer, M. Ledda et al., "Ejaculate collection efficiency and post-thaw semen quality in wild-caught Griffon vultures from the Sardinian population," *Reproductive Biology and Endocrinology*, vol. 7, article 18, 2009.

[15] K. J. Mattson, A. De Vries, S. M. McGuire, J. Krebs, E. E. Louis, and N. M. Loskutoff, "Successful artificial insemination in the corn snake, *Elaphe guttata*, using fresh and cooled semen," *Zoo Biology*, vol. 26, no. 5, pp. 363–369, 2007.

[16] R. Shine, "Reproductive strategies in snakes," *Proceedings of the Royal Society B: Biological Sciences*, vol. 270, no. 1519, pp. 995–1004, 2003.

[17] D. M. Sever and W. C. Hamlett, "Female sperm storage in reptiles," *Journal of Experimental Zoology*, vol. 292, no. 2, pp. 187–199, 2002.

[18] R. L. Zacariotti, K. F. Grego, W. Fernandas, S. S. Sant'Anna, and M. A. de Barras Vaz Guimarães, "Semen collection and evaluation in free-ranging Brazilian rattlesnakes (Crotalus durissus terrificus)," *Zoo Biology*, vol. 26, no. 2, pp. 155–160, 2007.

[19] M. Moshiri, F. Todehdehghan, and A. Shiravi, "Study of sperm reproductive parameters in mature zanjani viper," *Cell Journal*, vol. 16, no. 2, pp. 111–116, 2014.

[20] B. M. Fahrig, M. A. Mitchell, B. E. Eilts, and D. L. Paccamonti, "Characterization and cooled storage of semen from corn snakes (*Elaphe guttata*)," *Journal of Zoo and Wildlife Medicine*, vol. 38, no. 1, pp. 7–12, 2007.

[21] D. E. Bassett Jr., M. S. Boguski, F. Spencer, R. Reeves, M. Goebl, and P. Hieter, "Comparative genomics, genome cross-referencing and XREFdb," *Trends in Genetics*, vol. 11, no. 9, pp. 372–373, 1995.

[22] D. E. Bassett Jr., M. S. Boguski, F. Spencer et al., "Genome cross-referencing and XREFdb: implications for the identification and analysis of genes mutated in human disease," *Nature Genetics*, vol. 15, no. 4, pp. 339–344, 1997.

[23] C. Suter-Crazzolara and G. Kurapkat, "An infrastructure for comparative genomics to functionally characterize genes and proteins," *Genome Inform Ser Workshop Genome Inform*, vol. 11, pp. 24–32, 2000.

[24] S. Oduru, J. L. Campbell, S. Karri, W. J. Hendry, S. A. Khan, and S. C. Williams, "Gene discovery in the hamster: a comparative genomics approach for gene annotation by sequencing of hamster testis cDNAs," *BMC Genomics*, vol. 4, no. 1, article 22, 2003.

[25] S. K. Hawthorne, G. Goodarzi, J. Bagarova et al., "Comparative genomics of the sperm mitochondria-associated cysteine-rich protein gene," *Genomics*, vol. 87, no. 3, pp. 382–391, 2006.

[26] L. Yang, X. Zhang, J. Chen et al., "ReCGiP, a database of reproduction candidate genes in pigs based on bibliomics," *Reproductive Biology and Endocrinology*, vol. 8, article 96, 2010.

[27] K. J. Irizarry, S. B. Malladi, X. Gao et al., "Sequencing and comparative genomic analysis of 1227 Felis catus cDNA sequences enriched for developmental, clinical and nutritional phenotypes," *BMC Genomics*, vol. 13, no. 1, article 31, 2012.

[28] P. G. Baker, C. A. Goble, S. Bechhofer, N. W. Paton, R. Stevens, and A. Brass, "An ontology for bioinformatics applications," *Bioinformatics*, vol. 15, no. 6, pp. 510–520, 1999.

[29] M. Ashburner, C. A. Ball, J. A. Blake et al., "Gene ontology: tool for the unification of biology. The Gene Ontology Consortium," *Nature Genetics*, vol. 25, no. 1, pp. 25–29, 2000.

[30] N. H. Shah and N. V. Fedoroff, "CLENCH: a program for calculating CLuster ENriCHment using the gene ontology," *Bioinformatics*, vol. 20, no. 7, pp. 1196–1197, 2004.

[31] X. Wang, S. Pyne, and I. Dinu, "Gene set enrichment analysis for multiple continuous phenotypes," *BMC Bioinformatics*, vol. 15, article 260, 2014.

[32] C. L. Smith and J. T. Eppig, "The mammalian phenotype ontology: enabling robust annotation and comparative analysis," *Wiley Interdisciplinary Reviews: Systems Biology and Medicine*, vol. 1, no. 3, pp. 390–399, 2009.

[33] P. N. Robinson and S. Mundlos, "The human phenotype ontology," *Clinical Genetics*, vol. 77, no. 6, pp. 525–534, 2010.

[34] T. F. Hayamizu, S. de Coronado, G. Fragoso, N. Sioutos, J. A. Kadin, and M. Ringwald, "The mouse-human anatomy ontology mapping project," *Database*, vol. 2012, Article ID bar066, 2012.

[35] M. Kanehisa, "The KEGG database," *Novartis Foundation Symposium*, vol. 247, pp. 91–252, 2002.

[36] K. F. Aoki-Kinoshita and M. Kanehisa, "Gene annotation and pathway mapping in KEGG," *Methods in Molecular Biology*, vol. 396, pp. 71–91, 2007.

[37] F. Hahne, A. Mehrle, D. Arlt, A. Poustka, S. Wiemann, and T. Beissbarth, "Extending pathways based on gene lists using InterPro domain signatures," *BMC Bioinformatics*, vol. 9, article 3, 2008.

[38] D. W. Huang, B. T. Sherman, and R. A. Lempicki, "Systematic and integrative analysis of large gene lists using DAVID bioinformatics resources," *Nature Protocols*, vol. 4, no. 1, pp. 44–57, 2009.

[39] V. González-Cadavid, J. A. M. Martins, F. B. Moreno et al., "Seminal plasma proteins of adult boars and correlations with sperm parameters," *Theriogenology*, vol. 82, no. 5, pp. 697–707, 2014.

[40] R. J. Llanos, C. M. Whitacre, and D. C. Miceli, "Potential involvement of C_3 complement factor in amphibian fertilization," *Comparative Biochemistry and Physiology—Part A: Molecular and Integrative Physiology*, vol. 127, no. 1, pp. 29–38, 2000.

[41] A. Asano, J. L. Nelson-Harrington, and A. J. Travis, "Phospholipase B is activated in response to sterol removal and stimulates acrosome exocytosis in murine sperm," *The Journal of Biological Chemistry*, vol. 288, no. 39, pp. 28104–28115, 2013.

[42] J. B. Aitken, N. Naumovski, B. Curry, C. G. Grupen, Z. Gibb, and R. J. Aitken, "Characterization of an L-amino acid oxidase in equine spermatozoa," *Biology of Reproduction*, vol. 92, no. 5, article 125, 2015.

[43] V. G. Da Ros, M. W. Muñoz, M. A. Battistone et al., "From the epididymis to the egg: participation of CRISP proteins in mammalian fertilization," *Asian Journal of Andrology*, vol. 17, no. 5, pp. 711–715, 2015.

[44] N. Emami, A. Scorilas, A. Soosaipillai, T. Earle, B. Mullen, and E. P. Diamandis, "Association between kallikrein-related peptidases (KLKs) and macroscopic indicators of semen analysis: their relation to sperm motility," *Biological Chemistry*, vol. 390, no. 9, pp. 921–929, 2009.

[45] S. D. Carson and C. J. De Jonge, "Activation of coagulation factor X in human semen," *Journal of Andrology*, vol. 19, no. 3, pp. 289–294, 1998.

[46] Y.-T. Tseng, J.-Y. Hsia, C.-Y. Chen, N.-T. Lin, P. C.-S. Chong, and C.-Y. Yang, "Expression of the sperm fibrous sheath protein CABYR in human cancers and identification of α-enolase as an interacting partner of CABYR-a," *Oncology Reports*, vol. 25, no. 4, pp. 1169–1175, 2011.

[47] A. Geadkaew, N. Kosa, S. Siricoon, S. V. Grams, and R. Grams, "A 170 kDa multi-domain cystatin of *Fasciola gigantica* is active in the male reproductive system," *Molecular and Biochemical Parasitology*, vol. 196, no. 2, pp. 100–107, 2014.

[48] S. Asuvapongpatana, A. Saewu, C. Chotwiwatthanakun, R. Vanichviriyakit, and W. Weerachatyanukul, "Localization of cathepsin D in mouse reproductive tissues and its acquisition onto sperm surface during epididymal sperm maturation," *Acta Histochemica*, vol. 115, no. 5, pp. 425–433, 2013.

[49] D. Selvi Rani, A. Vanniarajan, N. J. Gupta, B. Chakravarty, L. Singh, and K. Thangaraj, "A novel missense mutation C11994T in the mitochondrial ND4 gene as a cause of low sperm motility in the Indian subcontinent," *Fertility and Sterility*, vol. 86, no. 6, pp. 1783–1785, 2006.

[50] R. Zhou, R. Wang, Y. Qin et al., "Mitochondria-related miR-151a-5p reduces cellular ATP production by targeting CYTB in asthenozoospermia," *Scientific Reports*, vol. 5, Article ID 17743, 2015.

[51] G.-H. Mao, Y.-N. Wang, M. Xu, W.-L. Wang, L. Tan, and S.-B. Tao, "Polymorphisms in the MT-ATP6 and MT-CYB genes in

in vitro fertilization failure," *Mitochondrial DNA*, vol. 26, no. 1, pp. 20–24, 2015.

[52] D. W. Huang, B. T. Sherman, Q. Tan et al., "DAVID bioinformatics resources: expanded annotation database and novel algorithms to better extract biology from large gene lists," *Nucleic Acids Research*, vol. 35, no. 2, pp. W169–W175, 2007.

[53] G. Dennis Jr., B. T. Sherman, D. A. Hosack et al., "DAVID: database for annotation, visualization, and integrated discovery," *Genome Biology*, vol. 4, no. 5, article P3, 2003.

[54] M. Kanehisa, S. Goto, S. Kawashima, Y. Okuno, and M. Hattori, "The KEGG resource for deciphering the genome," *Nucleic Acids Research*, vol. 32, pp. D277–D280, 2004.

[55] B. T. Sherman, D. W. Huang, Q. Tan et al., "DAVID knowledgebase: a gene-centered database integrating heterogeneous gene annotation resources to facilitate high-throughput gene functional analysis," *BMC Bioinformatics*, vol. 8, article 426, 2007.

[56] X. Jiao, B. T. Sherman, D. W. Huang et al., "DAVID-WS: a stateful web service to facilitate gene/protein list analysis," *Bioinformatics*, vol. 28, no. 13, pp. 1805–1806, 2012.

[57] D. J. E. Housley, Z. A. Zalewski, S. E. Beckett, and P. J. Venta, "Design factors that influence PCR amplification success of cross-species primers among 1147 mammalian primer pairs," *BMC Genomics*, vol. 7, article 253, 2006.

[58] F. Montiel-Sosa, E. Ruiz-Pesini, J. A. Enríquez et al., "Differences of sperm motility in mitochondrial DNA haplogroup U sublineages," *Gene*, vol. 368, no. 1-2, pp. 21–27, 2006.

[59] C.-Q. Feng, Y.-B. Song, Y.-G. Zou, and X.-M. Mao, "Mutation of MTCYB and MTATP6 is associated with asthenospermia," *Zhonghua Nan Ke Xue*, vol. 14, no. 4, pp. 321–323, 2008.

[60] X. Chen, Y. Yue, Y. He et al., "Identification and characterization of genes differentially expressed in X and Y sperm using suppression subtractive hybridization and cDNA microarray," *Molecular Reproduction and Development*, vol. 81, no. 10, pp. 908–917, 2014.

[61] E. Giacomini, B. Ura, E. Giolo et al., "Comparative analysis of the seminal plasma proteomes of oligoasthenozoospermic and normozoospermic men," *Reproductive BioMedicine Online*, vol. 30, no. 5, pp. 522–531, 2015.

[62] J. P. A. Rego, J. M. Crisp, A. A. Moura et al., "Seminal plasma proteome of electroejaculated Bos indicus bulls," *Animal Reproduction Science*, vol. 148, no. 1-2, pp. 1–17, 2014.

[63] A. Bernardini, F. Hozbor, E. Sanchez, M. W. Fornés, R. H. Alberio, and A. Cesari, "Conserved ram seminal plasma proteins bind to the sperm membrane and repair cryopreservation damage," *Theriogenology*, vol. 76, no. 3, pp. 436–447, 2011.

[64] A. Saewu, S. Asuvapongpatana, C. Chotwiwatthanakun, A. Tantiwongse, W. Weerachatyanukul, and S. Thitilertdecha, "Cathepsin D in human reproductive tissues: cellular localization in testis and epididymis and surface distribution in different sperm conditions," *Journal of Andrology*, vol. 33, no. 4, pp. 726–734, 2012.

[65] M. A. Dietrich, G. J. Arnold, J. Nynca, T. Fröhlich, K. Otte, and A. Ciereszko, "Characterization of carp seminal plasma proteome in relation to blood plasma," *Journal of Proteomics*, vol. 98, pp. 218–232, 2014.

[66] F. De Smet, Y. Moreau, K. Engelen, D. Timmerman, I. Vergote, and B. De Moor, "Balancing false positives and false negatives for the detection of differential expression in malignancies," *British Journal of Cancer*, vol. 91, no. 6, pp. 1160–1165, 2004.

[67] J. Wu, N. I. Lenchik, and I. C. Gerling, "Approaches to reduce false positives and false negatives in the analysis of microarray data: applications in type 1 diabetes research," *BMC Genomics*, vol. 9, supplement 2, article S12, 2008.

3

SOD1 Gene +35A/C (exon3/intron3) Polymorphism in Type 2 Diabetes Mellitus among South Indian Population

K. Nithya,[1] T. Angeline,[1] W. Isabel,[1] and A. J. Asirvatham[2]

[1]*PG & Research Department of Zoology & Biotechnology, Lady Doak College, Madurai, Tamil Nadu 625 002, India*
[2]*Arthur Asirvatham Hospital, Madurai, Tamil Nadu 625 020, India*

Correspondence should be addressed to T. Angeline; angelanand@yahoo.co.in

Academic Editor: Norman A. Doggett

Superoxide dismutase is an antioxidant enzyme that is involved in defence mechanisms against oxidative stress. Cu/Zn SOD is a variant that is located in exon3/intron3 boundary. The aim of the present study was to investigate whether the Cu/Zn SOD (+35A/C) gene polymorphism is associated with the susceptibility to type 2 diabetes mellitus among south Indian population. The study included patients with type 2 diabetes mellitus ($n = 100$) and healthy controls ($n = 75$). DNA was isolated from the blood and genotyping of Cu/Zn SOD gene polymorphism was done by polymerase chain reaction based restriction fragment length polymorphism method. Occurrence of different genotypes and normal (A) and mutant (C) allele frequencies were determined. The frequency of the three genotypes of the total subjects was as follows: homozygous wild-type A/A (95%), heterozygous genotype A/C (3%), and homozygous mutant C/C (2%). The mutant (C) allele and the mutant genotypes (AC/CC) were found to be completely absent among the patients with type 2 diabetes mellitus. Absence of mutant genotype (CC) shows that the Cu/Zn SOD gene polymorphism may not be associated with the susceptibility to type 2 diabetes mellitus among south Indian population.

1. Introduction

Diabetes mellitus is a polygenic metabolic disease characterized by hyperglycemia resulting from defects in both insulin secretion and its action [1]. Both the genetic and the environmental factors are involved in the development of type 2 diabetes mellitus [2]. Oxidative stress and oxidative damage to the tissue are common end points of chronic diseases, such as atherosclerosis, diabetes, and rheumatoid arthritis [3]. It has been demonstrated that oxidative stress in diabetes can be accelerated not only due to increased production of reactive oxygen species (ROS) caused by hyperglycaemia but also by reduced ability of antioxidant defense system caused at least partly by polymorphisms of scavenger enzymes like superoxide dismutase, catalase, and reduced glutathione [4]. Superoxide dismutase (SOD) is considered as a primary enzyme since it is involved in the direct elimination of reactive oxygen species [5]. In mammals, there are 3 isoforms of SOD, copper-zinc SOD (SOD1), manganese SOD (SOD2), and an extracellular form of CuZn-SOD (EC-SOD or SOD3). Each superoxide dismutase is the product of distinct genes but catalyzes the same reaction [6]:

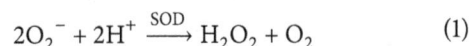

$$2O_2^- + 2H^+ \xrightarrow{\text{SOD}} H_2O_2 + O_2 \qquad (1)$$

Among the 3 isoforms, experimental evidences from mice indicate that activity of CuZn-SOD accounts for 50% to 80% of total SOD activity. Mn-SOD accounts for approximately 2% to 12% of total vascular SOD, and EC-SOD accounts for the remainder [7]. SOD1 [EC1.15.1.1] has a molecular mass of about 32,000 Da and has been found in the cytoplasm, nuclear compartments, and lysosomes of mammalian cells [8]. The SOD1 gene has been extensively studied and DNA sequence changes in the gene can be detected in about 0.3% of humans worldwide [9]. It has been hypothesized that mutations in the SOD1 gene may impair antioxidant enzyme activity thereby leading to accumulation of toxic superoxide anions [10].

SOD1 gene has five exons and the +35A/C polymorphism (rs2234694) is adjacent to the splicing point (exon3/intron3), being related to macrovascular complications in diabetes and having no association with microvascular complications [4]. While data regarding the SOD1 (+35A/C) gene polymorphism is available for other populations, including North Indians, the Bangladeshi, the Finns, the Romanians, the New Zealanders, and the Czechs [4, 11–15], it is lacking among South Indian population. Therefore, the present study has been conducted to determine the prevalence of +35A/C polymorphism in intron3/exon3 boundary of the SOD1 gene among South Indian population. In addition, the genetic association between polymorphism in the (+35A/C) SOD1 gene and type 2 diabetes mellitus has also been studied.

2. Materials and Methods

2.1. Study Design. The study population consisted of seventy-five controls with no history of cardiovascular disease, diabetes, hypertension, cancer, or any infectious diseases and one hundred patients with type 2 diabetes mellitus, belonging to South Indian population. The patients included were characterized based on the alteration in blood sugar level (>126 mg/dL). The samples from the subjects were collected into EDTA coated tubes and the informed consent was obtained. Clinical data including information on duration of diabetes, presence of any complication, history of other disorders, age, gender, lipid profile, blood sugar level, and systolic and diastolic blood pressure were collected using a questionnaire. Ethical clearance was obtained for this study.

2.1.1. DNA Extraction. Genomic DNA was extracted from the frozen blood by phenol-chloroform method [16]. For DNA extraction, $500 \mu L$ of the blood was used and the isolated DNA dissolved in TE was stored at $-20°C$. The quality of the DNA was checked in 0.7 percent agarose (Hi-Media, Mumbai) gel electrophoresis and quantified using UV spectrophotometry (Hitachi, Japan).

2.1.2. PCR Analysis of SOD1 (+35A/C) Gene [17]. PCR analysis was carried out using a thermal cycler (Eppendorf Mastercycler, Germany). Approximately 120 ng of genomic DNA was incubated in a total reaction mixture of $20 \mu L$ containing both the forward primer 5′ CTATCCAGAAAACACG-GTGGGCC 3′ and the reverse primer 5′ TCTATATTCAAT-CAAATGCTACAAAACC 3′ (~10 picomoles) (GenScript Corp., USA), $200 \mu M$ deoxynucleotide triphosphate, 10x PCR buffer pH-8.3 containing $MgCl_2$ 15 mM, and 5 units of *Taq* DNA polymerase (New England Biolabs, Beverly). DNA was initially denatured at $95°C$ for 8 min prior to amplification. The PCR amplification conditions were as follows: 30 cycles consisting of 1 min denaturation at $94°C$, 50 sec annealing at $64°C$, and 1 min extension at $72°C$. The final extension included 7 mins at $72°C$. The PCR product (278 bp) was confirmed by 2% agarose (Hi-Media, Mumbai) gel electrophoresis. The amplified product was used for further restriction fragment analysis.

2.1.3. Restriction Enzyme Analysis [17]. Restriction digestion was performed in a total volume of $20 \mu L$ consisting of $10 \mu L$

TABLE 1: Clinical data of the study subjects.

Parameters	Type 2 diabetic patients ($n = 100$)	Controls ($n = 75$)
Age (mean ± SD)	49.67 ± 10.24	35.94 ± 11.89
Sex (M/F)	50/50	32/43
Hypercholesterolemia	47	—
Family history	45	—
Hypertension	18	—
Other complications	15	—

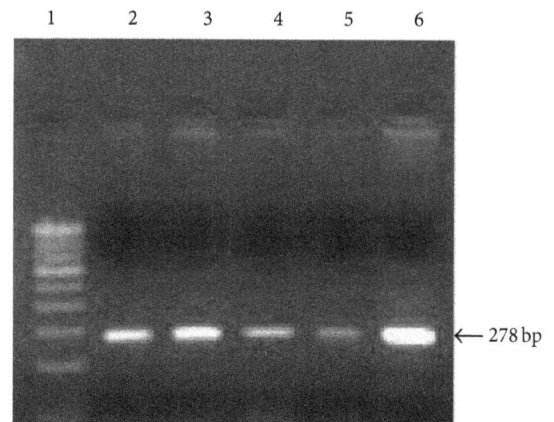

FIGURE 1: PCR analysis of SOD1 gene. Lane 1: 100 bp DNA ladder. Lanes 2–6: 278 bp PCR amplicon.

amplicon, $4 \mu L$ NE buffer (50 mM potassium acetate, 20 mM Tris-acetate, 10 mM magnesium acetate, and 1 mM dithiothreitol pH-7.9 at $25°C$), and 10 U of *HhaI* enzyme (Fermentas Life Sciences, Germany). Samples were incubated for 6-7 hrs at $37°C$ and the digested PCR products were resolved in 2 percent agarose gel electrophoresis stained with ethidium bromide and separated bands were observed using gel documentation system. PCR-RFLP analysis revealed the existence of the three genotypes (AA, AC, and CC) of +35A/C SOD1 gene polymorphism.

2.1.4. Sequencing of the PCR Amplified Product. The PCR product obtained was subjected to DNA sequencing which was carried out by Sanger's sequencing method (Synergy Scientific Services, Chennai) [18]. The DNA sequencing was done to check and confirm whether the amplified product was SOD1 gene sequence. The obtained sequence was then subjected to BLASTN analysis to study the homology sequence of the amplified product.

3. Results and Discussion

The clinical data including the risk factors of study subjects are shown in Table 1. A 278 bp fragment that resides in the SOD1 gene was amplified (Figure 1) and sequenced. When

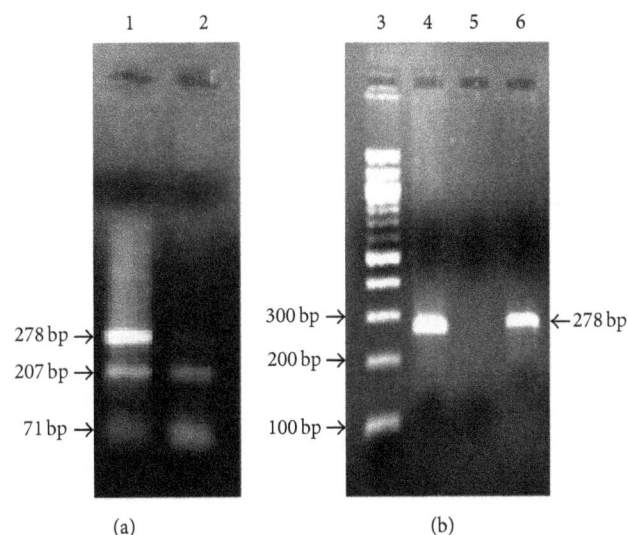

FIGURE 2: Restriction fragment length analysis of SOD1 gene +35A/C (exon3/intron3) polymorphism. Agarose gels (a) and (b) showing heterozygous (AC) genotype (Lane 1), homozygous mutant "CC" genotype (Lane 2), DNA marker-100 bp ladder (Lane 3), and homozygous wild-type "AA" genotype (Lanes 4 and 6).

TABLE 2: SOD1 gene polymorphism: genotype and allele frequency in patients and controls.

Genotype/allele	Patients ($n = 100$)	Controls ($n = 75$)
Genotypes (+35A/C)		
AA	100	65
AC	—	6
CC	—	4
Alleles		
A	1.0	0.91
C	—	0.09

BLASTN analysis was performed with the DNA sequence of PCR product (278 bp), 100 percent homology was found between the SOD1 gene and the submitted DNA sequence. After digesting the PCR amplicons with *HhaI*, the following fragment sizing patterns were observed by the gel electrophoresis: the SOD1 +35A/A genotype resulted in no cleavage of the amplified 278 bp fragments and the A/C genotype resulted in 3 fragments of 278, 207, and 71 bp. C/C genotype resulted in complete cleavage of the 278 bp fragment into 207 and 71 bp (Figure 2).

Genotype and allelic frequencies for the +35A/C SNP of the SOD1 gene in diabetic patients and nondiabetic healthy individuals were calculated. The results indicated that the occurrence of homozygous mutant (CC) and heterozygous mutant (AC) was completely absent in type 2 diabetes mellitus patients whereas such condition was observed in 4 (5%) and 6 (8%) of the controls (Table 2). As the two genotypes

were completely absent in type 2 diabetes mellitus patients, Hardy-Weinberg equilibrium was not tested. The A allele frequency was found to be 1.0 for diabetic patients and 0.91 for controls. The C allele frequency was completely absent in patients when compared to controls (0.09), which indicates that the Cu/Zn SOD gene polymorphism may not be associated with the susceptibility to type 2 diabetes mellitus among South Indian population.

Similar result was observed in a study conducted among Finnish population that there is no difference in the genotype distributions and allele frequencies of the CuZn-SOD gene polymorphism between type 2 diabetes mellitus patients and controls [13]. Another study has also reported that the (CC) genotype and C allele were completely absent among North Indian population and that there is no association between diabetes and SOD1 +35A/C gene polymorphism [11]. However, contradictory results indicating that there is an association between diabetes and SOD1 +35A/C gene polymorphism were obtained in a study conducted among Bangladeshi population [12]. Similarly, Panduru et al. provided evidences that there is an association of SOD1 +35A/C gene polymorphism with diabetic nephropathy in Romanian population [14]. When the prevalence of the SOD1 +35A/C gene polymorphism among South Indian Tamil population was determined and the association of the mutant allele with type 2 diabetes mellitus was evaluated, no association was found between SOD1 +35A/C gene polymorphism and type 2 diabetes mellitus.

As allele frequencies are population specific, there is variation in allelic frequencies in different population (Table 3). The highest mutant allele frequency of SOD1 +35A/C gene polymorphism was observed among Czechs (0.4) and the lowest (0.018) was observed among South Indian population.

Both genetic and biochemical characterization of SOD1 [19] demonstrates that SOD1 gains importance in the development of diseases such as heart failure [20], cancer [21], diabetes [4], Down's syndrome [22], and amyotrophic lateral sclerosis [23]. It has been reported that the SOD's involvement in the pathogenesis of diseases is due to altered SOD activity and ROS concentration [24]. It was also found that the SOD1 gene polymorphism influences SOD activity [25]. Decreased activity of the antioxidant enzymes and depletion of total antioxidant capacity may increase the susceptibility of diabetic patients to oxidative injury [26]. Similarly, studies conducted by Sayed et al. and Fujita et al. have reported that the enzymatic activity of SOD was significantly decreased in diabetic patients with retinopathy [27] and nephropathy [28]. Clinical studies conducted in different populations have shown a decrease in SOD activity in type 2 diabetic patients when compared to controls [29, 30].

4. Conclusions

As the mutant genotype and allele frequency were completely absent in type 2 diabetic patients among South Indian population, it was concluded that the CuZn-SOD (+35A/C) gene polymorphism may not be associated with the susceptibility to type 2 diabetes mellitus. However, this lack of association

TABLE 3: Prevalence of the SOD1+35A/C gene polymorphism in different populations.

Country	Ethnic population	Number of subjects	Disease risk	Mutant allele frequency	References
India	South Indian	P-100 C-75	T2DM	P-0 C-0.09	(Present study)
India	North Indian	P-207 C-210	T2DM	P-0 C-0	[11]
Bangladesh	Bangladeshi	P-109 C-144	T2DM	P-0.018 C-0.014	[12]
Finland	Finns	P-239 C-245	T2DM	P-0.09 C-0.10	[13]
Romania	Romanian	P-106 C-132	T1DM	P-0.06 C-0.01	[14]
New Zealand	European	P-230 C-210	COPD	P-0.05 C-0.06	[15]
Czech Republic	Czechs	P-306 C-140	T2DM	P-0.42 C-0.48	[4]

P: patients; C: controls.

might be due to smaller sample size and ethnic variation. Further study is needed to investigate the other SNPs (Mn-SOD and EC-SOD) in SOD gene to elicit the potential role of antioxidant defense and susceptibility in type 2 diabetes mellitus.

Acknowledgment

The authors thank the Science and Engineering Research Board (SERB), New Delhi, India, for funding to Dr. T. Angeline.

References

[1] J. R. Gavin, K. G. M. M. Alberti, M. B. Davidson et al., "Report of the expert committee on the diagnosis and classification of diabetes mellitus," *Diabetes Care*, vol. 26, no. 1, pp. 1183–1197, 2003.

[2] D. Giugliano, A. Ceriello, and K. Esposito, "Glucose metabolism and hyperglycemia," *The American Journal of Clinical Nutrition*, vol. 87, no. 1, pp. 217–222, 2008.

[3] J. W. Baynes and S. R. Thorpe, "Role of oxidative stress in diabetic complications: a new perspective on an old paradigm," *Diabetes*, vol. 48, no. 1, pp. 1–9, 1999.

[4] M. Flekac, J. Skrha, J. Hilgertova, Z. Lacinova, and M. Jarolimkova, "Gene polymorphisms of superoxide dismutases and catalase in diabetes mellitus," *BMC Medical Genetics*, vol. 9, article 30, 2008.

[5] J. M. McCord, B. B. Keele Jr., and I. Fridovich, "An enzyme-based theory of obligate anaerobiosis: the physiological function of superoxide dismutase," *Proceedings of the National Academy of Sciences of the United States of America*, vol. 68, no. 5, pp. 1024–1027, 1971.

[6] F. M. Faraci and S. P. Didion, "Vascular protection: superoxide dismutase isoforms in the vessel wall," *Arteriosclerosis, Throm-bosis, and Vascular Biology*, vol. 24, no. 8, pp. 1367–1373, 2004.

[7] P. Stralin, K. Karlsson, B. O. Johansson, and S. L. Marklund, "The interstitium of the human arterial wall contains very large amounts of extracellular superoxide dismutase," *Arteriosclerosis, Thrombosis, and Vascular Biology*, vol. 15, no. 11, pp. 2032–2036, 1995.

[8] W. Liou, L.-Y. Chang, H. J. Geuze, G. J. Strous, J. D. Crapo, and J. W. Slot, "Distribution of CuZn superoxide dismutase in rat liver," *Free Radical Biology and Medicine*, vol. 14, no. 2, pp. 201–207, 1993.

[9] P. M. Andersen, K. B. Sims, W. W. Xin et al., "Sixteen novel mutations in the Cu/Zn superoxide dismutase gene in amyotrophic lateral sclerosis: a decade of discoveries, defects and disputes," *Amyotrophic Lateral Sclerosis and Other Motor Neuron Disorders*, vol. 4, no. 2, pp. 62–73, 2003.

[10] M. E. Gurney, H. Pu, A. Y. Chiu et al., "Motor neuron degeneration in mice that express a human Cu,Zn superoxide dismutase mutation," *Science*, vol. 264, no. 5166, pp. 1772–1774, 1994.

[11] P. Vats, N. Sagar, T. P. Singh, and M. Banerjee, "Association of *Superoxide dismutases* (SOD1 and SOD2) and *Glutathione peroxidase* 1 (GPx1) gene polymorphisms with Type 2 diabetes mellitus," *Free Radical Research*, vol. 49, no. 1, pp. 17–24, 2015.

[12] L. A. Akhy, P. Deb, M. Das, L. Ali, M. Omar Faruque, and Z. Hassan, "Superoxide dismutase 1 gene +35A>C (intron3/exon3) polymorphism in diabetic nephropathy patients among Bangladeshi population," *Journal of Molecular Pathophysiology*, vol. 3, no. 4, pp. 52–57, 2014.

[13] O. Ukkola, P. H. Erkkilä, M. J. Savolainen, and Y. A. Kesäniemi, "Lack of association between polymorphisms of catalase, copper-zinc superoxide dismutase (SOD), extracellular SOD and endothelial nitric oxide synthase genes and macroangiopathy in patients with type 2 diabetes mellitus," *Journal of Internal Medicine*, vol. 249, no. 5, pp. 451–459, 2001.

[14] N. M. Panduru, E. Moţa, M. Moţa, D. Cimponeriu, C. Serafinceanu, and D. M. Cheţa, "Polymorphism of catalase gene promoter in Romanian patients with diabetic kidney disease

and type 1 diabetes," *Romanian Journal of Internal Medicine*, vol. 48, no. 1, pp. 81–88, 2010.

[15] R. P. Young, R. Hopkins, P. N. Black et al., "Functional variants of antioxidant genes in smokers with COPD and in those with normal lung function," *Thorax*, vol. 61, no. 5, pp. 394–399, 2006.

[16] V. M. Iranpur and A. K. Esmailizadeh, *Rapid Extraction of High Quality DNA from Whole Blood Stored at 4°C for Long Period*, Department of Animal Science, Faculty of Agriculture, Shahrekord University, Shahrekord, Iran, 2010.

[17] V. L. Kinnula, S. Lehtonen, P. Koistinen et al., "Two functional variants of the superoxide dismutase genes in Finnish families with asthma," *Thorax*, vol. 59, no. 2, pp. 116–119, 2004.

[18] F. Sanger and A. R. Coulson, "A rapid method for determining sequences in DNA by primed synthesis with DNA polymerase," *Journal of Molecular Biology*, vol. 94, no. 3, pp. 441–448, 1975.

[19] E. C. Chang, B. F. Crawford, Z. Hong, T. Bilinski, and D. J. Kosman, "Genetic and biochemical characterization of Cu,Zn superoxide dismutase mutants in *Saccharomyces cerevisiae*," *Journal of Biological Chemistry*, vol. 266, no. 7, pp. 4417–4424, 1991.

[20] F. M. F. Alameddine and A. M. Zafari, "Genetic polymorphisms and oxidative stress in heart failure," *Congestive Heart Failure*, vol. 8, no. 3, pp. 157–172, 2002.

[21] L. W. Oberley and G. R. Buettner, "Role of superoxide dismutase in cancer: a review," *Cancer Research*, vol. 39, no. 4, pp. 1141–1149, 1979.

[22] R. De la Torre, A. Casado, E. López-Fernández, D. Carrascosa, V. Ramírez, and J. Sáez, "Overexpression of copper-zinc super-oxide dismutase in trisomy 21," *Experientia*, vol. 52, no. 9, pp. 871–873, 1996.

[23] D. R. Rosen, T. Siddique, D. Patterson et al., "Mutations in Cu/Zn superoxide dismutase gene are associated with familial amy-otrophic lateral sclerosis," *Nature*, vol. 362, no. 6415, pp. 59–62, 1993.

[24] I. N. Zelko, T. J. Mariani, and R. J. Folz, "Superoxide dismutase multigene family: a comparison of the CuZn-SOD (SOD1), Mn-SOD (SOD2), and EC-SOD (SOD3) gene structures, evolution, and expression," *Free Radical Biology and Medicine*, vol. 33, no. 3, pp. 337–349, 2002.

[25] M. H. Ghattas and D. M. Abo-Elmatty, "Association of polymor-phic markers of the catalase and superoxide dismutase genes with type 2 diabetes mellitus," *DNA and Cell Biology*, vol. 31, no. 11, pp. 1598–1603, 2012.

[26] M. E. Rahbani-Nobar, A. Rahimi-Pour, M. Rahbani-Nobar, F. Adi-Beig, and S. M. Mirhashemi, "Total antioxidant capacity, superoxide dismutase and glutathion peroxidase in diabetic patients," *Medical Journal of Islamic Academy of Sciences*, vol. 12, no. 4, pp. 109–114, 1999.

[27] A. A. Sayed, Y. Aldebasi, S. O. Abd-Allah, S. M. El Gendy, A. S. Mohamed, and M. S. Abd El-Fattah, "Molecular and bio-chemical study of superoxide dismutase gene polymorphisms in egyptian patients with type 2 diabetes mellitus with and without retinopathy," *British Journal of Medicine and Medical Research*, vol. 3, no. 4, pp. 1258–1270, 2013.

[28] H. Fujita, H. Fujishima, S. Chida et al., "Reduction of renal superoxide dismutase in progressive diabetic nephropathy," *Journal of the American Society of Nephrology*, vol. 20, no. 6, pp. 1303–1313, 2009.

[29] K. Yamashita, K. Takahiro, F. Kamezaki, T. Adachi, and H. Tasaki, "Decreased plasma extracellular superoxide dismutase level in patients with vasospastic angina," *Atherosclerosis*, vol. 191, no. 1, pp. 147–152, 2007.

[30] M. Liao, Z. Liu, J. Bao et al., "A proteomic study of the aortic media in human thoracic aortic dissection: implication for oxidative stress," *Journal of Thoracic and Cardiovascular Surgery*, vol. 136, no. 1, pp. 65–72, 2008.

4

Genetic Variants in CSMD1 Gene are Associated with Cognitive Performance in Normal Elderly Population

Vadim Stepanov,[1,2] **Andrey Marusin,**[1] **Kseniya Vagaitseva,**[1,2] **Anna Bocharova,**[1] **and Oksana Makeeva**[1,3]

[1]*Institute of Medical Genetics, Tomsk National Medical Research Center, Tomsk, Russia*
[2]*Tomsk State University, Tomsk, Russia*
[3]*Nebbiolo Center for Clinical Trials, Tomsk, Russia*

Correspondence should be addressed to Vadim Stepanov; vadim.stepanov@medgenetics.ru

Academic Editor: Chao Zhao

Recently, genetic markers rs10503253 and rs2616984 in the CUB and Sushi multiple domains-1 (CSMD1) gene have been reported to be associated with schizophrenia and cognitive functions in genome-wide association studies. We examined the associations of the above SNPs with cognitive performance evaluated by the Montreal Cognitive Assessment (MoCA) tool in a cohort of the normal elderly from the Russian population. Significant association of rs2616984 genotypes with the MoCA scores was found using nonparametric analysis. No association of rs10503253 with MoCA scores was observed using both parametric and nonparametric statistics. Significant combined effect of two-locus CSMD1 genotypes on MoCA scores was demonstrated by median test. Allele "A" and genotype "AA" of rs2616984 were significantly associated with the lower MoCA scores in comparison of 1st and 4th quartiles of MoCA total score distribution. The results suggest that genetic variants in CSMD1 gene are likely a part of genetic component of cognitive performance in the elderly.

1. Introduction

The single-nucleotide polymorphisms (SNPs) in the CUB and Sushi multiple domains-1 (CSMD1) gene were recently identified in genome-wide association studies (GWAS) as significant genetic markers for schizophrenia (SZ) and cognitive performance. CSMD1 gene spanning over 2.6 Mb on chromosome 8p23.2 is highly expressed in the central nervous system and epithelial tissues [1] and encodes an important cell adhesion molecule involved in the development, connection, and plasticity of brain circuits. Despite the fact that the exact role of CSMD1 in neurodevelopmental process is not clear, murine models indicate that CSMD1 knockout induces behaviors reminiscent of blunted emotional responses, anxiety, and depression, suggesting an influence of the CSMD1 on psychopathology and endophenotypes of the negative symptom spectra [2]. A common intronic CSMD1 variant, rs10503253, was reported as genome-wide significant for SZ by Schizophrenia Psychiatric Genome-Wide Association Study Consortium [3] and was subsequently replicated in other GWA and meta-analysis studies [4–6]. A minor allele "A" of rs10503253 is associated with deleterious effects across a number of neurocognitive phenotypes, such as poorer performance on neuropsychological measures of general cognitive ability and memory function in SZ patients [7], and affects general cognitive ability and executive function in healthy individuals [8]. Another intronic variant in CSMD1, rs2616984, located 302 kb away, was found as genome-wide significant for the performance on standardized cognitive tests [9]. Recently we have demonstrated the association of the latter genetic variant with both SZ and Alzheimer's disease in a Russian population [10, 11]. These data suggested that neuropsychological effects of CSMD1 and the plausible role of its genetic variation in schizophrenia and other neuropsychiatric diseases are based on common underlying neurological mechanism developed via cognitive endophenotypes. In this study, we investigated the role of common genetic variation in CSMD1 gene in cognitive performance in normal elderly population.

TABLE 1: Demographic and health characteristics of the sample.

Characteristics	Mean ± SD or % where indicated
Age, years	70.9 ± 5.7
Gender, female	74%
Education	
Years of education	13.3 ± 3.1
Min education, years	4
Max education, years	20
Maximum achieved level of education	
Less than high school graduate	11%
High school graduate	8%
Some college or associate's degree	32%
Bachelor's degree	42%
Master's or higher professional degree	
Doctoral degree	7%
Memory testing history	7%
Clinical reasons	3%
Research participant	3%
Health characteristics	
Coronary artery disease	44%
Atrial fibrillation	24%
Stroke	7%
Congestive heart failure	17%
High blood pressure	81%
Obesity	26%
Diabetes, type 2	18%
Smoking	
Smoking during last 30 days	9%
Smoking more than 100 cigarettes over life	20%

2. Material and Methods

708 elderly individuals of European Russian descent without dementia and neurological diseases were recruited from a population-based cohort study on primary prevention of Alzheimer's disease in Tomsk, Russia [12, 13]. The sample consisted of 74% of females (see Table 1). Mean age in the sample was 70.9 (varying from 60 to 89) years. Level of education of the study participants was assessed with two measures: number of years of education and achieved level of education. Level of education was relatively high with 42% of study participants having bachelor's or master's degree and 7% having doctoral degree. Only 7% of the sample had been exposed to memory testing previously with clinical or research purpose reasons.

Health characteristics are presented in Table 1. Health conditions were self-reported with the notice from investigator to report only on diagnoses established by medical doctors and confirmed in medical records. The sample had high prevalence of cardiovascular conditions with high blood pressure indicated in 81% of study subjects.

In each participant, the cognitive performance was assessed using the Montreal Cognitive Assessment (MoCA). MoCA measures 8 cognitive domains: memory, attention, naming, visuospatial/executive, language, abstraction, delayed recall, and orientation domains. MoCA scores ranged between 0 and 30 points, and higher scores indicate better cognitive function.

Genotyping of 2 genetic variants in CSMD1 gene, rs10503253 and rs2616984, was performed by multiplex PCR with the following iPLEX primer extension reaction and detection of allele-specific extension products by matrix-assisted laser desorption/ionization time-of-flight (MALDI-TOF) mass spectrometry on Sequenom MassARRAY 4 platform. Details of the genotyping method have been previously described elsewhere [14].

Linkage disequilibrium (LD) between genetic variants and haplotype frequencies were estimated using Haploview 4.2 software. Statistical analysis of relationships between genetic markers and cognitive performance was performed in Statistica 7.0 (StatSoft Inc.) package using parametric (analysis of variance, ANOVA) and nonparametric (Kruskal-Wallis test and median test) statistics. For ANOVA analysis, MoCA scores were adjusted for age and education using linear regression model. Differences in allele and genotype frequencies between quartiles of MoCA distribution were estimated by chi-square test.

3. Results

Allele and haplotype frequencies of rs1050325 and rs2616984 in the total sample of 708 elderly subjects, as well as in the first (MoCa < 21) and fourth (MoCA > 24) quartiles of MoCA scores distribution, are presented in Table 2. Genotype distribution for both genetic variants in CSMD1 gene corresponded to Hardy-Weinberg equilibrium. No difference in the allele frequency between men and women was found. Minor alleles of both SNPs demonstrated very similar frequency in the total sample: 0.267 for "A" allele of rs10503253 and 0.272 for "G" allele of rs2616984. Two intronic variants in CSMD1 gene, located 302 kb apart, show very low level of linkage disequilibrium ($D' = 0.107$; LOD = 0.24).

Significantly higher frequency of the major allele of rs2616984 ("A") was observed in a subsample of individuals with MoCA score less than 21 compared to fourth quartile's subsample (0.754 versus 0.674; chi-square = 6.008; $p = 0.0142$), while no significant differences between lower and upper quartiles were found for rs10503253.

Odds ratio values for lower MoCA score associated with allele "A" and genotype "AA" of rs2616984 were 1.49 (95% CI 1.07–2.07, $p = 0.014$) and 1.70 (95% CI 1.11–2.61, $p = 0.009$), respectively. Genotype "AG" of rs2616984 and haplotype "CG" of rs105032/rs2616984 were associated with higher MoCA score (Table 3).

One-way ANOVA demonstrates no significant differences in the mean MoCA scores among genotypes of both genetic variants in CSMD1 gene, as well as among combinations of genotypes of two SNPs (Table 4). However, the effect of rs2616984 on MoCA scores was close to statistically significant ($F = 2.814$, $p = 0.060$).

TABLE 2: Allele and haplotype frequency of two CSMD1 genetic variants in the total sample and in the lower (Q1) and upper (Q4) quartiles of the MoCA distribution.

Allele, haplotype	Total sample, $N = 708$	Lower quartile (MoCA $\leq$ 20), $N = 188$	Upper quartile (MoCA $\geq$ 25), $N = 193$	Q1 versus Q4, p
		rs1050325		
C	0.733	0.741	0.749	0.799
A	0.267	0.259	0.251	
		rs2616984		
A	0.726	0.754	0.674	0.014
G	0.274	0.246	0.326	
		rs1050325/rs2616984 haplotypes		
CA	0.524	0.557	0.497	0.101
CG	0.209	0.184	0.251	0.024
AA	0.201	0.197	0.176	0.455
AG	0.065	0.062	0.075	0.478

TABLE 3: Odds ratio values for lower MoCA score for alleles, genotypes, and haplotypes of CSMD1 genetic variants in comparison of lower (Q1) and upper (Q4) quartiles of MoCA score distribution.

Allele, genotype, haplotype	OR	95% CI	Chi-square	p
		rs10503253		
A	1,04	0.74–1,46	0.06	0,799
C	0.96	0.68–1.35	0.06	0.799
AA	0.95	0.42–2.16	0.01	0.905
CA	1.13	0.73–1.76	0.34	0.562
CC	0.90	0.58–1.37	0.29	0.592
		rs2616984		
A	**1,49**	**1,07–2,07**	**6,01**	**0,014**
G	**0.67**	**0.48–0.94**	**6.01**	**0.0142**
AA	**1.70**	**1.11–2.61**	**6.64**	**0.0099**
AG	**0.65**	**0.42–1.00**	**4.25**	**0.039**
GG	0.70	0.32–1.51	0.96	0.326
		rs10503253/rs2616984 haplotypes		
CA	1,27	0,94–1,70	2,63	0,105
CG	0,67	0,47–0,97	4,97	0,026
AA	1,15	0,79–1,69	0,59	0,443
AG	0,81	0,44–1,47	0,55	0,457

Significant values are in bold.

In addition, less conservative nonparametric statistics indicate that genetic variation in rs2616984 locus significantly influenced the MoCA values in the normal elderly population. Under codominant model, p value for median test was 0.024 and Kruskal-Wallis p was 0.020. No significant association of rs10503253 with MoCA scores was observed using both parametric and nonparametric statistics. Significant combined effect of two-locus CSMD1 genotypes on MoCA scores was demonstrated by median test (chi-square = 16.19; df = 8; $p = 0.039$).

4. Discussion

Current data on contribution of CSMD1 genetic variation to neuropsychiatric diseases and to cognitive performance are quite controversial. More evidence is accumulated on rs10503253, which is genome-wide significant marker for SZ in Europeans according to initial GWAS and subsequent replicative studies (see above), but is not replicated for Japanese [15] and Han Chinese [16]. Association of this genetic variant with cognitive and memory functions initially reported for European SZ patients [7] was confirmed for normal Greek population, but was not replicated in Norwegian healthy cohort [17]. Another intronic variant in CSMD1, rs2616984, was associated with cognitive performance in European GWAS [9] and recently was found as susceptibility marker for both SZ and Alzheimer's disease in Russians [10, 11], but not in Central Asian Kazakh population [18]. Other CSMD1 SNPs were also reported to be associated with cognitive and memory functions in healthy Norwegians [17],

TABLE 4: One-way ANOVA analysis of MoCA scores among genotypes of genetic variants in CSMD1 gene.

Genotype	N	MoCA mean	MoCA std. deviation
rs1050325, $F = 0.105$, $p = 0.899$			
CA	247	22.08502	3.747744
AA	57	22.35088	3.319269
CC	370	22.14595	4.159463
All	674	22.14095	3.942305
rs2616984, $F = 2.814$, $p = 0.060$			
AA	361	21.81717	3.966503
AG	255	22.57255	3.928699
GG	57	22.31579	3.728069
All	673	22.14562	3.943373
rs1050325/rs2616984, $F = 0.860$, $p = 0.549$			
AA/AA	37	22.43243	3.484414
CA/AA	126	21.70635	3.655281
CC/AA	198	21.77273	4.239512
AA/AG	16	22.25000	3.193744
CA/AG	100	22.55000	3.957795
CC/AG	139	22.62590	4.005926
AA/GG	3	23.00000	2.645751
CA/GG	21	22.14286	3.119066
CC/GG	33	22.36364	4.211726
All	673	22.14562	3.943373

with cognitive decline in Alzheimer's disease in patients of European descent [19], as well as with clinical outcomes of SZ in Japanese [20].

Between-population variability in CSMD1 effects on neurocognitive phenotypes may reflect population-specific composition of genetic risk factors and/or population-specific LD patterns of associated markers with unknown functional variant(s) within CSMD1 or in adjacent genetic loci. But, nevertheless, growing amount of data on CSMD1 genetic association with various neuropsychological traits clearly indicates that structural variability at this part of the genome is likely a part of the common neurological mechanism, underlying various diseases and normal variability in neurocognitive traits. Genetic overlapping over these traits may be mediated by cognitive endophenotypes, such as variability in normal cognitive performance due to normal functional differences in brain structure and functions. This point finds support in several functional studies, demonstrating that that SZ risk variants show greater effects at the level of imaging based metrics of brain structure and function than at the level of behavior [21]. Particularly, CSMD1 rs10503253 "A" allele was associated with comparatively reduced cortical activation in the middle occipital gyrus and cuneus during performance of a spatial working memory task [22].

Along with population- and genetic background-related effects, age-related differences also may play a substantial role in the variability in genetic patterns of neurocognitive traits. Association of CSMD1 variants with Alzheimer's disease probably suggests that some part of genetic susceptibility at this locus may appear in the older age, compared to SZ susceptibility variants, but may be expressed in earlier preclinical stages as markers of cognitive and memory performance. However currently we cannot delineate possible molecular mechanisms underlying the age-related difference in CSMD1 effects on neurocognitive functions. The cohort tested in the present study is, to the best of our knowledge, the oldest sample ever used for CSMD1 genetic associations with neurocognitive phenotypes. We observed the association with MoCA performance for the genetic variant previously found to be associated with cognitive functions in younger people [9], as well as with late-onset Alzheimer's disease in our previous work [10]. Thus the findings reported in the current paper support the growing amount of evidence for significant role of CSMD1 in normal and pathological cognitive phenotypes.

5. Conclusions

Our study suggests that genetic variants in CSMD1 gene are likely a part of genetic component of cognitive performance in the elderly.

Acknowledgments

This work was funded by the Russian Scientific Foundation (Project no. 16-15-00020).

References

[1] D. M. Kraus, G. S. Elliott, H. Chute et al., "CSMD1 is a novel

multiple domain complement-regulatory protein highly expressed in the central nervous system and epithelial tissues," *The Journal of Immunology*, vol. 176, no. 7, pp. 4419–4430, 2006.

[2] V. M. Steen, C. Nepal, K. M. Ersland et al., "Neuropsychological deficits in mice depleted of the schizophrenia susceptibility gene CSMD1," *PLoS ONE*, vol. 8, no. 11, Article ID e79501, 2013.

[3] Schizophrenia Psychiatric Genome-Wide Association Study (GWAS) Consortium, "Genome-wide association study identifies five new schizophrenia loci," *Nature Medicine*, vol. 18, no. 43, pp. 969–976, 2011.

[4] Cross-Disorder Group of the Psychiatric Genomics Consortium, "Identification of risk loci with shared effects on five major psychiatric disorders: a genome-wide analysis," *The Lancet*, vol. 381, no. 9875, pp. 1371–1379, 2013.

[5] Schizophrenia Working Group of the Psychiatric Genomics Consortium, "Biological insights from 108 schizophrenia-associated genetic loci," *Nature*, vol. 511, pp. 421–427, 2014.

[6] F. S. Goes, J. Mcgrath, D. Avramopoulos et al., "Genome-wide association study of schizophrenia in Ashkenazi Jews," *American Journal of Medical Genetics Part B: Neuropsychiatric Genetics*, vol. 168, no. 8, pp. 649–659, 2015.

[7] G. Donohoe, J. Walters, A. Hargreaves et al., "Neuropsychological effects of the CSMD1 genome-wide associated schizophrenia risk variant rs10503253," *Genes, Brain and Behavior*, vol. 12, no. 2, pp. 203–209, 2013.

[8] E. Koiliari, P. Roussos, E. Pasparakis et al., "The CSMD1 genome-wide associated schizophrenia risk variant rs10503253 affects general cognitive ability and executive function in healthy males," *Schizophrenia Research*, vol. 154, no. 1-3, pp. 42–47, 2014.

[9] E. T. Cirulli, D. Kasperavičiūt, D. K. Attix et al., "Common genetic variation and performance on standardized cognitive tests," *European Journal of Human Genetics*, vol. 18, no. 7, pp. 815–820, 2010.

[10] V. A. Stepanov, A. V. Bocharova, A. V. Marusin, N. G. Zhukova, V. M. Alifirova, and I. A. Zhukova, "Replicative association analysis of genetic markers of cognitive traits with Alzheimer's disease in the Russian population," *Journal of Molecular Biology*, vol. 48, no. 6, pp. 835–844, 2014.

[11] A. V. Bocharova, V. A. Stepanov, A. V. Marusin et al., "Association study of genetic markers of schizophrenia and its cognitive endophenotypes," *Russian Journal of Genetics*, vol. 53, no. 1, pp. 139–146, 2017.

[12] K. M. Hayden, O. A. Makeeva, L. K. Newby et al., "A comparison of neuropsychological performance between US and Russia: preparing for a global clinical trial," *Alzheimer's & Dementia*, vol. 10, no. 6, pp. 760–768, 2014.

[13] O. A. Makeeva, H. R. Romero, V. V. Markova et al., "Vascular risk factors confer domain-specific deficits in cognitive performance within an elderly russian population," *Alzheimer's & Dementia*, vol. 11, no. 7, pp. P894–P895, 2015.

[14] V. A. Stepanov and E. A. Trifonova, "Multiplex genotyping of single nucleotide polymorphisms by MALDI-TOF mass-spectrometry: ferequencies of 56 SNP in immune response genes in human populations," *Molekularna Biologija*, vol. 47, no. 6, pp. 976–986, 2013.

[15] K. Ohi, R. Hashimoto, H. Yamamori et al., "The impact of the genome-wide supported variant in the cyclin M2 gene on gray matter morphology in schizophrenia," *Behavioral and Brain Functions*, vol. 9, no. 1, article 40, 2013.

[16] Y. Liu, Z. Cheng, J. Wang et al., "No association between the rs10503253 polymorphism in the CSMD1 gene and schizophrenia in a Han Chinese population," *BMC Psychiatry*, vol. 16, no. 1, article 206, 2016.

[17] L. Athanasiu, S. Giddaluru, C. Fernandes et al., "A genetic association study of CSMD1 and CSMD2 with cognitive function," *Brain, Behavior, and Immunity*, vol. 61, pp. 209–216, 2017.

[18] V. A. Stepanov, A. V. Bocharova, K. Z. Saduakassova et al., "Replicative study of susceptibility to childhood-onset schizophrenia in Kazakhs," *Russian Journal of Genetics*, vol. 51, no. 2, pp. 185–192, 2015.

[19] R. Sherva, Y. Tripodis, D. A. Bennett et al., "Genome-wide association study of the rate of cognitive decline in Alzheimer's disease," *Alzheimer's & Dementia*, vol. 10, no. 1, pp. 45–52, 2014.

[20] S. Sakamoto, M. Takaki, Y. Okahisa et al., "Individual risk alleles of susceptibility to schizophrenia are associated with poor clinical and social outcomes," *Journal of Human Genetics*, vol. 61, no. 4, pp. 329–334, 2016.

[21] E. J. Rose and G. Donohoe, "Brain vs behavior: an effect size comparison of neuroimaging and cognitive studies of genetic risk for schizophrenia," *Schizophrenia Bulletin*, vol. 39, no. 3, pp. 518–526, 2013.

[22] E. J. Rose, D. W. Morris, A. Hargreaves et al., "Neural effects of the CSMD1 genome-wide associated schizophrenia risk variant rs10503253," *American Journal of Medical Genetics Part B: Neuropsychiatric Genetics*, vol. 162, no. 6, pp. 530–537, 2013.

5

Phenotypic Nonspecificity as the Result of Limited Specificity of Transcription Factor Function

Anthony Percival-Smith ⓘ

Department of Biology, The University of Western Ontario, London, ON, Canada N6A 1B7

Correspondence should be addressed to Anthony Percival-Smith; aperciva@uwo.ca

Academic Editor: Martin Kupiec

Drosophila transcription factor (TF) function is phenotypically nonspecific. Phenotypic nonspecificity is defined as one phenotype being induced or rescued by multiple TFs. To explain this unexpected result, a hypothetical world of limited specificity is explored where all TFs have unique random distributions along the genome due to low information content of DNA sequence recognition and somewhat promiscuous cooperative interactions with other TFs. Transcription is an emergent property of these two conditions. From this model, explicit predictions are made. First, many more cases of TF nonspecificity are expected when examined. Second, the genetic analysis of regulatory sequences should uncover *cis*-element bypass and, third, genetic analysis of TF function should generally uncover differential pleiotropy. In addition, limited specificity provides evolutionary opportunity and explains the inefficiency of expression analysis in identifying genes required for biological processes.

1. Introduction: The Specific World of TFs

In Biology gene expression is a major concern and is focused on how a single genotype exhibits multiple phenotypes. The classical example is an *E. coli* cell grown in glucose has very low levels of beta-galactosidase activity, whereas, the same cell will have high levels when grown in lactose [1]. Developmental biology provides nice examples of cells with the same DNA sequence, or genotype, giving rise to specialized transparent lens cells and red blood cells packed with hemoglobin. These distinct phenotypes are the result of differential programs of gene expression. The regulation of gene expression occurs at many different points along the flow of genetic information. One major mechanism of regulation is the control of the initiation of transcription, which is mediated by TFs. Eukaryotic TFs are generally thought to be composed of multiple functional domains: a DNA binding domain that recognizes a specific DNA sequence, a transcriptional regulation domain, and allosteric regulation domain(s) [2, 3]. This multipartite organization confers multiple levels for the regulation of TF specificity. Mechanisms of specificity include specific DNA sequence recognition mediated by the DNA binding domain and regulation of the activity of the TF

prior to binding DNA or after binding DNA. In this paper, I will assume that the initial binding of TFs sets in motion the recruitment of RNA polymerase and all subsequent gene and chromatin modifications and, therefore, although important, these epigenetic mechanisms are not discussed.

The first mechanism regulating TF specificity is specific DNA sequence recognition by the DNA binding domain. The evidence for this mechanism is reflected in the thousand or so structures of protein DNA complexes solved to date. Visualizing amino acid base interactions of a protein DNA complex was a watershed moment in the study of gene expression [4]. From the first structures of lambda repressors to the structure of many more protein DNA complexes, clear, specific amino acid base interactions were observed (Figure 1(a)). Perhaps the most beautiful mechanism of specific DNA recognition is the plant virulence TFs, TAL effectors [5]. The DNA binding domain of TAL TFs is a series of repeats of 34 amino acids with each repeat recognizing a specific base (Figure 1(b)). This one to one repeat to base recognition allows the engineering of proteins that can specifically recognize any DNA sequence in a genome [6]. Indeed, the idea that specific amino acid base contacts were made in the protein DNA complex leads to the isolation or design of change of specificity mutants [7–11]. The

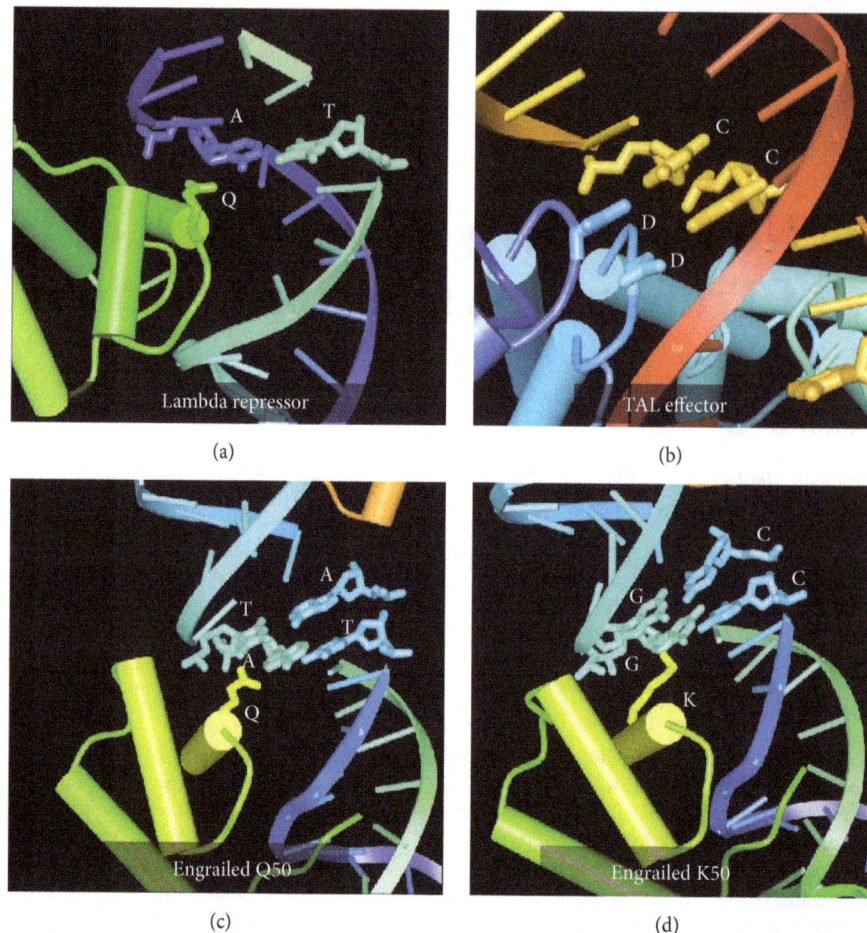

FIGURE 1: The specific amino acid base interactions of DNA binding protein domains with their recognition site. The interaction of a glutamine (green) of the 434 lambda repressor with a specific A:T base pair (blue:teal) (a) [4]. The interaction of the aspartic acids (blue) of two HD TAL repeats with cytosine bases (brown) (b) [5]. The interaction of the glutamine (yellow) of the wild-type Engrailed Q50 homeodomain with TA:AT base pairs (teal:blue) (c). The interaction of the lysine (yellow) of the change of specificity mutant Engrailed K50 homeodomain with GG:CC base pairs (teal:blue) (d) [12]. All images were generated with the Cn3D rendering program from the coordinates in the NCBI database [13].

structural basis for a change of specificity in a homeodomain (HD) has been determined [12] (Figures 1(c) and 1(d)). The studies of change of specificity mutants reinforce the idea that there are mechanisms of specificity required for the regulation of gene expression. Zooming out from the specific interactions at the interface of a TF DNA complex reveals the TF DNA complex on the promoter regulating the expression of a gene. Important for genetic analysis of gene expression, mutations in the DNA recognition site result in misregulation of gene expression identifying the sequence as a *cis*-acting regulatory element.

Although the structural studies beautifully illustrate how TFs recognize specific DNA sequences, there are long-standing questions about how TFs find the specific targets in complex genomes. Mathematical analysis of the kinetics of the lac repressor operator interaction in the *E. coli* was a major topic 40 years ago. One hypothesis proposed for the lac repressor system is that the lac repressor slides on one-dimensional DNA to find the recognition site, and this hypothesis is supported by image analysis in living cells [16].

In addition, analysis of the binding constants and kinetics of the protein DNA interactions point out two interesting properties. First the difference between the dissociation constant (K_d) for recognition of specific DNA sequences and the nonspecific sequences can be large in the case of the lac repressor (10^{-7}-10^{-3}) or small in the case of the HD (10^{-2}) [17, 18]. In cases of small differences, it does not take much nonspecific DNA to compete for binding with the recognition site. Second, for a dynamic system the dissociation constant needs to be large enough such that the TF dissociates from the DNA recognition site; otherwise, the TF is bound irreversibly. The half-life of the lac repressor DNA and HD DNA complexes are in the order of 20-30 minutes [18, 19]. In order to have high specificity for binding site recognition and a dynamic system, weak interactions between TFs in prokaryotes and eukaryotes occur and are a mechanism of cooperativity [20]. In prokaryotes many TFs oligomerize and the complex recognizes a sequence of 6-8 nucleotides and these oligomers interact on DNA using weak interactions to effectively increase the size of the target site recognized to

12-16 nucleotides. This is also observed in eukaryotes with the interactions between the Homeotic selector (HOX) proteins and Extradenticle (EXD) [21]. The cooperative interaction between HOX proteins and EXD increases the target size recognized and confers distinct target site recognition to many HOX proteins when in the complex with EXD. With the addition of Homothorax to the HOX EXD complex the target site becomes larger. However, in vitro analyses identifying the sequence of high affinity sites identifies many sequences that are not identified in the genome as occupied in a ChIP seq analysis [22]. Despite decades of work, how TFs find target sequences is still an interesting and open issue.

The application of information theory to the sequence information recognized by the DNA-binding domain of TFs establishes a clear difference between bacterial and eukaryotic TFs [23]. The DNA binding sites recognized by bacterial transcription factors have higher information content than the bacterial genome to be searched harboring these sequences; whereas, the DNA binding sites recognized by eukaryotic transcription factors have much lower information than the genome being searched and significantly lower information than bacterial TF DNA binding sites. Bacterial TFs recognize specific DNA elements in the genome because the DNA binding sites present enough information for specific binding; whereas, eukaryotic TFs do not recognize enough information in the DNA binding sites resulting in a large amount of spurious binding throughout the genome. This problem with the information present in the interaction between a eukaryotic transcription factor and its DNA binding sites is the heart of the Futility Theorem which asserts that essentially all predicted TF binding sites generated with models for the binding of individual TFs have no functional role [24]. This Futility Theorem still holds today for the analysis of eukaryotic TFs and particularly human TFs [25].

A second mechanism of specificity is the regulation of TF activity. TFs are regulated either prior to the formation of a DNA protein complex or after. During development, a simple mechanism of regulation prior to complex formation is whether the TF is expressed in the specific cell or not. Expression is not the only mechanism of regulation, and other mechanisms that regulate TF factor activity prior to DNA binding include regulation of subcellular localization. Nuclear receptor activity is regulated by whether it is in the nucleus or not, a mechanism shared by the NFkB family of TFs [26, 27]. In addition, TF activity is regulated after the formation of the DNA protein complex. An example of this is the yeast Gal4p transcription factor being bound to the *cis*-regulatory sequences in the presence or absence of galactose [28]. Gal4p activity is controlled by whether it is bound to Gal80p or not. The analysis of the last two mechanisms often involves the genetic dissection of the functional domains of the TF to identify the domains required for specific subcellular localization, binding specific regulatory proteins or allosteric effectors. The best examples of the effects of allosteric effectors on TF function are allolactose on the lac repressor interaction with DNA and steroids on nuclear factors. The specific world of transcription factor function is presently the major model for describing TF function. In the proceeding section, I propose an alternate model for

eukaryotic TF function. Although in describing this model, I do not incorporate or discuss much from the specific world; I do acknowledge that there are specific mechanisms regulating TF function, which would have to be incorporated into a more complete model for the regulation of the initiation of transcription of all genes in the genome.

2. A World of Limited Specificity

2.1. Hypothesis: Limited Specificity of TFs. Extensive phenotypic nonspecificity of TF function has been observed in Drosophila [14, 29–34]. In these observations, phenotypic nonspecificity is defined as one phenotype being induced or rescued by multiple TFs (Figure 2). Phenotypic nonspecificity is observed for HD containing TFs in the induction of wingless, eyeless, arista to tarsus transformations, ectopic thoracic beards, *Hox*-mediated control of autophagy, suppression of *spalt* expression phenotypes following ectopic expression and the rescue of neuromere development and mesoderm formation. This phenotypic nonspecificity was somewhat expected because the HOX DNA binding domain (the HD) recognizes the same recognition sequence. But surprisingly, the phenotypic nonspecificity is not just restricted to HD containing TFs. Both HD containing and non-HD containing TFs when ectopically expressed induce wingless, eyeless, ectopic thoracic beard phenotypes, and the reduced maxillary palp phenotype of *proboscipedia* is rescued by expression of the TF Doublesex male [14]. In the examples of phenotypic nonspecificity (Figure 2), four TFs that induce wingless and eyeless phenotypes have distinct DNA recognition sites of low information content and three are HD TFs and one is a Zn-finger TF. These observations by themselves do not prove a particular mechanism. Here I propose a mechanism and use this model to make explicit testable predictions that would support the hypothesis.

Three different models for TF function can be imagined. A true nonspecific model where every TF has no DNA sequence preference for binding which would lead to an absence of differential gene expression, and an extreme model of specificity where TFs find the promoters of specific sets of genes relying on a high degree of specific DNA sequence recognition which is the combined result of the DNA recognition of binding sites by individual DNA binding domains of TFs and coordinated with very specific protein::protein interactions between only a few TFs for cooperative binding to increase the information content of the DNA protein complex interaction. In the third model, I propose that the DNA sequence recognition and cooperative interactions are limited in specificity and not sufficient to target the expression of a very limited set of genes required for a specific phenotype. This model of limited specificity of TFs is supported by TF occupancy observed in a genome browser (ChIP seq) [35] at two different scales: at the small scale of a few kilobases TFs look to recognize specific regions (TF binding sites) of a promoter, but at a genome scale of megabases each TF is seemingly randomly distributed, and each random distribution of a TF is specific to each individual TF. I propose the distribution of TFs is dependent

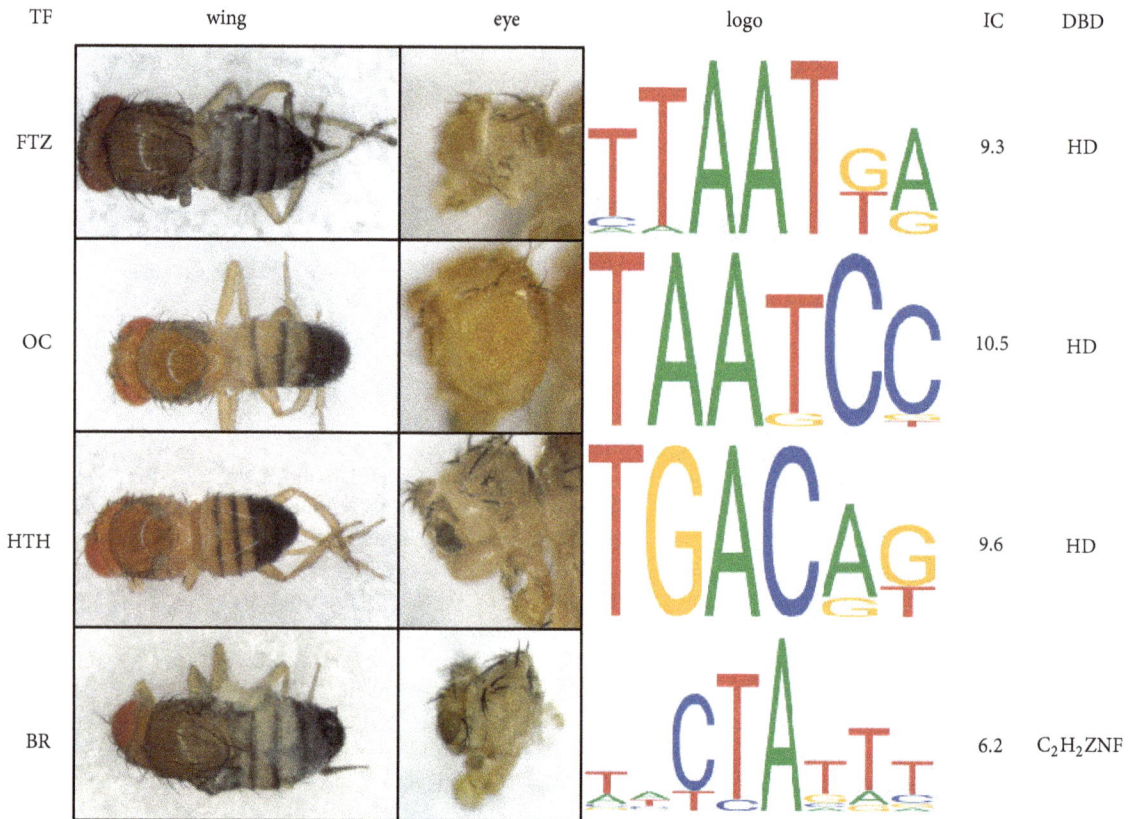

FIGURE 2: Phenotypic nonspecificity in Drosophila. The transcription factors are Fushi tarazu (FTZ), Ocelliless (OC), Homothorax (HTH), and Broad (BR). Wing and eye refer to the wingless and eyeless phenotypes induced by ectopic expression of the respective TF by either the rhomboid GAL4 or eyeless GAL4 driver [14]. The sequence logos for the recognition sequences of the four TFs are from the JASPAR database [15]. The information content (IC) for the recognition site is listed along with the DNA binding domain (DBD) of the TFs either the homeodomain (HD) or the C2H2 Zn fingers (ZNF).

on the recognition of DNA sequences by the DNA binding domain and cooperative interactions between TFs. The low information content of the DNA sites leads to multiple binding of TFs all along the genome (for a TF that recognizes five bases, this would be a binding site on average every kilobase), and these distributions are refined by cooperative interactions. The cooperative interactions between TFs are also of limited specificity: some TFs interact homophilically, some TFs do not interact homophilically, and all TFs have a very large set of heterophilic interactions. Figure 3 shows the distribution of three TFs, the proportion of genes that have 0, 1, 2,... TFs conforms to a Poisson distribution. (For simplicity's sake I have assumed each TF recognizes a binding site of the same length, each TF has the same ability to recruit RNA polymerase, and all genes are the same size). Transcription is an emergent property of these random distributions of TFs. To keep it simple in Figure 3, I assume that a gene with two TFs or more is expressed; however, there is not a good estimate of the real number required. Later I will modify and extend the model to explain default repression. The major point is that random distributions of TFs result in large sets of genes being transcribed. In the model of limited specificity, the genes required for completion of a particular process are a small proportion of the genes regulated by the TF and most genes have no role.

Limited specificity explains phenotypic nonspecificity because it allows the substitution of one TF with another. For example, a process that requires the expression of four genes, and the four genes, plus a hundred more, are regulated by TFa. All that is required for phenotypic nonspecificity is that the expression of TFb in place of TFa is able to result in the expression of the four genes required for the process; even though, the total set of regulated genes are quite distinct between the two TFs apart from these four genes.

I predict that there are many more cases of phenotypic nonspecificity. Indeed, there may be cases reported in the literature that have not been recognized as phenotypic nonspecificity. One interesting potential observation of phenotypic nonspecificity is that the initial set of four TFs identified that transform fibroblasts to pluripotent stem cells is not the only set of four TFs that induce pluripotency [36, 37]. Future analysis of the induction of pluripotency may uncover an even larger set of TFs capable of inducing this phenotype. In addition, in an analysis of the neuron cell fates of the Drosophila optic lobe, phenotypic convergence was observed where a neural trait is regulated independently by different combinations of transcription factors [38]. Phenotypic convergence like phenotypic nonspecificity may be a consequence of limited specificity. Because phenotypic nonspecificity is not often tested directly, the following set

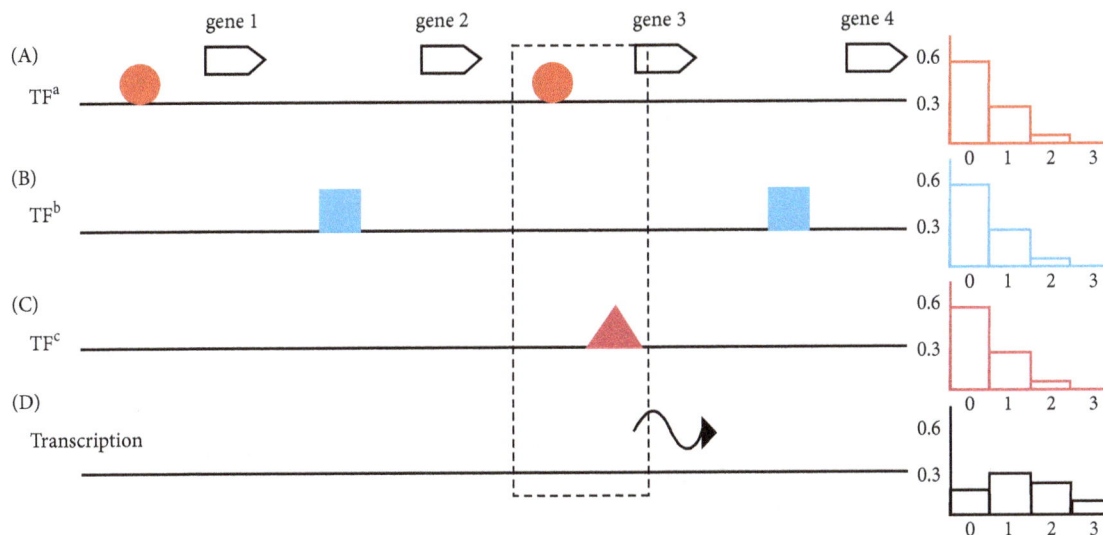

FIGURE 3: A simplified representation for the nonspecific model for transcriptional regulation. The pattern of binding of three TFs (TFa red circle, TFb blue square, and TFc magenta triangle) to a region of a genome with four genes (A-C) is shown. To the right of each distribution is the expected proportion of genes in the genome with no, 1, 2, and 3 TFs bound. The dotted box indicates a gene with two TFs bound, which in this simplified representation, is sufficient to recruit RNAP for transcription. Transcription (wiggly line) is an emergent property of random TF distributions. To the right is the expected proportion of genes in the genome with no, 1, 2, and 3 of these three TFs bound.

of predictions is made. First, I predict that many more TF combinations induce pluripotency. Second, the hierarchy of segmentation along the anterior posterior axis of Drosophila uses two large sets of TFs, the Gap proteins and the Pair-rule proteins [39]. Based on my previous observations of frequencies of phenotypic nonspecificity in Drosophila [14], I predict in 30-50% of these genes that the coding region can be substituted by 10-20% of Drosophila TFs and result in normal segmentation. In addition, this pattern of phenotypic rescue will also apply to the set of TFs expressed along the Dorsal ventral axis of the nerve cord in vertebrates, and the set of TFs temporally expressed during the development of the Drosophila brain [40, 41]. Based on the small size of the regulatory regions of yeast genes relative to Drosophila genes, I also predict that 10-20% of yeast transcription factor loci can have their coding regions substituted by 5-10% of yeast transcription factors. If any of these are shown, then these processes require the expression of a set of TFs rather than just the specific set identified, suggesting that TFs do not have intrinsic properties associated with a specific process. Basically, apart from some DNA sequence binding preferences and some preferences for TF partners to interact with cooperatively, TFs are very similar functionally, that is they lack unique and identifying leg, heart or muscle determining protein domains and functions.

2.2. A Language Change Associated with Phenotypic Nonspecificity. The language for discussing mechanisms of transcriptional regulation is biased. For example, the formulation of a common question in the study of HOX proteins "Given the high degree of sequence conservation of the HD between HOX proteins and therefore conservation of the DNA binding sites recognized by these HDs, how do the HOX proteins function to specifically direct specific segmental identities during development?" asserts that there is phenotypic specificity of TF function; a mechanism for specificity is assumed and sought at the outset. The question of whether the phenotype regulated by a TF is phenotypically specific or nonspecific is neither discussed nor tested at the outset. The assumption of phenotypic specificity is down to the level of the syntactic articles used. For example, prior to my observation of extensive phenotypic nonspecificity in Drosophila [14], I would have confidently and definitively written that the TF Proboscipedia directs maxillary palp development; however, now I write a TF directs maxillary palp development, because the TF Doublesex Male rescues the proboscipedia phenotype. Phenotypic specificity uses the definite article "the", and phenotypic nonspecificity uses the indefinite article "a". In a further example of bias, in the discussion of multiple genes regulated by a TF, the genes are often divided into specific targets and nonspecific, off-targets suggesting that there is an intrinsic difference between these genes regulated by a TF and further implying that off-targets are safe to ignore. This use of language is arbitrarily dismissing a set of regulated genes as unimportant. In the model of limited specificity, more genes regulated by a TF are not required than are required for a process, and therefore, there are no such things as a limited set of confounding off-targets.

2.3. Promoter Analysis in a World of Limited Specificity: cis-Element Bypass. The mutational analysis of a promoter is a common approach for identifying *cis*-acting regulatory sequences controlling expression of a gene. In a world with limited specificity this approach is still valid, but the interpretation of the results is restricted to an analysis of the present wild-type promoter and not an analysis of the potential of the promoter. When the promoter analysis is being performed, the set of TFs required for regulation of the gene does not change, and hence, does not address the potential of the promoter. If TFa can be substituted for another TFb then the phenomenon of cis-element bypass is expected, where the promoter with the mutation in the binding site for TFa is now expressed in the presence of TFb because TFb binds to its own distinct cis-element. Conversely, a promoter with a mutation in the cis-element binding TFb would be expressed in the presence of TFa, but not when TFa is substituted by TFb. Cis-element bypass is an explicit prediction of genetic analysis in the world of limited specificity and would suggest that the cis-elements are sufficient but not always necessary. The substitution of one TF for another uncovers the potential of the regulatory region. In a world of limited specificity, arguments about the importance of the organization of cis-elements in the promoter for regulation of gene expression are not very important.

2.4. Analysis of the Functional Domains of TFs: Differential Pleiotrophy.

The aim of mutational analysis of TF loci is determining the functional organization of the proteins by identifying functional domains and motifs important for establishing the wild-type pattern of gene expression. This approach is particularly important in the determination of the modular structure of TFs like Gal4p and nuclear receptors [2, 3]. However, the genetic analysis of other TFs like the Drosophila HD containing TFs Ultrabithorax, Sex combs reduced and Antennapedia, the yeast HD containing TF Pho2p, and Human HOX proteins has uncovered differential pleiotrophy, the nonuniform behavior of mutant alleles across phenotypes suggesting motifs make small tissue specific contributions to overall TF activity [42–48]. The differential pleiotrophy of mutations in TF loci has been attributed to an ensemble nature to allostery of intrinsically disordered proteins [49], but alternatively differential pleiotrophy may also be an expected outcome in the model of limited specificity. In the model, the TF is randomly distributed, and each TF is not much different from another apart from its distribution. The TFs also interact cooperatively with limited specificity resulting in tissue specific TF distributions and activation of transcription. The mechanism of cooperativity is not restricted to two small highly structured protein domains making very specific amino acid interactions with each other such that only a limited number of TF::TF interactions occur. Rather the TF::TF interactions are mediated by more diffuse, nonspecific protein surfaces, and the way the surfaces of TFs interact may change from tissue to tissue and gene to gene. Therefore, the genetic analysis may not identify separate, specific interaction surfaces indicating a high degree of modularity. Rather the diffuse nonspecific surfaces and tissue specific and gene specific deployment of these surfaces show up in genetic analysis as differential pleiotrophy where specific regions of the TF seem to make small tissue specific contributions to overall TF activity. Differential pleiotrophy may be evidence supporting limited specificity.

2.5. Phenotypic Nonspecificity as an Evolutionary Opportunity.

The study of evolution and development has identified Toolkit genes [50]. Toolkit genes share four major conserved characteristics: structure (basically amino acid sequence), expression, requirement, and function. Many of the Toolkit genes encode TFs. Unfortunately, in the experimental set up testing conservation of function, it is assumed that the function of the proteins is highly specific that is only that specific protein/function can generate or rescue the phenotype. This is particularly true in the analysis of Toolkit TFs. The experimental test is to substitute a TF with an orthologous TF from another organism and determine whether the ortholog can induce or rescue a particular developmental phenotype [51–56]. Observation of phenotypic nonspecificity renders this test uninformative because multiple nonorthologous/nonparalogous TFs can induce or rescue the phenotype [14, 29–34]. Therefore, some toolkit functions are now not functionally conserved, with structurally unrelated TFs able to substitute, severely weakening the Toolkit gene hypothesis at both the levels of conservation of function and structure leaving just conservation of expression and requirement.

In the evolutionary history of animals and plants it is hard to explain how a complex multicellular organism of multiple cell types arises as a result of development. A common assertion is that animals and plants have acquired the genes during evolution to accomplish this feat. A significant proportion of the genes required for determination of the body plan encodes TFs, and hypotheses have proposed that the increase in complexity is associated with the increase in the number of TFs and families of TFs [57]. In a sense, it is proposed that the acquisition of TFs recapitulates phylogeny [58]. This model has two problems. First, phenotypic nonspecificity suggests that TFs are not tailored by evolution to have a restricted role in a specific process. Second, the multiple body plan simplifications observed during evolution confounds the march to complexity. Phylogenomic analysis shows that between divergence of ctenophores and cnidarians, the simple porifera and placozoa phyla branch off; this body plan simplification is also observed in divergence of bilaterian phyla [59–61]. Phenotypic nonspecificity and body plan simplification suggests that a simple addition of genes does not explain or can be used to track the rise of complexity.

Also, an interesting consideration is the effect of the proliferation of transcription factors during evolution in the specific and limited specificity models. In the specific model, a transcription factor like Eyeless/PAX6 becomes associated with the regulation of genes required for eye development and once established is conserved during evolution. The origin of this specificity is difficult to explain. For example, when the gene encoding the TF arose what was the function of the TF at that point? Does it have no function and spends time this way, even though it can bind to DNA and affect transcription, until during evolution it becomes employed for a specific role? Does the TF have an initial function that during evolution is moved to another process and the original role replaced by another TF that has somehow changed its specificity? These are not considerations encountered with limited specificity. From the first time it is expressed, the TF is functional and as long as it does not result in a pattern of gene expression that is detrimental to survival, it is maintained in the genome. During evolution expression of the TF or the genes it regulates may change such at upon genetic analysis in the extant organism it seems to be the essential TF regulator of a process but in reality, it is one of many TFs that have the potential to regulate the process. An expectation of limited specificity is the existence of TF loci that when mutant result in phenotypically viable organisms with no difference in fitness; however, analysis of differential gene expression shows large changes in the pattern of gene expression.

The phenomenon of phenotypic nonspecificity may be good for evolution and development, as phenotypic specificity is an evolutionary constraint, whereas, phenotypic nonspecificity is an evolutionary opportunity. In recently evolved systems involving regulation of multiple genes like the Bicoid and Dorsal morphogens, it would take a large number of random sequence changes in cis-elements during evolution in a specific world to bring about the specific pattern of gene expression required. Whereas, in a world of limited specificity, all that is required is a sequence change that results in a change in the expression pattern of a TF that has the potential of bringing about the pattern of gene expression required. Large-scale changes in gene expression can occur in the mechanism of limited specificity with one mutational change because the recognition sites (cis-elements) for the TF already exist in the genes required for that phenotypic change. Distalless normally required for limb development can be seconded into a different situation to pattern eyespots in butterfly wings because the cis-elements for Distalless already exist in the genes required for eye spot formation [62]. The set of Dorsal regulated genes, which pattern the early dorsal ventral axis of Drosophila could be a consequence of a simple change in Dorsal expression [63]. Cis-element bypass uncovers the potential of regulatory regions, which is why it is an important prediction with ramifications for evolution. The world of limited specificity may be a powerful model to explain large-scale changes in gene expression that may have occurred during evolution.

2.6. Phenotypic Nonspecificity and the Analysis of Patterns of Gene Expression. In the world of limited specificity, a TF factor is required for the regulation of large sets of genes, the size of which depends on how widely the TF is expressed. In addition, as long as the genes whose products are required for the development of a specific organ are expressed at the correct time, place, and level whether tissue specifically or not is all that matters. These two considerations explain why expression analysis is inefficient at identifying the components required for specific developmental/biological processes. The example I will use is the simple developmental program of sporulation in *S. cerevisiae*, which includes meiosis and the encapsulation of the four haploid products in a spore wall. The general belief in the early 1980s was that genes required for sporulation would be expressed specifically during sporulation; indeed, a few genes specifically expressed during sporulation are required but a large majority of genes specifically expressed during sporulation are not required for sporulation [64–67]. In addition, a systematic screen of genes not required for yeast viability showed that there are more genes required for sporulation that are not specifically expressed during sporulation than genes that are specifically expressed during sporulation, and more surprisingly genes downregulated during sporulation are also required for sporulation [68, 69]. Therefore, identifying genes required for sporulation using an expression approach was inefficient due to two reasons: first a minority of genes specifically expressed during sporulation were required for sporulation, and second the set of genes required for sporulation but not preferentially expressed during sporulation is larger than the set specifically expressed during sporulation and required. This result is expected in a world of limited specificity as the pattern of gene expression is less important than whether the gene is just expressed at the correct time and place irrespective of other considerations and, therefore, questioning the value of preforming expression analysis to screen for genes required for biological processes. And conversely, the expression pattern of a gene may also not provide much information on how it is required. For example, in Drosophila String (Cdc-25) expression is restricted to mitotically dividing cells of the embryo potentially suggesting an important regulatory role for the cell cycle to occur or not. However, expression of String in all cells rescues the string phenotype and does not induce ectopic mitosis, suggesting that String does not have an important role in initiating mitosis [70]. In addition, the TF Fushi tarazu is expressed and required in the even-number parasegments but low-level expression in all cells partially rescues the fushi tarazu phenotype [71]. In the world of limited specificity, the relationships between expression pattern and requirement are not straightforward and do not generally support "guilt by association". Differential gene expression is a well-described phenomenon; however, limited specificity may change how differential expression is primarily viewed: from a mechanism that promotes the expression of genes that initiate a process to a mechanism that inhibits the expression of genes that might disrupt a process, which would be under selection.

2.7. Default Repression. When an experimental perturbation is applied, which includes the alteration in the expression of a TF(s), gene expression analysis identifies genes that are both activated and repressed [72]. Interestingly the random distribution of TFs in the genome proposed in the model of limited specificity results in some promoters with few TFs bound and other promoters with many TFs bound. Using this characteristic of the model of limited specificity, I explain the activation/repression phenomenon with the incorporation of the three habits model and poised RNA polymerase II (RNAP) [73–75]. The three habits model explains the behavior of the terminal TFs of major cell-cell signaling pathways. The three habits are: activator insufficiency, where a ligand activated signaling pathway response element binding TFs (SPRE TF) is insufficient for activating transcription of a target gene alone: cooperativity, where a ligand activated SPRE TF with other local activators (TFs) cooperates and activates transcription; and default repression, where in the absence of a ligand the SPRE TF and local TFs repress transcription of the target gene. Although the three habits

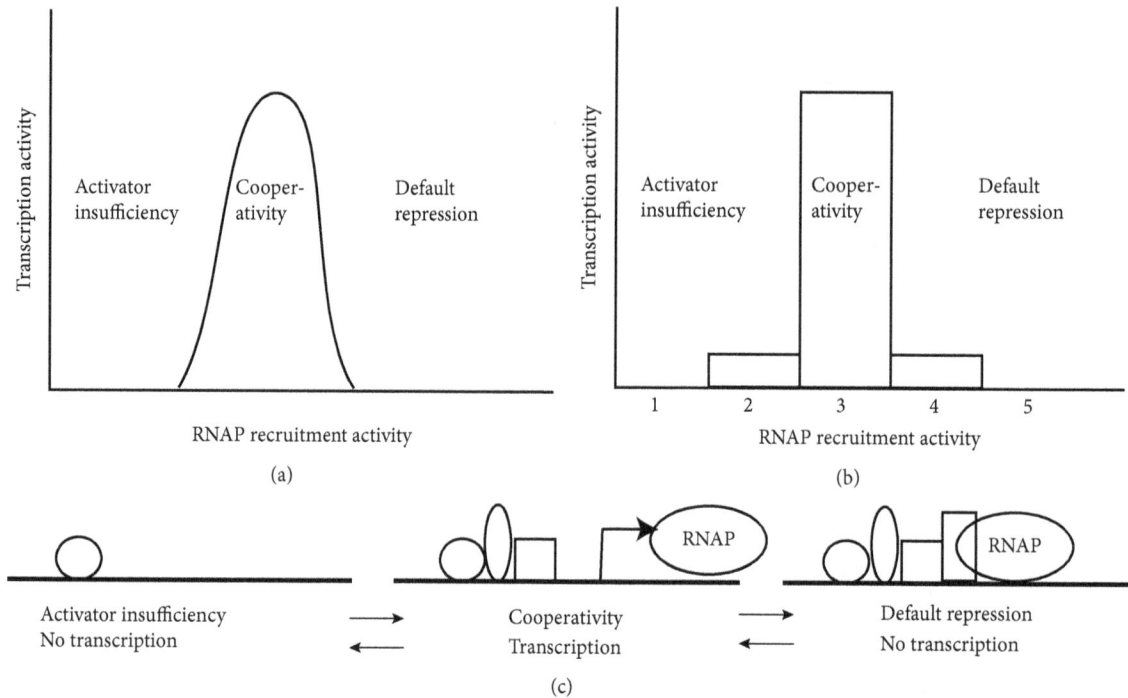

FIGURE 4: The model for default repression. Transcription activity is the rate of transcription initiation of a gene and RNAP recruitment activity is related to the amount of time RNAP is around the start of transcription of a gene (a and b). In (a) RNAP recruitment activity is assumed to be continuous, but in (b) it is assumed that each TF bound contributes an integer amount of TF recruitment activity that interacts cooperatively to recruit RNAP. It is assumed for simplicity of representation that when three TFs are bound, transcription is at its maximum. The three habits of the gene expression are shown as three interchangeable states (c). When one TF is bound transcription does not occur because of activator insufficiency. When three TFs are bound they cooperate and transcription occurs at the maximum rate. When four TFs are bound RNAP is strongly recruited to the promoter and does not engage in transcription and the gene is in default repression.

were proposed to explain observations on the expression of target genes at the end of cell-cell signaling pathways, I will extend it to all genes. I propose that transcriptionally active promoters are in a sweet spot or Goldilocks zone for the number of TFs bound. Starting from no bound TFs (Figure 4), as more TFs bind the amount of RNAP recruitment activity (the term RNAP recruitment is more vague than how it is defined by [76] which is the formation of a closed complex and more reflects time spent around the start of transcription), increase using cooperative interactions and transcription proceeds; however, there can be too much of a good thing where the RNAP recruiting activity is too high and RNAP is conformationally stuck or poised at the promoter. In Figure 4(b), I have shown a simplified example where all TFs have the same recruitment activity. At 1 or 2 TFs bound there is no or little expression (activator insufficiency), at three TFs bound there is the highest level of transcription (cooperativity), but above that number, RNAP is poised on the promoter and less transcription occurs (default repression). This model proposes two mechanisms for the activation of transcription: the gain of RNAP recruitment activity by the addition of TFs and the loss of RNAP recruitment activity by the loss or inactivation of TFs. Most models of transcription put transcription as the end point of RNAP recruitment activity such that poised RNAPs are a prelude to transcription; here poised RNAPs are the consequence of

too much RNAP recruitment activity. This explains nicely why in a genome wide analysis of expression, genes are both repressed and activated when a transcription factor is either added (ectopic expression) or taken away (loss of function). When a TF is added many genes now have enough RNAP recruitment activity to be transcribed and, in addition, now other genes have too much RNAP recruitment activity and are not transcribed as highly. Likewise, when a TF is taken away many genes now do not have enough RNAP recruitment activity to be transcribed while other genes have lost enough RNAP recruitment activity to now be transcribed. In this model, the TF is neither a repressor nor an activator, only the number of TFs bound at the promoter sets whether a gene is transcribed or not.

One prediction of this model is the existence of TF loci that reciprocally oscillate between high levels of transcription and high levels of TF accumulation due to both an autoregulatory element with many binding sites for the TF and the TF having a short half-life. The oscillation occurs because at low TF concentration the gene is highly transcribed but as the TF concentration increases and more TFs are bound to the autoregulatory element transcription deceases due to default repression resulting in a subsequent decrease in the concentration of the TF as it is rapidly degraded. New techniques in the live imaging of TF accumulation and active transcription may uncover examples of this [77].

I have no knowledge of experimental examples in eukaryotes of the phenomenon that reduction in RNAP recruitment results in the expression of a gene. However, there may be a bacterial example in the most classical of systems, the lac repressor [78, 79]. The lac repressor interacts with RNAP when not bound to allolactose. The RNAP in this complex with the lac repressor cycles through abortive initiation making short transcripts. The addition of allolactose to the lac repressor RNAP complex results in transient activation of transcription. The exact mechanism of this has been debated and investigated extensively [80]. The heat shock factor and RNAP are bound to (poised) the heat shock promoter in non-heat shocked cells. Poised RNAP II in eukaryotic cells also initiate short transcripts [81]. It is not difficult to imagine that the heat shock response is just temperature sensitive RNAP recruitment, as temperature increases recruitment deceases and transcription is initiated. The model also explains why the mutation of a silencer site results in activation of gene transcription, as the mutation just reduces the number of TFs that bind. In addition, the model explains how a TF can be both an activator and repressor in the same cell as each promoter has a different number of TFs bound. Another prediction of this model is that genes with poised RNA polymerases have more TFs bound than when they are transcribed. To test this prediction, the unbiased detection of the number of TFs bound to a promoter when transcribed or not is the major technical issue that needs to be overcome.

2.8. Limited Specificity and the Mathematical Modeling of Gene Expression. One of the interesting aspects of limited specificity is the potential to simplify mathematical modeling of eukaryotic gene expression. In the world of limited specificity all TFs have a similar intrinsic potential, differing only in the relative random distributions of binding in the genome. By applying a few rules to TF behavior, such as the DNA-binding preference and set of interacting TFs, the transcription profile of a cell expressing a defined set of transcription factors could be determined. It is easy to imagine doing this theoretically in a simulation, but it may be possible to take known transcription profiles and TF occupancy profiles to inform such a model and predict the outcome of a perturbation in vivo. The gene expression profile would be represented as a product of a matrix of genes, DNA-binding sites and TFs. Mathematical modeling is made easier in the world of limited specificity because TFs have a smaller number of specific attributes (variables).

3. Conclusions

To explain phenotypic nonspecificity of TFs in the induction or rescue of a phenotype, I propose a model where TFs are not inherently much different from one another in terms of intrinsic function and where transcription is an emergent property of the random distributions of these TFs which is influenced by DNA sequence binding preference and somewhat promiscuous TF::TF interactions. Using this model, I make three predictions. First further examination will uncover many more cases of phenotypic nonspecificity.

Second, in cases where multiple TFs can function to determine a phenotype, *cis*-element bypass will be observed. To my knowledge, no cases of *cis*-element bypass have been reported because it has yet to be directly tested. Third, genetic analysis of TFs in the nonspecific world will uncover differential pleiotropy. Differential pleiotropy is observed for the HD containing Drosophila, Human and yeast TFs: Ultrabithorax, Fushi tarazu, Sex combs reduced, Antennapedia, HOXA7, HOXB3, and Pho2p; therefore, in order to make a more general statement, observation of differential pleiotropy needs to be extended to other TF families. If the world of limited specificity exists, it has far reaching ramifications for evolution, expression analysis, and mathematical modeling of gene expression.

Although I have been outlining genetic analysis strictly in a world of limited specificity, highly specific mechanisms exist. One example is the specificity conferred to nuclear receptors by the ligand binding domains [82], and another example is the specific and modular interaction between Gal4p with Gal80p [83]. Therefore, if limited specificity exists, the next big problem would be what proportion of the observed patterns of gene expression does it explain? For example, if limited specificity explains 90% of gene expression where specificity only explains 10% then it would have to be considered the dominant model.

Acknowledgments

This work was supported by the National Science and Engineering Research Council (NSERC) Discovery and Accelerator grants to Anthony Percival-Smith.

References

[1] F. Jacob and J. Monod, "Genetic regulatory mechanisms in the synthesis of proteins," *Journal of Molecular Biology*, vol. 3, pp. 318–356, 1961.

[2] R. Brent and M. Ptashne, "A eukaryotic transcriptional activator bearing the DNA specificity of a prokaryotic repressor," *Cell*, vol. 43, no. 3, pp. 729–736, 1985.

[3] S. Rusconi and K. R. Yamamoto, "Functional dissection of the hormone and DNA binding activities of the glucocorticoid receptor," *EMBO Journal*, vol. 6, no. 5, pp. 1309–1315, 1987.

[4] A. K. Aggarwal, D. W. Rodgers, M. Drottar, M. Ptashne, and S. C. Harrison, "Recognition of a DNA operator by the repressor of phage 434: A view at high resolution," *Science*, vol. 242, no. 4880, pp. 899–907, 1988.

[5] D. Deng, C. Yan, X. Pan et al., "Structural basis for sequence-specific recognition of DNA by TAL effectors," *Science*, vol. 335, no. 6069, pp. 720–723, 2012.

[6] M. Cristian, T. Cermak, EL. Doyle, C. Schmidt, and etal., "DNA double-strand breaks with TAL effectors," *Genetics*, vol. 186, pp. 757–761, 2010.

[7] P. Youderian, A. Vershon, S. Bouvier, R. T. Sauer, and M. M. Susskind, "Changing the DNA-binding specificity of a repressor," *Cell*, vol. 35, no. 3, pp. 777–783, 1983.

[8] M. Hollis, D. Valenzuela, D. Pioli, and R. Wharton, "Specific recognition of a chimeric operator by a repressor heterodimer," *Proc. Natl. Acad. Sci. USA*, vol. 85, pp. 5834–5838, 1988.

[9] S. D. Hanes and R. Brent, "DNA specificity of the bicoid activator protein is determined by homeodomain recognition helix residue 9," *Cell*, vol. 57, no. 7, pp. 1275–1283, 1989.

[10] A. Percival-Smith, M. Muller, M. Affolter, and W. J. Gehring, "The interaction with DNA of wild-type and mutant fushi tarazu homeodomains," *EMBO Journal*, vol. 9, no. 12, pp. 3967–3974, 1990.

[11] S. D. Hanes and R. Brent, "A genetic model for interaction of the homeodomain recognition helix with DNA," *Science*, vol. 251, no. 4992, pp. 426–430, 1991.

[12] L. Tucker-Kellogg, M. A. Rould, K. A. Chambers, S. E. Ades, R. T. Sauer, and C. O. Pabo, "Engrailed (Gln50→Lys) homeodomain-DNA complex at 1.9 Å resolution: Structural basis for enhanced affinity and altered specificity," *Structure*, vol. 5, no. 8, pp. 1047–1054, 1997.

[13] Y. Wang, L. Y. Geer, C. Chappey, J. A. Kans, and S. H. Bryant, "Cn3D: sequence and structure views for Entrez," *Trends in Biochemical Sciences*, vol. 25, no. 6, pp. 300–302, 2000.

[14] A. Percival-Smith, "Non-specificity of transcription factor function in Drosophila melanogaster," *Development Genes and Evolution*, vol. 227, no. 1, pp. 25–39, 2017.

[15] A. Khan, O. Fornes, A. Stigliani et al., "JASPAR 2018: Update of the open-access database of transcription factor binding profiles and its web framework," *Nucleic Acids Research*, vol. 46, no. 1, pp. D260–D266, 2018.

[16] P. Hammar, P. Leroy, A. Mahmutovic, E. G. Marklund, O. G. Berg, and J. Elf, "The lac repressor displays facilitated diffusion in living cells," *Science*, vol. 336, no. 6088, pp. 1595–1598, 2012.

[17] S.-Y. Lin and A. D. Riggs, "Lac repressor binding to DNA not containing the lac operator and to synthetic poly dAT," *Nature*, vol. 228, no. 5277, pp. 1184–1186, 1970.

[18] M. Affolter, A. Percival-Smith, M. Muller, W. Leupin, and W. J. Gehring, "DNA binding properties of the purified Antennapedia homeodomain," *Proceedings of the National Acadamy of Sciences of the United States of America*, vol. 87, no. 11, pp. 4093–4097, 1990.

[19] A. D. Riggs, R. F. Newby, and S. Bourgeois, "lac repressor-Operator interaction. II. Effect of galactosides and other ligands," *Journal of Molecular Biology*, vol. 51, no. 2, pp. 303–314, 1970.

[20] A. D. Johnson, B. J. Meyer, and M. Ptashne, "Interactions between DNA-bound repressors govern regulation by the λ phage repressor," *Proceedings of the National Acadamy of Sciences of the United States of America*, vol. 76, no. 10, pp. 5061–5065, 1979.

[21] H. D. Ryoo, T. Marty, F. Casares, M. Affolter, and R. S. Mann, "Regulation of Hox target genes by a DNA bound homothorax/Hox/extradenticle complex," *Development*, vol. 126, no. 22, pp. 5137–5148, 1999.

[22] M. Slattery, T. Riley, P. Liu et al., "Cofactor Binding Evokes Latent Differences in DNA Binding Specificity between Hox Proteins," *Cell*, vol. 147, no. 6, pp. 1270–1282, 2011.

[23] Z. Wunderlich and L. A. Mirny, "Different gene regulation strategies revealed by analysis of binding motifs," *Trends in Genetics*, vol. 25, no. 10, pp. 434–440, 2009.

[24] W. W. Wasserman and A. Sandelin, "Applied bioinformatics for the identification of regulatory elements," *Nature Reviews Genetics*, vol. 5, no. 4, pp. 276–287, 2004.

[25] S. A. Lambert, A. Jolma, L. F. Campitelli et al., "The Human Transcription Factors," *Cell*, vol. 175, no. 2, pp. 598–599, 2018.

[26] D. Picard and K. R. Yamamoto, "Two signals mediate hormone-dependent nuclear localization of the glucocorticoid receptor.," *EMBO Journal*, vol. 6, no. 11, pp. 3333–3340, 1987.

[27] P. A. Baeuerle and D. Baltimore, "IkB: A specific inhibitor of NFkB transcription factor," in *Science*, vol. 242, pp. 540–546, 1988.

[28] E. Giniger, S. M. Varnum, and M. Ptashne, "Specific DNA binding of GAL4, a positive regulatory protein of yeast," *Cell*, vol. 40, no. 4, pp. 767–774, 1985.

[29] S. Greig and M. Akam, "The role of homeotic genes in the specification of the Drosophila gonad," *Current Biology*, vol. 5, no. 9, pp. 1057–1062, 1995.

[30] F. Hirth, T. Loop, B. Egger, D. F. B. Miller, T. C. Kaufman, and H. Reichert, "Functional equivalence of Hox gene products in the specification of the tritocerebrum during embryonic brain development of Drosophila," *Development*, vol. 128, no. 23, pp. 4781–4788, 2001.

[31] A. Percival-Smith, W. A. Teft, and J. L. Barta, "Tarsus determination in Drosophila melanogaster," *Genome*, vol. 48, no. 4, pp. 712–721, 2005.

[32] A. Banreti, B. Hudry, M. Sass, A. Saurin, and Y. Graba, "Hox Proteins Mediate Developmental and Environmental Control of Autophagy," *Developmental Cell*, vol. 28, no. 1, pp. 56–69, 2014.

[33] K. M. Lelli, B. Noro, and R. S. Mann, "Variable motif utilization in homeotic selector (Hox)-cofactor complex formation controls specificity," *Proceedings of the National Acadamy of Sciences of the United States of America*, vol. 108, no. 52, pp. 21122–21127, 2011.

[34] A. Percival-Smith, L. Sivanantharajah, J. J. H. Pelling, and W. A. Teft, "Developmental competence and the induction of ectopic proboscises in Drosophila melanogaster," *Development Genes and Evolution*, vol. 223, no. 6, pp. 375–387, 2013.

[35] D. Karolchik, R. M. Kuhn, R. Baertsch et al., "The UCSC genome browser database: 2008 update," *Nucleic Acids Research*, vol. 36, no. 1, pp. D773–D779, 2008.

[36] K. Takahashi and S. Yamanaka, "Induction of pluripotent stem cells from mouse embryonic and adult fibroblast cultures by defined factors," *Cell*, vol. 126, no. 4, pp. 663–676, 2006.

[37] M. Maekawa, K. Yamaguchi, T. Nakamura, and R. Shibukawa, "Direct reprogaramming of somatic cells is promoted by maternal transcription factor Glis1," *Nature*, vol. 474, pp. 225–229, 2011.

[38] N. Konstantinides, K. Kapuralin, C. Fadil, and Barboza., "Phenotypic convergence: distinct transcription factors regulate common terminal features," *Cell*, vol. 174, pp. 622–635, 2018.

[39] M. Akam, "The molecular basis for metameric pattern in the Drosophila embryo," *Development*, vol. 101, no. 1, pp. 1–22, 1987.

[40] G. Le Dréau and E. Martí, "Dorsal-ventral patterning of the neural tube: a tale of three signals," *Developmental Neurobiology*, vol. 72, no. 12, pp. 1471–1481, 2012.

[41] X. Li, T. Erclik, C. Bertet et al., "Temporal patterning of Drosophila medulla neuroblasts controls neural fates," *Nature*, vol. 498, no. 7455, pp. 456–462, 2013.

[42] L. T. Bhoite, J. M. Allen, E. Garcia et al., "Mutations in the Pho2 (Bas2) transcription factor that differentially affect activation with its partner proteins Bas1, Pho4, and Swi5," *The Journal of Biological Chemistry*, vol. 277, no. 40, pp. 37612–37618, 2002.

[43] E. Tour, C. T. Hittinger, and W. McGinnis, "Evolutionarily conserved domains required for activation and repression functions of the Drosophila Hox protein Ultrabithorax," *Development*, vol. 132, no. 23, pp. 5271–5281, 2005.

[44] C. T. Hittinger, "Pleiotropic functions of a conserved insect-specific Hox peptide motif," *Development*, vol. 132, no. 23, pp. 5261–5270, 2005.

[45] F. Prince, T. Katsuyama, Y. Oshima et al., "The YPWM motif links Antennapedia to the basal transcriptional machinery," *Development*, vol. 135, no. 9, pp. 1669–1679, 2008.

[46] L. Sivanantharajah and A. Percival-Smith, "Analysis of the sequence and phenotype of *Drosophila* Sex combs reduced alleles reveals potential functions of conserved protein motifs of the Sex combs reduced protein," *Genetics*, vol. 182, no. 1, pp. 191–203, 2009.

[47] L. Sivanantharajah and A. Percival-Smith, "Acquisition of a leucine zipper motif as a mechanism of antimorphy for an allele of the Drosophila Hox gene sex combs reduced," *G3: Genes, Genomes, Genetics*, vol. 4, no. 5, pp. 829–838, 2014.

[48] A. Dard, J. Reboulet, F. JiaY Bleicher, and M. Duffraisse, "Human HOX protein use diverse and context-dependent motifs to interact with TALE calss cofactors," *Cell Reports*, vol. 22, pp. 3058–3071, 2018.

[49] L. Sivanantharajah and A. Percival-Smith, "Differential pleiotropy and HOX functional organization," *Developmental Biology*, vol. 398, no. 1, pp. 1–10, 2015.

[50] S. B. Carroll, "Evo-devo and an expanding evolutionary synthesis: a genetic theory of morphological evolution," *Cell*, vol. 134, no. 1, pp. 25–36, 2008.

[51] J. Malicki, K. Schughart, and W. McGinnis, "Mouse Hox-2.2 specifies thoracic segmental identity in Drosophila embryos and larvae," *Cell*, vol. 63, no. 5, pp. 961–967, 1990.

[52] J. J. Zhao, R. A. Lazzarini, and L. Pick, "The mouse Hox-1.3 gene is functionally equivalent to the Drosophila Sex combs reduced gene," *Genes & Development*, vol. 7, no. 3, pp. 343–354, 1993.

[53] C. Hunter and C. Keynon, "Specification of anteroposterior cell fates in Caenorabditis elegans by Drosophila Hox proteins," in *Nature*, vol. 377, pp. 229–232, 1995.

[54] G. Halder, P. Callaerts, and W. J. Gehring, "Induction of ectopic eyes by targeted expression of the eyeless gene in Drosophila," *Science*, vol. 267, no. 5205, pp. 1788–1792, 1995.

[55] B. Lutz, H.-C. Lu, G. Eichele, D. Miller, and T. C. Kaufman, "Rescue of Drosophila labial null mutant by the chicken ortholog Hoxb-1 demonstrates that the function of Hox genes is phylogenetically conserved," *Genes & Development*, vol. 10, no. 2, pp. 176–184, 1996.

[56] A. Percival-Smith and J. A. Laing Bondy, "Analysis of murine HOXA-2 activity in Drosophila melanogaster," *Developmental Genetics*, vol. 24, no. 3-4, pp. 336–344, 1999.

[57] J. F. Ryan, K. Pang, J. C. Mullikin, M. Q. Martindale, and A. D. Baxevanis, "The homeodomain complement of the ctenophore Mnemiopsis leidyi suggests that Ctenophora and Porifera diverged prior to the ParaHoxozoa," *EvoDevo*, vol. 1, no. 1, 2010.

[58] RA. Raff and TC. Kaufman, "Embryos, genes, and evolution," in *pages 251-261*, pp. 251–261, Indiana University Press, Bloomington, IN, USA, 1991.

[59] J. F. Ryan, K. Pang, C. E. Schnitzler, A. D. Nguyen, and R. T. Moreland, "The geneome of the ctenophore Mnemiopsis leidyi and its implications for cell type evolution," *Science*, vol. 342, Article ID 1232592, 2013.

[60] L. L. Moroz, K. M. Kocot, M. R. Citarella, S. Dosung, and T. P. Norekian, "The ctenophore genome and the evolutionary origins of neural systems," *Nature*, vol. 510, pp. 109–114, 2014.

[61] J. Aruga, Y. S. Odaka, A. Kamiya, and H. Furuya, "Dicyema Pax6 and Zic: Tool-kit genes in a highly simplified bilaterian," *BMC Evolutionary Biology*, vol. 7, no. 1, article no. 201, 2007.

[62] B. Dhungel, Y. Ohno, R. Matayoshi et al., "Distal-less induces elemental color patterns in Junonia butterfly wings," *Zoological Letters*, vol. 2, no. 1, 2016.

[63] A. Stathopoulos, M. Van Drenth, A. Erives, M. Markstein, and M. Levine, "Whole-genome analysis of Dorsal-ventral patterning in the Drosophila embryo," *Cell*, vol. 111, no. 5, pp. 687–701, 2002.

[64] A. Percival-Smith and J. Segall, "Isolation of DNA sequences preferentially expressed during sporulation in Saccharomyces cerevisiae.," *Molecular and Cellular Biology*, vol. 4, no. 1, pp. 142–150, 1984.

[65] A. Percival-Smith and J. Segall, "Characterization and mutational analysis of a cluster of three genes expressed preferentially during sporulation of Saccharomyces cerevisiae.," *Molecular and Cellular Biology*, vol. 6, no. 7, pp. 2443–2451, 1986.

[66] A. T. Garber and J. Segall, "The SPS4 gene of Saccharomyces cerevisiae encodes a major sporulation-specific mRNA.," *Molecular and Cellular Biology*, vol. 6, no. 12, pp. 4478–4485, 1986.

[67] S. Chu, J. DeRisi, M. Eisen et al., "The transcriptional program of sporulation in budding yeast," *Science*, vol. 282, no. 5389, pp. 699–705, 1998.

[68] A. M. Deutschbauer, R. M. Williams, A. M. Chu, and R. W. Davis, "Parallel phenotypic analysis of sporulation and postgermination growth in Saccharomyces cerevisiae," *Proceedings of the National Acadamy of Sciences of the United States of America*, vol. 99, no. 24, pp. 15530–15535, 2002.

[69] A. H. Enyenihi and W. S. Saunders, "Large-scale functional genomic analysis of sporulation and meiosis in Saccharomyces cerevisiae," *Genetics*, vol. 163, no. 1, pp. 47–54, 2003.

[70] B. A. Edgar and P. H. O'Farrell, "The three postblastoderm cell cycles of Drosophila embryogenesis are regulated in G2 by string," *Cell*, vol. 62, no. 3, pp. 469–480, 1990.

[71] B. Argiropoulos, J. Ho, B. J. Blachuta, I. Tayyab, and A. Percival-Smith, "Low-level ectopic expression of Fushi tarazu in Drosophila melanogaster results in ftzUal/Rpl-like phenotypes and rescues ftz phenotypes," *Mechanisms of Development*, vol. 120, no. 12, pp. 1443–1453, 2003.

[72] S. D. Hueber, D. Bezdan, S. R. Henz, M. Blank, H. Wu, and I. Lohmann, "Comparative analysis of Hox downstream genes in Drosophila," *Development*, vol. 134, no. 2, pp. 381–392, 2007.

[73] D. S. Gilmour and J. T. Lis, "RNA polymerase II interacts with the promoter region of the noninduced hsp70 gene in Drosophila melanogaster cells.," *Molecular and Cellular Biology*, vol. 6, no. 11, pp. 3984–3989, 1986.

[74] S. Barolo and J. W. Posakony, "Three habits of highly effective signaling pathways: Principles of transcriptional control by developmental cell signaling," *Genes & Development*, vol. 16, no. 10, pp. 1167–1181, 2002.

[75] J. Zeitlinger, A. Stark, M. Kellis et al., "RNA polymerase stalling at developmental control genes in the Drosophila melanogaster embryo," *Nature Genetics*, vol. 39, no. 12, pp. 1512–1516, 2007.

[76] B. Krummel and M. J. Chamberlin, "RNA Chain Initiation by Escherichia coli RNA Polymerase. Structural Transitions of the Enzyme in Early Ternary Complexes," *Biochemistry*, vol. 28, no. 19, pp. 7829–7842, 1989.

[77] J. P. Bothma, M. R. Norstad, S. Alamos, and H. G. Garcia, "LlamaTags: A Versatile Tool to Image Transcription Factor Dynamics in Live Embryos," *Cell*, vol. 173, no. 7, pp. 1810–1822.e16, 2018.

[78] A. Schmitz and D. J. Galas, "The interaction of RNA polymerase and lac repressor with the lac control region," *Nucleic Acids Research*, vol. 6, no. 1, pp. 111–137, 1979.

[79] S. B. Straney and D. M. Crothers, "Lac repressor is a transient gene-activating protein," *Cell*, vol. 51, no. 5, pp. 699–707, 1987.

[80] A. Sanchez, M. L. Osborne, L. J. Friedman, J. Kondev, and J. Gelles, "Mechanism of transcriptional repression at a bacterial promoter by analysis of single molecules," *EMBO Journal*, vol. 30, no. 19, pp. 3940–3946, 2011.

[81] I. Jonkers and J. T. Lis, "Getting up to speed with transcription elongation by RNA polymerase II," *Nature Reviews Molecular Cell Biology*, vol. 16, no. 3, pp. 167–177, 2015.

[82] S. Green and P. Chambon, "Oestradiol induction of a glucocorticoid-responsive gene by a chimeric receptor," *Nature*, pp. 75–78, 1987.

[83] J. B. Thoden, L. A. Ryan, R. J. Reece, and H. M. Holden, "The interaction between an acidic transcriptional activator and its inhibitor: The molecular basis of Gal4p recognition by Gal80p," *The Journal of Biological Chemistry*, vol. 283, no. 44, pp. 30266–30272, 2008.

6

Lack of Association between Variant rs7916697 in *ATOH7* and Primary Open Angle Glaucoma in a Saudi Cohort

Altaf A. Kondkar [ID],[1] **Taif A. Azad** [ID],[1] **Faisal A. Almobarak** [ID],[1] **Ibrahim M. Bahabri**,[2]
Hatem Kalantan,[1] **Khaled K. Abu-Amero** [ID],[1] and **Saleh A. Al-Obeidan** [ID][1]

[1]*Glaucoma Research Chair, Department of Ophthalmology, College of Medicine, King Saud University, Riyadh, Saudi Arabia*
[2]*King Khaled University Hospital, King Saud University, Riyadh, Saudi Arabia*

Correspondence should be addressed to Altaf A. Kondkar; akondkar@gmail.com

Academic Editor: Francine Durocher

A case-control genetic association study was performed to investigate whether variant rs7916697 in *atonal bHLH transcription factor 7* (*ATOH7*), which has been previously reported to be associated with optic disc parameters and primary open angle glaucoma (POAG) in different ethnic groups, is a risk factor for POAG or any of its clinical phenotypes in a Saudi cohort. Genotyping of rs7916697 (G>A) variant was performed in 186 unrelated POAG cases and 171 unrelated nonglaucomatous controls of Saudi origin using real-time Taq-Man® assay. Genotypic and allelic association with POAG and its related clinical indices were evaluated. Demographic and systemic disease status did not differ significantly between POAG cases and controls. Association analysis between POAG cases and controls showed no significant genotype effect under additive (p=0.707), dominant (p=0.458), and recessive (p=0.554) models. Besides, the minor 'A' allele frequency was 0.39 in POAG cases and 0.36 in controls with no significant distribution (p=0.406). In addition, there was no significant difference between genotypes and clinical phenotypes such as intraocular pressure and cup/disc ratio within the POAG group, or any age and sex adjusted genotype effect on the disease outcome in regression analysis. Variant rs7916697 in *ATOH7* is not associated with POAG or its clinical indices such as IOP and cup/disc ratio in a Saudi cohort.

1. Introduction

With an estimated heritability of 0.81 [1], primary open angle glaucoma (POAG) follows a complex multi-factorial inheritance pattern involving both genetics and environmental factors [2]. *Atonal bHLH transcription factor 7* (*ATOH7*), located on cytogenic band 10q21.3-22.1, is a single exon gene that encodes transcription factor known to play a central role in differentiation of retinal ganglion cells (RGCs) and optic nerve formation [3]. Studies in animal model of glaucoma have demonstrated that abnormal ATOH7 expression results in an increased number of differentiated RGCs [4, 5]. Mice retinal progenitor cells have been demonstrated to express *Atoh7* [6] and its regulation by *Pax6* in embryonic retina [7] is clinically relevant because *PAX6* gene mutations have been associated with various optic nerve abnormalities in humans [8]. Genome-wide association studies (GWAS)

have previously reported strong association between optic disk parameters and variant rs7916697 near *ATOH7* in an Australian twin cohort [6], the Rotterdam study [9], and Singapore Asians [10] and very recently in Latino population [11]. A recent meta-analysis of GWAS within the International Glaucoma Genetics Consortium revealed that rs7916697, among others, significantly affected at least one of the optic disc parameters [12]. In addition, a suggestive protective association was also noted for rs7916697 in Afro-Caribbean Barbados population in POAG [13] indicating *ATOH7* as an important susceptibility gene associated with glaucoma and optic disc parameters. Despite strong evidence for involvement of *ATOH7* in pathogenesis of POAG the exact molecular mechanism(s) leading to glaucomatous damage of the optic nerve and the possibility of any interaction of *ATOH7* with other risk factors is still largely unknown.

Our previous study has shown that another polymorphism rs1900004 in *ATOH7*, also reported to be strongly associated with optic disk parameters and POAG [6, 9], was not associated with POAG in a small number of Saudi patients that we investigated [14]. The present study was performed in a different and almost double the number of sample cohorts of Saudi origin to investigate any association between variant rs7916697 in *ATOH7* and POAG or any of its clinical indices.

2. Material and Methods

2.1. Study Population. A case-control study was performed according to the principles of the Declaration of Helsinki for human research and all of the study participants provided a written informed consent. The study received ethical clearance from College of Medicine institutional review board committee at King Saud University (approval number # 08-657). Saudi patients with clinically confirmed diagnosis of POAG (n=186) and nonglaucomatous healthy controls (n=171) of the same ethnicity were recruited into the study at King Abdulaziz University Hospital, King Saud University, Riyadh, Saudi Arabia. The criteria for selection of patients and controls have been detailed elsewhere [15, 16]. A standardized ophthalmic examination was performed in all the participating patients that included measurement of intraocular pressure (IOP) by Goldmann applanation tonometry mounted at the slit lamp, examination of anterior chamber angles by gonioscopy, dilated pupil examination of the lens and fundus, and visual field testing by Humphrey automated field analyzer. POAG patients were diagnosed by glaucoma specialist ophthalmologist and satisfied the following diagnostic criteria: (1) presence of progressive glaucomatous damage at the optic disk or retinal nerve fiber layer changes, such as narrowing of the neuroretinal rim, diffuse thinning of retinal nerve fibre layer, or localized defects; (2) presence of visual field defects which are typical of glaucoma such as nasal step defect, arcuate or paracentral scotomata, or generalized tunnel vision; (3) bilaterally open anterior chamber angles as examined by gonioscopy; and (4) adult onset. Any secondary form of glaucoma cases such as pigmentary glaucoma, uveitic, pseudoexfoliation, and history of steroid use or ocular trauma was excluded. Ethically matched control subjects were selected from ophthalmology screening clinics and subjects with normal IOP (without anti-glaucoma medicine), open anterior chamber angles, and healthy optic disk on examination with no previous history of ocular disease(s) or ophthalmic surgeries. Patient or control subjects refusing to participate were excluded. Other details on history of systemic diseases, health awareness, and smoking habits were procured through medical records of the patients or personal interviews for controls.

2.2. Genotyping of rs7916697 in ATOH7 Gene. DNA samples were obtained from peripheral EDTA blood using the illustra blood genomicPrep Mini Spin kit (GE Healthcare, Buckinghamshire, UK) and genotyped for rs7916697 G>A polymorphism near *ATOH7* gene (NG_031934.1) using the TaqMan®

SNP Genotyping Assay (assay ID: C_27850155_10; Applied Biosystems Inc., Foster City, CA, USA) on ABI 7500 Real-Time PCR System (Applied Biosystems) using recommended cycling conditions as described previously [17]. Fluorescence was measured at annealing step and genotype calling was performed using the automated 2-color allele discrimination software on ABI 7500.

2.3. Statistical Analyses. SPSS version 22 (IBM Inc., Chicago, Illinois, USA) was used to perform the analyses. Chi-square analysis was used to test Hardy-Weinberg Equilibrium (HWE) and allelic/genotype association. Independent samples t-test, one-way ANOVA, and Kruskal-Wallis tests were used to detect the mean difference across genotypes/groups. Regression analysis was done to determine the effect of age, sex, and genotype on the disease (POAG) outcome. Odds ratios (OR) were calculated, confidence interval (CI) level was set to 95%, and a p value less than 0.05 was considered statistically significant.

3. Results

3.1. Demographic and Clinical Characteristics of the Participants. A total of 357 participants consisting of 186 POAG cases and 171 nonglaucomatous controls were included in this study. Except for family history of glaucoma (p=0.039), the mean age, gender distribution, smoking habits, and status of systemic diseases such as diabetes, hypertension, coronary artery disease, and hypercholesterolemia showed no significant difference between the two study groups (Table 1).

3.2. Genotype and Allelic Association with POAG. The genotypes did not deviate significantly from the HWE (p>0.05). Both the genotype and allele frequency of rs7916697 in *ATOH7* did not differ significantly between POAG cases and controls (Table 2). The minor 'A' allele frequency was 0.39 in POAG cases and 0.36 in controls with no significant distribution between cases and controls (p=0.406). There was no significant genotype association with POAG as assessed by additive (p=0.707), dominant (p=0.458), and recessive models (p=0.554). In addition, there was no significant difference between genotypes and clinical phenotypes such as intraocular pressure and cup/disc ratio within the POAG group (Table 3), or any age and sex adjusted genotype effect on the disease outcome in regression analysis (Table 4).

4. Discussion

Variant rs7916697 is located in the 5′ untranslated region of *ATOH7*, an important candidate for human optic nerve aplasia and related clinical syndromes [3]. A genetic association between rs7916697 and risk of POAG in a Saudi cohort was investigated.

The minor allele frequency (MAF) of rs7916697'A' in *ATOH7* varies across different population ranging from 0.25 to 0.82 with an overall MAF of 0.44 (1000 Genomes Project, Ensembl database). The MAF in our Saudi cohort was 0.36 in controls and 0.39 in POAG cases. The observed MAF was

TABLE 1: Demographic and clinical characteristics of POAG cases and controls genotyped for SNP rs7916697.

Variables	Controls (n = 171) No. (%)	Cases (n = 186) No. (%)	p value[a]
Demographic Characteristics			
Age in years, mean (±SD)	60.9 (10.6)	58.9 (11.5)	0.096*
Male	98 (57.3)	101 (55.2)	0.560
Female	73 (42.7)	85 (45.7)	-
Systemic Diseases			
Diabetes mellitus	65 (38.0)	75 (40.3)	0.654
Coronary artery disease	4 (2.3)	6 (3.2)	0.612
Hypertension	56 (32.7)	71 (38.1)	0.285
Hypercholesterolemia	8 (4.6)	14 (7.5)	0.263
Health Awareness / Behavior			
Family history of glaucoma	7 (4.1)	18 (9.6)	0.039
Smoking	15 (8.7)	20 (10.7)	0.527

[a]Pearson Chi^2 test, *t-test.

TABLE 2: Association analysis of allele frequency and genotype distribution for SNP rs7916697 in POAG patients and controls.

SNP (Gene)	rs7916697 (*ATOH7*)				
	Controls (n = 171) No. (%)	POAG (n = 186) No. (%)	Odds ratio	95% confidence interval	p value[a]
Allelic analysis					
G	219 (64.0)	227 (61.0)	1	Reference	-
A*	123 (36.0)	145 (39.0)	0.88	0.65 – 1.19	0.406
HWE P	0.968	0.936	-	-	-
Genotype and Model analysis					
G/G	70 (40.9)	69 (37.1)	1	Reference	-
G/A	79 (46.2)	89 (47.8)	0.87	0.55 – 1.37	0.560
A/A	22 (12.8)	28 (15.0)	0.77	0.40 – 1.48	0.438
Additive	-	-	-	-	0.707§
Dominant	-	-	0.85	0.55 – 1.30	0.458
Recessive	-	-	0.83	0.45 – 1.52	0.554

[a]Pearson Chi^2 test. *Risk variant. HWE P, Hardy-Weinberg equilibrium p value. §Fisher exact test.

higher than 0.27 reported in US Caucasians [18], similar to 0.38 reported in Latino study [11] and much lower than 0.76 observed in Afro-Caribbeans where 'G' was reported to be the minor allele [13], highlighting an ethnic specific complex genetic etiology in POAG.

Genetic studies have reported both positive and negative association of variants in *ATOH7* and risk of POAG in different ethnic groups. Polymorphism rs7916697 in *ATOH7* was reported to be an important genetic determinant of optic disc size in a meta-analysis consisting of UK and Australian cohort (p=1.3x10^{-10}) that explained 1.7% variation in the optic disc area in the UK cohort [6], implicating its pathophysiological role in POAG. Studies in Rotterdam and Singapore had identified significant association between rs7916697 and optic disc area (p=2x10^{-15}) in Asians [10].

The Blue Mountains Eye Study and the Twins study also replicated a strong association at rs7916697 for optic disc size that was dependent on vertical cup/disc ratio (VCDR) [19]. Similarly, consistent with other findings, rs7916697 showed a borderline genome-wide significance (p=5.44x10^{-8}) and was associated with decrease in VCDR in the Latino population [11]. In contrast, we did not find any significant association between rs7916697 and POAG or any of its clinical indices such as IOP, cup/disc ratio, and number of anti-glaucoma medications that serve as clinical markers of disease severity/progression. Consistent with our findings, rs7916697 was also not found to be significantly associated with POAG and other clinical indices in US Caucasians, where another variant rs1900004 in *ATOH7* was reported to influence optic disc area [18]. Similarly, a modest protective effect was

TABLE 3: Analysis of genotype effect on demographic and clinical characteristics within PAOG group.

| Characteristics | Genotypes | | | p value[a] |
	G/G (n= 69) No. (%)	G/A (n= 89) No. (%)	A/A (n= 28) No. (%)	
DEMOGRAPHIC				
Age in years, mean (SD)	62.2 (9.7)	60.0 (11.0)	60.4 (11.4)	0.413*
Male	35 (50.7)	50 (56.1)	16 (57.1)	0.750
Female	34 (49.3)	39 (43.8)	12 (42.8)	-
MEDICAL HISTORY				
Family history of glaucoma	6 (8.7)	7 (7.8)	5 (17.8)	0.278
Smoking	5 (7.2)	12 (13.5)	3 (10.7)	0.455
Diabetes mellitus	22 (31.9)	38 (42.7)	15 (53.5)	0.116
Hypertension	24 (34.8)	33(37.0)	14 (50.0)	0.360
Coronary artery disease	1 (1.4)	4 (4.5)	1 (3.5)	0.557
Hypercholesterolemia	3 (4.3)	7 (7.8)	4 (14.3)	0.240
GLAUCOMA INDICES				
Intraocular pressure in mmHg, mean (SD)	22.5 (8.9)	24.0 (9.7)	22.6 (7.9)	0.444**
Cup/disc ratio	0.84 (0.6)	0.75 (0.2)	0.82 (0.1)	0.259**
Number of anti-glaucoma medications	1.7 (1.1)	2.0 (1.0)	1.9 (1.1)	0.325**

[a]Pearson Chi^2 test. *One-way ANOVA. **Kruskal-Wallis test.

TABLE 4: Binary logistic regression analysis to assess the effect of age, sex, and genotype on disease outcome.

Variables	Odds ratio	95% confidence interval	p value
Age	1.01	0.99 – 1.03	0.096
Sex[a]	0.91	0.59 – 1.39	0.676
Genotype[b]	-	-	0.627
G/A	1.17	0.74 – 1.83	0.499
A/A	1.34	0.70 – 2.59	0.371

[a]Female as reference, [b]G/G as reference.

reported for rs7916697 (allelic p=0.0096, genotypic p=0.01) for the 'G' allele (OR=0.67; 95% CI=0.50–0.91) in Afro-Caribbean subjects. Though this finding failed to withstand correction for multiple testing, there was significant evidence for an interactive effect with rs1063192 (near *CDKN2B/AS1* on chromosome 9) [13].

The molecular mechanisms leading to the development and progression of optic nerve defect in POAG could be IOP-and/or non-IOP-dependent (RGC/optic nerve vulnerability-related). Genes affecting optic nerve quantitative traits such as optic disc area and VCDR may have a plausible role in pathogenesis of POAG by affecting developmental-related pathways [20]. *ATOH7* gene encodes Math5, a protein that plays a central role in RGC differentiation [4] and thus may not be a major risk modulator or causal factor in IOP-related pathogenesis of late-onset POAG. A complex disease such as glaucoma can result from interactions between several genes. An additive effect of genetic variants associated with IOP, VCDR, and high or normal tension glaucoma [13, 21] cannot be ruled out in this study. In addition, we have previously shown that polymorphism rs1900004 in *ATOH7* was also not

associated with POAG [14], indicating that specific genetic variants may be more enriched in one ethnic population than another, highlighting racial differences, and that *ATOH7* may not have a major role in POAG development/progression in this population. Nonetheless, the findings of our study require cautious interpretation. Considering the global MAF (from 1000 Genomes database) or the MAF observed in our Saudi cohort, the current sample size of the study has an estimated power of >80% (with alpha risk of 5%) to detect an odds ratio of 2.0. However, it will certainly require a much larger sample size to detect a 1.5-fold relative risk. As with other complex diseases, large sample sizes are required to ensure sufficient power to fully define the underlying genetic causal effect which may be a major limitation in this study. Besides, the role of other variants and gene-gene interaction also cannot be ruled out.

5. Conclusion

In an attempt to link this polymorphism with POAG among the Saudi cohort our study shows that variant rs7916697 in the *ATOH7* gene lacks significant association with POAG or related phenotypes such as IOP and cup/disc ratio. This observation needs further validation in a much larger sample population of clinically well-defined POAG patients potentially with age, gender, and ethnicity matched controls to assess the risk it may contribute to the development or progression of the disease in this population.

Acknowledgments

The authors would like to thank the Deanship of Scientific Research and Glaucoma Research Chair of Department of Ophthalmology, College of Medicine, King Saud University, for their support and use of laboratory facilities.

References

[1] J. Charlesworth, P. L. Kramer, T. Dyer et al., "The path to open-angle glaucoma gene discovery: Endophenotypic status of intraocular pressure, cup-to-disc ratio, and central corneal thickness," *Investigative Ophthalmology & Visual Science*, vol. 51, no. 7, pp. 3509–3514, 2010.

[2] K. Abu-Amero, A. A. Kondkar, and K. V. Chalam, "An updated review on the genetics of primary open angle glaucoma," *International Journal of Molecular Sciences*, vol. 16, no. 12, pp. 28886–28911, 2015.

[3] N. L. Brown, S. L. Dagenais, C.-M. Chen, and T. Glaser, "Molecular characterization and mapping of ATOH7, a human atonal homolog with a predicted role in retinal ganglion cell development," *Mammalian Genome*, vol. 13, no. 2, pp. 95–101, 2002.

[4] Z. Yang, K. Ding, L. Pan, M. Deng, and L. Gan, "Math5 determines the competence state of retinal ganglion cell progenitors," *Developmental Biology*, vol. 264, no. 1, pp. 240–254, 2003.

[5] W.-T. Song, X.-Y. Zhang, and X.-B. Xia, "Atoh7 promotes the differentiation of Müller cells-derived retinal stem cells into retinal ganglion cells in a rat model of glaucoma," *Experimental Biology and Medicine*, vol. 240, no. 5, pp. 682–690, 2015.

[6] S. Macgregor, A. W. Hewitt, P. G. Hysi et al., "Genome-wide association identifies ATOH7 as a major gene determining human optic disc size," *Human Molecular Genetics*, vol. 19, no. 13, pp. 2716–2724, 2010.

[7] A. N. Riesenberg, T. T. Le, M. I. Willardsen, D. C. Blackburn, M. L. Vetter, and N. L. Brown, "Pax6 regulation of Math5 during mouse retinal neurogenesis," *Genesis*, vol. 47, no. 3, pp. 175–187, 2009.

[8] N. Azuma, Y. Yamaguchi, H. Handa et al., "Mutations of the PAX6 gene detected in patients with a variety of optic-nerve malformations," *American Journal of Human Genetics*, vol. 72, no. 6, pp. 1565–1570, 2003.

[9] W. D. Ramdas, L. M. van Koolwijk, M. K. Ikram et al., "A genome-wide association study of optic disc parameters," *PLoS Genetics*, vol. 6, no. 6, 2010.

[10] A. Broeks, M. K. Schmidt, and M. E. Sherman, "Low penetrance breast cancer susceptibility loci are associated with specific breast tumor subtypes: findings from the Breast Cancer Association Consortium," *Human Molecular Genetics*, vol. 20, no. 16, pp. 3289–3303, 2011.

[11] D. R. Nannini, M. Torres, Y. I. Chen et al., "A genome-wide association study of vertical cup-disc ratio in a latino population," *Investigative Opthalmology & Visual Science*, vol. 58, no. 1, pp. 87–95, 2017.

[12] H. Springelkamp, A. I. Iglesias, and A. Mishra, "New insights into the genetics of primary open-angle glaucoma based on meta-analyses of intraocular pressure and optic disc characteristics," *Human Molecular Genetics*, vol. 26, no. 2, pp. 438–453, 2017.

[13] D. Cao, X. Jiao, X. Liu et al., "CDKN2B polymorphism is associated with primary open-angle glaucoma (POAG) in the Afro-Caribbean population of Barbados, West Indies," *PLoS ONE*, vol. 7, no. 6, 2012.

[14] A. A. Kondkar, A. Mousa, T. A. Azad et al., "Analysis of Polymorphism rs1900004 in Atonal bHLH Transcription Factor 7 in Saudi Patients with Primary Open Angle Glaucoma," *Genetic Testing and Molecular Biomarkers*, vol. 20, no. 11, pp. 715–718, 2016.

[15] K. Abu-Amero, T. Sultan, S. Al-Obeidan, and A. Kondkar, "Analysis of CYP1B1 sequence alterations in patients with primary open-angle glaucoma of Saudi origin," *Clinical Ophthalmology*, vol. 12, pp. 1413–1416, 2018.

[16] A. A. Kondkar, N. B. Edward, H. Kalantan et al., "Lack of association between polymorphism rs540782 and primary open angle glaucoma in Saudi patients," *Journal of Negative Results in BioMedicine*, vol. 16, no. 1, 2017.

[17] A. A. Kondkar, T. A. Azad, F. A. Almobarak et al., "Polymorphism rs10483727 in the SIX1/SIX6 gene locus is a risk factor for primary open angle glaucoma in a saudi cohort," *Genetic Testing and Molecular Biomarkers*, vol. 22, no. 1, pp. 74–78, 2018.

[18] B. J. Fan, D. Y. Wang, L. R. Pasquale, J. L. Haines, and J. L. Wiggs, "Genetic variants associated with optic nerve vertical cup-to-disc ratio are risk factors for primary open angle glaucoma in a US Caucasian population," *Investigative Ophthalmology & Visual Science*, vol. 52, no. 3, pp. 1788–1792, 2011.

[19] C. Venturini, A. Nag, P. G. Hysi et al., "Clarifying the role of ATOH7 in glaucoma endophenotypes," *British Journal of Ophthalmology*, vol. 98, no. 4, pp. 562–566, 2014.

[20] W. D. Ramdas, L. M. van Koolwijk, H. G. Lemij et al., "Common genetic variants associated with open-angle glaucoma," *Human Molecular Genetics*, vol. 20, no. 12, pp. 2464–2471, 2011.

[21] F. Mabuchi, N. Mabuchi, Y. Sakurada et al., "Additive effects of genetic variants associated with intraocular pressure in primary open-angle glaucoma," *PLoS ONE*, vol. 12, no. 8, 2017.

7

Genetic Variants Associated with Hyperandrogenemia in PCOS Pathophysiology

Roshan Dadachanji ⓘ**, Nuzhat Shaikh** ⓘ**, and Srabani Mukherjee** ⓘ

Department of Molecular Endocrinology, National Institute for Research in Reproductive Health, J.M. Street, Parel, Mumbai 400012, India

Correspondence should be addressed to Srabani Mukherjee; mukherjees@nirrh.res.in

Academic Editor: Fabio M. Macciardi

Polycystic ovary syndrome is a multifactorial endocrine disorder whose pathophysiology baffles many researchers till today. This syndrome is typically characterized by anovulatory cycles and infertility, altered gonadotropin levels, obesity, and bulky multifollicular ovaries on ultrasound. Hyperandrogenism and insulin resistance are hallmark features of its complex pathophysiology. Hyperandrogenemia is a salient feature of PCOS and a major contributor to cosmetic anomalies including hirsutism, acne, and male pattern alopecia in affected women. Increased androgen levels may be intrinsic or aggravated by preexisting insulin resistance in women with PCOS. Studies have reported augmented ovarian steroidogenesis patterns attributed mainly to theca cell hypertrophy and altered expression of key enzymes in the steroidogenic pathway. Candidate gene studies have been performed in order to delineate the association of polymorphisms in genes, which encode enzymes in the intricate cascade of steroidogenesis or modulate the levels and action of circulating androgens, with risk of PCOS development and its related traits. However, inconsistent findings have impacted the emergence of a unanimously accepted genetic marker for PCOS susceptibility. In the current review, we have summarized the influence of polymorphisms in important androgen related genes in governing genetic predisposition to PCOS and its related metabolic and reproductive traits.

1. Introduction

Polycystic ovary syndrome affects scores of women worldwide with a prevalence of nearly 6–10% of premenopausal women [1]. Typical features of PCOS comprise distorted gonadotropin ratios, chronic anovulation, and subsequently irregular menstrual cycles, insulin resistance, increased androgen levels, and appearance of polycystic ovarian morphology upon ultrasound imaging [2]. Besides reproductive anomalies, these women are at an increased risk of developing type II diabetes, metabolic syndrome, and cardiovascular diseases (CVD), with subclinical markers being detected at earlier ages. While a clear-cut origin of PCOS has not emerged to explain its underlying pathophysiology, androgen excess and insulin resistance are reported to be the pivotal pathogenic drivers which extend reproductive, metabolic, and cosmetic consequences to affected women. The ovary remains the primary source of hyperandrogenism in women with PCOS which is mainly attributed to thecal

cell hyperplasia leading to intense ovarian steroidogenesis. Evidence suggests that hyperandrogenism is an important factor in promoting anovulation due to follicular arrest [3] and high androgen level has been linked to reduced oocyte developmental competence and maturation rates. Further testosterone has been correlated with fertilization rates, embryo development, and miscarriage rates in women with PCOS [4]. Adrenal androgen excess has been reported in 20–30% of women with PCOS possibly due to defects in cortisol metabolism or common steroid pathway biosynthesis enzymes [5]. Apart from modifying reproductive outcomes, hyperandrogenemia also predicts severity of cardiometabolic profiles and CVD risk [6]. A recent meta-analysis has accentuated that hyperandrogenemia unfavourably influences the incidence of dyslipidemia, indices of insulin resistance, and metabolic syndrome risk [7]. The possibility of a genetic basis of hyperandrogenemia in PCOS has been recognized early [8] and candidate gene studies investigating the association of genes involved in androgen synthesis and action have strengthened this concept further [9, 10]. In the present review, we have outlined studies detailing the association of polymorphisms in these genes with PCOS susceptibility and its related traits.

2. General Steroid Metabolism

Ovary, the chief organ of interest is endowed with important functions of maintaining the female reproductive physiology. These include timely development of ovarian follicles and production of mature oocytes as well as the steroid hormones synthesis [11]. Steroidogenesis comprises processes by which the precursor cholesterol is converted to biologically active steroid hormones. Steroidogenic enzymes are responsible for the biosynthesis of various steroid hormones including glucocorticoids, mineralocorticoids, progestins, androgens, and estrogens. They consist of several specific cytochrome P450 enzymes (CYPs), hydroxysteroid dehydrogenases (HSDs), and steroid reductases [12]. De novo synthesis of all steroid hormones starts with the conversion of cholesterol to pregnenolone by CYP11A (cholesterol side-chain cleavage) [13]. CYP11A is bound to the inner membrane of the mitochondrion and is found in all steroid producing organs and tissues [12]. Pregnenolone is converted to progesterone by 3β-hydroxysteroid dehydrogenase (3β-HSD), found in both mitochondria and smooth endoplasmic reticulum. 3β-HSD is widely distributed in steroidogenic and nonsteroidogenic tissues and consists of two isoenzymes, which are regulated in a tissue-specific manner [14–17]. The type 2 3β-HSD is predominantly expressed in steroidogenic tissues such as adrenal, testis, and ovary, whereas type 1 is found in placenta and in nonsteroidogenic tissues such as liver, kidney, and skin. Pregnenolone and progesterone form the precursors for all other steroid hormones.

In the ovary, steroidogenesis is a well-regulated process governed by the gonadotropins and signaling mechanisms occurring in the ovarian cells. Androgen synthesis predominantly takes place in thecal cells which have LH receptors and subsequent signaling and activation of CYP17 enzyme convert pregnenolone and progesterone to dehydroepiandrosterone (DHEA) and androstenedione, respectively. These androgens are further acted upon by CYP19 aromatase enzymes present in the FSH stimulated granulosa cells to estrogens which are essential for normal physiological functions of the human ovary.

3. Hyperandrogenemia and PCOS

The most common clinical manifestation of hyperandrogenism in women is hirsutism and excessive terminal hair growth in androgen-dependent areas of the body. Other clinical manifestations of hyperandrogenism include acne vulgaris, weight gain, menstrual irregularities, and acanthosis nigricans [1]. Hyperandrogenemia has been the common feature included in all three mainly proposed and employed diagnostic criteria put forward by the National Institute of Health in 1990, consensus criteria by the American Society for Reproductive Medicine (ASRM) and the European Society of Human Reproduction and Embryology (ESHRE)

at Rotterdam in 2003, and more recently the Androgen Excess Society in 2006, which has asserted the inclusion of presence of clinical and/or biochemical hyperandrogenism to be imperative in diagnosis of PCOS. A controversial opinion regarding the inclusion of hyperandrogenism was debated at the 2012 joint meeting of ASRM and ESHRE whereupon it was noted that it was a significant predictor for diagnosis and prognosis of the syndrome and its accompanying metabolic maladies, thus forming an important criterion for inclusion into multicentric studies in PCOS [18].

PCOS is now considered as a disorder of androgen excess [19, 20]. In women with PCOS, there is increased gonadotropin-releasing hormone (GnRH) pulse frequency which favours increased LH secretion over that of follicle stimulating hormone (FSH) [21]. Under control of high pulsatile release of LH, the theca cells upregulate the expression of steroidogenic acute regulatory protein (StAR), P450 side-chain cleavage (P450scc), 3β-hydroxysteroid dehydrogenase (3β-HSD), and cytochrome P450c17 (CYP17) and increase steroidogenic activity in the theca cells [22], thereby producing androstenedione. This action is further enhanced in a synergistic fashion by the high levels of insulin commonly observed in PCOS women. Androstenedione is then converted by aromatase to estrogen in the granulosa cells under the influence of pituitary FSH. However, there is relative deficit in FSH secretion which often results in impaired and arrested follicular development and reduced aromatase activity, thereby resulting in excess androgen accumulation and hyperandrogenemia in PCOS women.

Polycystic ovaries typically consist of numerous follicles arrested primarily in the preantral and antral stages with thecal hyperplasia and follicular fluid accumulation subsequently forming cyst-like structures which line the periphery of the ovary giving it a string of pearls-like appearance. Increased ovarian stromal volume along with many fluid filled follicles make these ovaries enlarged, a common morphological feature observed in PCOS women. In addition to thickened thecal layers, these follicles show increased steroidogenic activity. Insulin resistance, another major player in PCOS pathophysiology, intensifies the steroid inducing action of LH and indirectly increases LH pulse amplitude and progressively worsens this hyperandrogenemia. Insulin may also act directly via the insulin receptors on the ovary to augment ovarian steroidogenesis [23] and may also stimulate P450c17α activity in ovary and adrenal glands of PCOS women [24]. Insulin indirectly exacerbates hyperandrogenemia by reducing hepatic biosynthesis of sex hormone binding globulin and increasing the free and bioavailable testosterone levels. This creates a precarious physiologic environment of hormonal imbalance promoted by a sequence of hyperinsulinemia followed by hyperandrogenemia [23]. Coupled with lowered aromatase activity and diminished conversion of testosterone to estrogen, the circulating androgen pool continues to grow. Long-term cultures of theca and granulosa cells demonstrated significantly increased enzymatic activities of P450c17α and 3βHSD in PCOS theca cells with subsequent increased synthesis of testosterone precursors compared to normal cells [25, 26]. Follicular hyperandrogenemia induces marked changes in methylation status of essential genes for reproduction and development such as *PPARγ*, *HDAC3*, and *NCOR1* [27]. Increased activity of both 17 and 20 lyase in the Δ4 pathway and 3β-hydroxysteroid dehydrogenase II combined with low aromatase activity was documented in hyperandrogenic PCOS women. Thus, the heterogeneity in androgenic phenotype may be attributed to differential activity of important enzymes in the steroidogenic pathway [28].

While the metabolic derangements of PCOS have mainly been attributed to the insulin resistance and obesity frequently present in these women, increased abdominal adiposity develops as a consequence of hyperandrogenemia. Sensitive indicators of hyperandrogenism include total and free testosterone and androstenedione levels, and free androgen index (FAI) which are capable of predicting phenotype heterogeneity and severity [29]. Recently use of mass spectrometry based techniques such as liquid chromatography- and gas chromatography-MS/MS for serum and urinary steroid hormone profiling have led to more accurate measurements of androgen levels in women with PCOS and have been hailed as the gold standard for testosterone assay by the Endocrine Society [30]. These techniques have enabled researchers to accurately measure low concentrations of testosterone (<5 nmol/l) and prevent inaccurate measurement due to cross-reactivity with DHEAS as seen in immunoassay methods [31]. This has led to assignment of lower cut-off values for total and free testosterone [32], thereby improving diagnosis and subclassification of PCOS women [33]. However hirsutism scores do not necessarily correlate with steroid concentrations measured in affected women [29]. Further, PCOS women showing biochemical androgen excess are more prone to reduced insulin sensitivity and dysglycemia [34], number of menstrual cycles in a year, elevated cardiovascular disease markers, dyslipidemia, and heightened predisposition to metabolic syndrome [35].

Experimental evidence strongly implicates the role of hyperandrogenic intrauterine hormonal milieu in influencing the development of PCOS-like reproductive and metabolic features in monkey and sheep animal models. Prenatally androgenized ewes showed early increase in LH secretion, coupled with progressive loss in reproductive cyclicity and ovulatory failure [36], increased number of primary follicles [37], and persistent follicular cysts [38] along with upregulation of steroidogenic genes such as *StAR*, *CYP11A*, and *CYP17* in thecal cells of female offspring [39]. Similarly, female rhesus monkeys prenatally exposed to testosterone demonstrate LH hypersecretion, anovulation, and polyfollicular ovaries with increased follicular recruitment with impaired oocyte developmental competence [40]. Moreover, maternal or fetal exposure to high doses of androgens has been reported to alter pancreatic morphology, particularly beta cell development, which may contribute to insulin resistance and consequent hyperinsulinemia in rhesus monkeys and sheep models [41, 42]. Adrenal androgen excess has also been reported in prenatally androgenized nonhuman primate models for PCOS as indicated by elevated basal circulating levels of DHEA and DHEAS. Furthermore, treatment with thiazolidinedione-based insulin sensitizers ameliorates adrenal steroidogenesis along with reducing insulin

resistance [43]. Women with PCOS were more likely to give birth to small for gestational age infants which has been associated with increased maternal testosterone levels [44]. Findings of high AMH levels in daughters born to women with PCOS suggest altered follicular development from an early age [45]. Familial prevalence of PCOS and its associated phenotypes provides evidence of possible maternal transmission and genetic inheritance of this disorder [10]. Maternal heritability had significant effects on the prevalence of fasting dysglycemia in women with PCOS [46]. These studies support prenatal activation and fetal programming which passes PCOS-like traits in subsequent generations. Xita et al. have elegantly hypothesized that exposure to androgen excess encourages fetal programming of PCOS by altering phenotypic expression of reproductive and metabolic tissues and results in altered differentiation of thecal cells, LH hypersecretion, and male-type fat distribution in female offspring. Moreover genotypes related to regulating androgen levels, activity and bioavailability in PCOS mothers can modulate the extent of androgen exposure in utero [47]. Epigenetic changes and subsequently altered gene expression and maternal nutrition have also been found to influence fetal programming [47–49]. On the other hand, it was also shown that increased maternal androgen levels may not induce PCOS in female fetus provided normal placental aromatization activity is maintained [49]. Altogether, these findings highlight the role of hyperandrogenism in critical windows of fetal development in modifying PCOS susceptibility.

PCOS is also thus regarded as a form of functional ovarian hyperandrogenism, where all the above-mentioned aberrations contribute towards not only the reproductive dysfunctions but also the metabolic anomalies observed in these women (Figure 1).

4. Genetics of PCOS

PCOS has a strong genetic component as evidenced by clustering of PCOS in families as well as PCOS-like features in both male and female relatives of affected women [1]. Approaches such as twin studies and linkage studies have been employed in order to decipher the contribution of heritability in this multifactorial disorder. Linkage studies of 37 candidate genes predicted strong association of follistatin and nominal association of *CYP11A1* gene in affected siblings with hyperandrogenemia and PCOS related traits. This same study established strong genetic association of *D19S884* allelic marker near *INSR* gene with PCOS, by transmission disequilibrium test [50]. Association studies of candidate genes involved in pathways related to the etiology of the syndrome and its associated anomalies have garnered interest from the research community to try and pinpoint the significance of genetic predisposition in manifestation of this syndrome [10]. Polymorphisms of genes involved in pathways including insulin signaling, gonadotropin regulation, chronic inflammation, and energy homeostasis have been studied [10]; however, the exact role of these susceptibility genes has not yet been established. Single nucleotide polymorphisms (SNPs) reveal functional changes due to

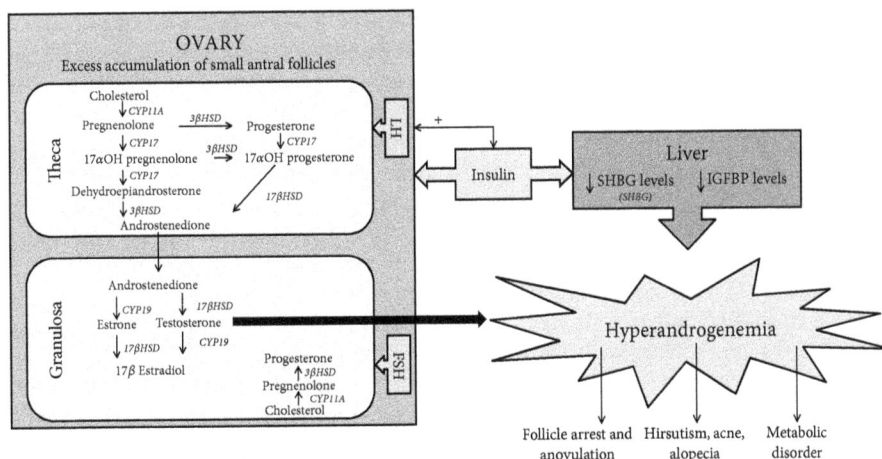

FIGURE 1: Overview of pathophysiology of PCOS. Androgen biosynthesis is a well-orchestrated process occurring in the ovary mediated by an enzymatic cascade under stimulation by pituitary LH. In PCOS, accumulation of small antral follicles with thecal hyperplasia along with overexpression of steroidogenic enzymes results in elevated testosterone levels. In contrast, downregulation of aromatase enzymes decreases testosterone to estradiol conversion, leading to release of large amounts of circulating testosterone. In addition, women with PCOS display insulin resistance coupled with compensatory hyperinsulinemia. Insulin acts directly on the ovary, via its receptors, as well as synergistically with LH to enhance androgen production. On the other hand, insulin acts indirectly via decreasing hepatic biosynthesis of sex hormone binding globulin, thereby raising biologically available testosterone levels. The hyperandrogenic phenotype is typically characterized by arrest in folliculogenesis and consequent anovulatory infertility and cosmetic problems such as hirsutism, acne, and androgenic alopecia. It also contributes to increased incidence of metabolic disorders including insulin resistance, dyslipidemia, metabolic syndrome, and cardiovascular disease.

the fact that amino acid variations or modulation of gene expression and candidate gene approaches are helpful in deciphering the impact of differential frequency distribution in healthy and diseased population. Simultaneously, while candidate gene approaches have been studied in relatively smaller populations, genome-wide association studies have revolutionized the study of PCOS genetics. Previously we have reviewed the genes involved in insulin action and regulation with PCOS susceptibility and related traits [51]. Given the importance of androgens in female reproductive health and PCOS development, in the current review, we will be concentrating on polymorphisms in genes involved in androgen synthesis, action, and bioavailability.

CYP11A1 Gene. CYP11A1 on 15q23-24 encodes the enzyme P450 cholesterol side-chain cleavage that catalyzes the rate limiting step of ovarian steroidogenesis, that is, the conversion of cholesterol to pregnenolone [52, 53]. Theca cells derived from PCOS ovaries and propagated in long-term culture demonstrate increased CYP11A expression compared to normal theca cells [54, 55]. An early linkage study carried out in 20 families showed involvement of *CYP11A* locus in PCOS development and subsequently association of 5'UTR (TTTTA)$_n$ pentanucleotide repeats in hirsute PCOS women [56]. Positive association of pentanucleotide repeat alleles with PCOS susceptibility were confirmed subsequently in women from United States [57], South India [58], and Greece [59] and nominally in women from United Kingdom [60]. Wang et al. have demonstrated that different allele combinations may increase or decrease the risk of PCOS in Chinese women [61]. In contrast to earlier findings, no association was reported in Spanish [62], Chinese [63, 64],

Argentinian [65], Indian [53], and Czech [66] women with PCOS. A recent meta-analysis confirmed strong association of this (TTTTA)$_n$ repeat polymorphism of *CYP11A* with increased risk of PCOS in Caucasian population [67]. Furthermore, another meta-analysis indicated that carriers of 4 repeats had increased risk considering the recessive model while carriers of 6 repeats showed decreased risk of PCOS considering the dominant model [68]. Conflicting reports regarding the association of these pentanucleotide repeats with PCOS related traits are available. Increased testosterone levels have been reported in carriers of short alleles in women with PCOS [53, 59] while no effect of allele dose was seen on CYP11A transcription [57] or serum androgen levels in another studies [57, 60, 63]. What is more this repeat polymorphism shows significant relationship with metabolic traits including obesity [61], higher waist-hip ratio, decreased AUC glucose values [64], alleviated dyslipidemia [66], and decreased FSH levels [66]. Another polymorphism, rs4077582, showed significant association in Chinese women with PCOS [69, 70] as well as altered testosterone and LH levels [70]. One more polymorphism, namely, rs11632698, showed both positive [69] and negative [70] association with PCOS risk in Chinese women. Together, these studies imply *CYP11A* to be a promising genetic biomarker for PCOS.

CYP17 Gene. The *CYP17* gene at 10q24.3 encodes cytochrome P450 enzyme with 17-hydroxylase activity, which converts pregnenolone and progesterone into 17-hydroxypregnenolone and 17-hydroxyprogesterone, respectively. The 17,20-lyase activity subsequently converts these steroids to dehydroepiandrosterone (DHEA) and 4-androstenedione [10]. The vast majority of studies have focused on a widely studied polymorphism at −34 position (−34 T/C) in the promoter, which creates an additional Sp1 transcription factor binding site, thereby regulating expression of CYP17 and consequently androgen levels [71]. In 1994, Carey et al. showed significant association of this polymorphism with PCO and male pattern baldness in a family-based study [72]; however, these findings were not persistent when they increased the sample size [73]. On similar lines, this polymorphism was not found to be a significant factor for PCOS development in British [74], Slovenian [75], Polish [76], American [77, 78], Korean [79], Chilean [80], Chinese [81], Thai [82], and Indian [83] women with PCOS or even in Turkish adolescents [84]. In contrast, Indian women with PCOS showed significantly increased frequency of C allele [53]. This polymorphism impacts the hyperandrogenic phenotype in women with PCOS [53, 65, 81]. Interestingly, this polymorphism negatively influenced metabolic traits including obesity [80, 82], waist circumference [80], and insulin resistance [80]. A meticulous meta-analysis taking into consideration all studies revealed that this variant was not associated with risk of PCOS development when considering any genetic model or even after stratification by country and ethnicity. Furthermore, amongst studies which were in Hardy-Weinberg equilibrium, significantly increased risk was seen considering the dominant genetic model. However, they suggest sample size may also influence these associations as shown by increased risk in small sample compared to large sample studies [85].

CYP19 Gene. The aromatase p450 enzyme, essential for synthesis of estrogen from androgens, is encoded by *CYP19* gene on chromosome 15q21.2 [86]. Reduced aromatase activity in both lean and obese women with PCOS has reported [87] and activity is further inhibited by hyperandrogenemia [28]. Yang et al. have demonstrated decreased aromatase expression concomitant with increased levels of testosterone in follicular fluid derived from PCOS women [88]. Promoter hypermethylation and reduced CYP19A1 mRNA and protein levels were evident in PCOS ovaries, suggesting repressed aromatase expression [89]. An intronic variant rs2414096 was shown to be significantly associated with increased risk of PCOS development and with increased estradiol to testosterone ratio (E2/T), FSH levels, and age of menarche in Han Chinese women [90]. Raised PCOS symptom score and changes in circulating estradiol and testosterone concentrations were observed in adolescent girls in the UK carrying this polymorphism [91]. Additionally, certain promoter variants were independently associated with PCOS symptom score in UK adolescents [92]. The rs2414096 polymorphism lacked association with PCOS or with alterations in hormonal and metabolic variables after undergoing a 6-month treatment regime of oral contraceptives in both anovulatory and ovulatory PCOS women [93]. Another common polymorphism, a tetranucleotide repeat polymorphism (TTTA)$_n$ in the fourth intron related to suboptimal aromatase activity [94], has been investigated. Reports are available indicating that short allele repeats, predominantly consisting of seven repeats, are prevalent in Greek [94–96] and Han Chinese [97] women with PCOS compared to controls. These short repeat

alleles were associated with hormonal parameters including increased testosterone levels, high LH : FSH ratios [94], and reduced reproductive markers such as number of large follicles and total oocyte count [95]. Interestingly, these alleles predicted successful pregnancy following assisted reproductive technique intervention [95]. Carriers of 11 repeat alleles are also commonly found in Chinese women with PCOS [97, 98] which influence lipid metabolism [97]. Another polymorphism, rs2470152, did not affect PCOS risk but the heterozygous TC genotype was found to be significantly associated with increased testosterone levels with decreased E2/T ratio, suggesting role of this polymorphism in regulating aromatase activity [99]. A missense polymorphism, Arg264Cys, increases aromatase activity and affects PCOS susceptibility [100]. However, the above findings indicate a definite role of this gene in PCOS outcome.

AR Gene. The androgen receptor (AR) gene located on the X chromosome encodes the AR, which consists of a poorly conserved N terminal domain containing highly polymorphic CAG repeats [101]. An inverse correlation has been demonstrated between CAG repeat number and AR transactivation efficiency [102]. An interesting case study reported that a woman carrying a heterozygous *AR* gene mutation gave birth to a baby with androgen insensitivity syndrome suggesting plausible repercussions on reproductive outcomes associated with *AR* gene mutations [103] and not only with repeat lengths. AR has been primarily localized in the theca interna cells of preantral follicles, granulosa cells of preantral and antral follicles, and both theca and granulosa cells of dominant follicles [104]. Inconsistent associations of the differences in number of CAG repeats in exon 1 have been reported with PCOS prevalence. It has later been ascertained that short *AR* CAG repeats were more frequent in PCOS cases and may possibly be linked to PCOS onset in both Chinese and Caucasian populations [100, 105–108]. This may contribute to the inherent hyperandrogenic phenotype commonly seen in women with PCOS by increasing AR activity and enhancing androgen sensitivity to even low circulating levels of testosterone, thereby promoting hirsutism, acne, and irregular cycles [106, 109]. Anovulatory normoandrogenic PCOS women showed a significant trend towards short CAG repeat length indicating increased intrinsic androgen sensitivity [110]. Furthermore, they found that Indian women showed comparatively shorter repeat lengths compared to Chinese women, indicating possible role of ethnic variation [110]. No association of *AR* CAG repeat lengths with PCOS was reported in Indian [111], Slovene [112], Korean [113], Croatian, [114] and Finnish women [115]. A few studies have indicated that CAG repeat lengths may also modify both testosterone and insulin resistance parameters in women with PCOS despite failing to show association with PCOS risk. This CAG repeat polymorphism was found to be a significant predictor of serum circulating testosterone levels in Croatian [114], Brazilian [116], Chinese [101] and Korean [113] women with PCOS. German women carrying short CAG repeats presented with increased testosterone which in turn aggravated insulin resistance in these women suggesting a putative effect of CAG repeats as an underlying mechanism of hyperandrogenemia induced insulin resistance [117]. In contrast, infertile Australian PCOS women showed preferential expression of long CAG repeat alleles compared to fertile PCOS women [118]. Meta-analyses examining the relationship between CAG repeat lengths at *AR* and PCOS risk have concluded that they may not be major determining factors in PCOS etiology [111, 119, 120]. Apart from CAG repeat, other groups have concluded that a GGN repeat polymorphism and rs6152G/A polymorphism were also significantly associated with PCOS in Chinese women [121, 122]. Thus, *AR* polymorphisms may exacerbate the hyperandrogenic phenotype of women with PCOS.

SHBG Gene. Sex hormone binding globulin (SHBG) is primarily synthesized in the liver, binds androgens, and estrogen with high affinity, thereby lowering circulating steroid hormones and rendering them biologically unavailable to target tissues [123]. Several polymorphisms in the *SHBG* gene located on chromosome 17 have been shown to alter hepatic biosynthesis, plasma levels, and plasma clearance efficiency of SHBG, thereby regulating the distribution of sex steroid hormones [123]. Two novel coding region mutations were discovered in a woman showing severe SHBG deficiency, one which resulted in abnormal glycosylation and the other to truncated SHBG synthesis. This led to remarkably low SHBG levels with elevated circulating free testosterone concentrations [124]. The putative genetic contribution of *SHBG* polymorphisms was further supported by evidence of association of longer TAAAA repeats with late onset of menarche [125] and decreased SHBG levels in hirsute French women [126]. Long TAAAA repeat alleles failed to show association with PCOS risk in Croatian [127], Slovenian [128], French [126], and Chinese [129] women. However Greek women with PCOS had significantly greater frequency of long repeat alleles compared to controls [130]. An inverse association between TAAAA repeat polymorphism alone [126–128, 130] or coupled with short *AR* CAG repeats [96] and SHBG serum levels has been established. Greek women with PCOS having long *SHBG* alleles coupled with short *CYP19* alleles demonstrated low SHBG levels and increased testosterone levels with raised FAI, DHEAS and T/E2 ratios [96]. A meta-analysis was unable to draw a conclusive association between the TAAAA repeat polymorphism with PCOS risk indicating that it may not be a reliable predictor of PCOS onset [131]. A functional missense polymorphism in exon 8 causes an amino acid change from aspartic acid to asparagine (D327N), delays SHBG half-life, and influences the metabolism of SHBG [126]. Another missense polymorphism, E326K lowered SHBG levels in women with PCOS independently of BMI, androgen, and insulin related traits [132]. Family-based and case-control association studies have found that rs1799941 and rs727428 in *SHBG* gene influenced SHBG metabolism in American and Mediterranean women with PCOS [133, 134], but not PCOS risk [133]. A recent study in Bahraini women has concluded that haplotypes spanning six polymorphisms were associated with either increased or decreased PCOS susceptibility [135] rekindling interest in *SHBG* gene polymorphisms in PCOS susceptibility.

StAR Gene. The *StAR* gene located on chromosome 8p11.2 encodes the steroidogenic acute regulatory protein which binds to and facilitates uptake of cholesterol into mitochondria of cells for steroidogenesis. However a pilot study carried out in Iranian women investigating seven known polymorphisms showed no significant association with PCOS risk [136].

HSD17B5 Gene. The enzyme type 17β-hydroxysteroid dehydrogenase type 5 (HSD17B5) is instrumental in converting androstenedione to testosterone in theca cells and adrenal glands [137]. The −71A/G polymorphism in the promoter region was revealed for the first time by Qin et al., who also investigated its prevalence in a population of ethnically diverse PCOS women. Here they found that this variant was associated with PCOS susceptibility in Caucasian but not in African American women with PCOS [138]. It also modulates testosterone biosynthesis and thereby plasma testosterone levels [138]. Subsequent studies failed to find this association in Greek [139] and Caucasian [140] women with PCOS. Intronic polymorphism rs12529 affected testosterone levels but PCOS risk remained unchanged in Chinese women. On the other hand, rs1937845 not only increased risk of PCOS development but also increased homeostasis model assessment of β-cell function (HOMA-B) index and testosterone levels in these women [137]. In Brazilian women with PCOS, improvement in hyperandrogenic phenotype could be attributed to treatment regimen with oral contraceptive pills but not *HSD17B5* polymorphisms [141].

INSL3 Gene. Insulin-like factor 3 (INSL3) is localized in the thecal cells and corpus luteum of the ovary. A pioneering study by Glister et al. established the role of INSL3-RXFP2 signaling in maintaining androgen production by the ovarian theca cells [142]. Recently, women with PCOS were reported to have increased serum INSL3 levels [143–145]. *INSL3* polymorphisms may have an important role in modulating ovarian steroidogenesis and hence contribute to the pathogenesis of PCOS. To the best of our knowledge, our group has conducted the first case-control association study investigating relationship between *INSL3* polymorphisms and its haplotypes with PCOS susceptibility and its related traits in a well characterized cohort of Indian women with PCOS [146]. Our study showed that the A/G *rs6523* polymorphism present in exon 1 of *INSL3* was significantly associated with PCOS susceptibility. Other coding region polymorphisms along with the *rs6543* SNP affect both the metabolic and hyperandrogenemia related traits of PCOS in both controls and women with PCOS. These polymorphisms have differential influence depending on the physiological state present [146]. No other studies have been attempted to replicate this association till date.

5. Conclusion

PCOS remains an endocrine enigma even today characterized by adverse hormonal perturbations raising metabolic and gynecological concerns in affected women. Genetic factors work in tandem with environmental signals contributing to its pathogenesis. A hallmark feature of PCOS remains augmented androgen synthesis and consequent circulating levels which is frequently associated with cosmetic complaints including hirsutism, acne, and alopecia. The ovary remains the primary source of hyperandrogenism in women with PCOS. Thecal cell hyperplasia coupled with enhanced steroidogenic potential of androgen pathway enzymes may contribute to excess androgen production in ovaries of affected women. The current review has encapsulated salient findings from candidate gene based association studies of polymorphisms in genes involved in steroidogenesis as well as androgen levels and action which are presumed to govern PCOS susceptibility and phenotypic heterogeneity of the disorder. However, candidate gene studies have not provided conclusive results due to different diagnostic criteria, the likely contribution of multiple genes, differences in lifestyle, environmental factors, and the sample size studied. On the other hand, genome-wide association studies (GWAS) empower researchers with the capacity to explore thousands of variants across the entire genome in both case and control participants to uncover association of genetic variants with complex disease in an unbiased manner. The notable GWAS studies in Chinese populations have essentially offered several loci mapping to *DENND1A*, *THADA*, *LHCGR*, *FSHR*, *INSR*, *TOX3*, *YAP1*, *RAB5B*, *c9orf3*, *HMGA2*, and *SUMO1P1/ZNF217* involved in steroidogenesis, gonadotropin action and regulation, follicular development, insulin signaling and type 2 diabetes mellitus (T2DM), calcium signaling, and endocytosis [147, 148]. Of these loci, *DENND1A* has been implicated as a driving force for PCOS hyperandrogenemia. Overexpression in normal ovaries was found to upregulate ovarian steroidogenesis whereas knockdown decreases steroid synthesis by reducing transcription of CYP11A1 and CYP17 [149]. Interestingly, alternative splicing of DENND1A to produce v.2 variant is supposed to be important in PCOS development [150] and DENND1A v.2 was highly concentrated in theca cells of ovaries of women with PCOS [149]. On the other hand, although DENDD1A v.1 was abundant in the NCI-H295 adrenal steroidogenic cell line, overexpression of v.2 increased expression of CYP17 and CYP11A enzymes [150]. The association of the *LHCGR* locus with PCOS in GWAS [148] strengthens the rationale that alteration in receptor expression could contribute to LH hyperstimulation, thereby enhancing steroidogenesis. Findings from GWAS in European population has highlighted the significant association of gene polymorphisms in *FDFT1* and *GATA4* involved in cholesterol synthesis and a potent regulator of steroidogenic gene transcription, respectively, suggesting altered androgen synthesis [151]. Thus while candidate genes have offered substantial evidence to strengthen the role of genetic variants in modulating PCOS hyperandrogenism, GWAS has provided new clues which need to be explored in greater detail in different ethnic populations. Selection of suitable candidate genes should continue in order to successfully delineate the genetic underpinnings of a multigenic complex disorder like PCOS. These would pave the way for establishing genetic predisposition profiles which could be harnessed for designing therapeutic management strategies in future.

Acknowledgments

The authors acknowledge the financial assistance provided by University Grants Commission to Roshan Dadachanji for pursuing her doctoral studies. The authors gratefully acknowledge NIRRH (REV/526/08-2017) for providing necessary support.

References

[1] M. O. Goodarzi, D. A. Dumesic, G. Chazenbalk, and R. Azziz, "Polycystic ovary syndrome: etiology, pathogenesis and diagnosis," *Nature Reviews Endocrinology*, vol. 7, no. 4, pp. 219–231, 2011.

[2] H. Teede, A. Deeks, and L. Moran, "Polycystic ovary syndrome: a complex condition with psychological, reproductive and metabolic manifestations that impacts on health across the lifespan," *BMC Medicine*, vol. 8, article 41, 2010.

[3] S. Jonard and D. Dewailly, "The follicular excess in polycystic ovaries, due to intra-ovarian hyperandrogenism, may be the main culprit for the follicular arrest," *Human Reproduction Update*, vol. 10, no. 2, pp. 107–117, 2004.

[4] J. Qiao and H. L. Feng, "Extra- and intra-ovarian factors in polycystic ovary syndrome: Impact on oocyte maturation and embryo developmental competence," *Human Reproduction Update*, vol. 17, no. 1, pp. 17–33, 2011.

[5] M. O. Goodarzi, E. Carmina, and R. Azziz, "DHEA, DHEAS and PCOS," *The Journal of Steroid Biochemistry and Molecular Biology*, vol. 145, pp. 213–225, 2015.

[6] N. M. P. Daan, Y. V. Louwers, M. P. H. Koster et al., "Cardiovascular and metabolic profiles amongst different polycystic ovary syndrome phenotypes: who is really at risk?" *Fertility and Sterility*, vol. 102, no. 5, pp. 1444.e3–1451.e3, 2014.

[7] R. Yang, S. Yang, R. Li, P. Liu, J. Qiao, and Y. Zhang, "Effects of hyperandrogenism on metabolic abnormalities in patients with polycystic ovary syndrome: A meta-analysis," *Reproductive Biology and Endocrinology*, vol. 14, no. 1, article no. 67, 2016.

[8] R. S. Legro, D. Driscoll, J. F. Strauss III, J. Fox, and A. Dunaif, "Evidence for a genetic basis for hyperandrogenemia in polycystic ovary syndrome," *Proceedings of the National Acadamy of Sciences of the United States of America*, vol. 95, no. 25, pp. 14956–14960, 1998.

[9] N. Xita, I. Georgiou, and A. Tsatsoulis, "The genetic basis of polycystic ovary syndrome," *European Journal of Endocrinology*, vol. 147, no. 6, pp. 717–725, 2002.

[10] N. Prapas, A. Karkanaki, I. Prapas, I. Kalogiannidis, I. Katsikis, and D. Panidis, "Genetics of polycystic ovary syndrome," *Hippokratia*, vol. 13, no. 4, pp. 216–223, 2009.

[11] E. A. McGee and A. J. W. Hsueh, "Initial and cyclic recruitment of ovarian follicles," *Endocrine Reviews*, vol. 21, no. 2, pp. 200–214, 2000.

[12] W. L. Miller, "Molecular biology of steroid hormone synthesis," *Endocrine Reviews*, vol. 9, no. 3, pp. 295–318, 1988.

[13] K. L. Parker and B. P. Schimmer, "Transcriptional regulation of the genes encoding the cytochrome P-450 steroid hydroxylases," in *Vitamins and Hormones*, vol. 51, pp. 339–370, 1995.

[14] S. Gingras, S. Côté, and J. Simard, "Multiple signal transduction pathways mediate interleukin-4-induced 3β-hydroxysteroid dehydrogenase/Δ5-Δ4 isomerase in normal and tumoral target tissues," *The Journal of Steroid Biochemistry and Molecular Biology*, vol. 76, no. 1-5, pp. 213–225, 2001.

[15] S. Leers-Sucheta, K.-I. Morohashi, J. I. Mason, and M. H. Melner, "Synergistic activation of the human type II 3β-hydroxysteroid dehydrogenase/Δ5-Δ4 isomerase promoter by the transcription factor steroidogenic factor-1/adrenal 4-binding protein and phorbol ester," *The Journal of Biological Chemistry*, vol. 272, no. 12, pp. 7960–7967, 1997.

[16] J. I. Mason, D. S. Keeney, I. M. Bird et al., "The regulation of 3β-hydroxysteroid dehydrogenase expression," *Steroids*, vol. 62, no. 1, pp. 164–168, 1997.

[17] J. Simard, M.-L. Ricketts, S. Gingras, P. Soucy, F. A. Feltus, and M. H. Melner, "Molecular biology of the 3β-hydroxysteroid dehydrogenase/Δ5-Δ4 isomerase gene family," *Endocrine Reviews*, vol. 26, no. 4, pp. 525–582, 2005.

[18] L. Gianaroli, C. Racowsky, J. Geraedts, M. Cedars, A. Makrigiannakis, and R. A. Lobo, "Best practices of ASRM and ESHRE: A journey through reproductive medicine," *Fertility and Sterility*, vol. 98, no. 6, pp. 1380–1394, 2012.

[19] R. Azziz, "Androgen excess is the key element in polycystic ovary syndrome," *Fertility and Sterility*, vol. 80, no. 2, pp. 252–254, 2003.

[20] E. Diamanti-Kandarakis, J. Papailiou, and S. Palimeri, "Hyperandrogenemia: pathophysiology and its role in ovulatory dysfunction in PCOS," *Pediatric Endocrinology Reviews*, vol. 3, supplement 1, pp. 198–204, 2006.

[21] S. K. Blank, C. R. McCartney, K. D. Helm, and J. C. Marshall, "Neuroendocrine effects of androgens in adult polycystic ovary syndrome and female puberty," *Seminars in Reproductive Medicine*, vol. 25, no. 5, pp. 352–359, 2007.

[22] E. Diamanti-Kandarakis, G. Argyrakopoulou, F. Economou, E. Kandaraki, and M. Koutsilieris, "Defects in insulin signaling pathways in ovarian steroidogenesis and other tissues in polycystic ovary syndrome (PCOS)," *The Journal of Steroid Biochemistry and Molecular Biology*, vol. 109, no. 3–5, pp. 242–246, 2008.

[23] S. Mukherjee and A. Maitra, "Molecular & genetic factors contributing to insulin resistance in polycystic ovary syndrome," *Indian Journal of Medical Research*, vol. 131, pp. 743–760, 2010.

[24] G. N. Allahbadia and R. Merchant, "Polycystic ovary syndrome and impact on health," *Middle East Fertility Society Journal*, vol. 16, no. 1, pp. 19–37, 2011.

[25] V. L. Nelson, K. Qin, R. L. Rosenfield et al., "The biochemical basis for increased testosterone production in theca cells propagated from patients with polycystic ovary syndrome," *The Journal of Clinical Endocrinology & Metabolism*, vol. 86, no. 12, pp. 5925–5933, 2001.

[26] K. Takayama, T. Fukaya, H. Sasano et al., "Immunohistochemical study of steroidogenesis and cell proliferation in polycystic ovarian syndrome," *Human Reproduction*, vol. 11, no. 7, pp. 1387–1392, 1996.

[27] F. Qu, F.-F. Wang, R. Yin et al., "A molecular mechanism underlying ovarian dysfunction of polycystic ovary syndrome: Hyperandrogenism induces epigenetic alterations in the granulosa cells," *Journal of Molecular Medicine*, vol. 90, no. 8, pp. 911–923, 2012.

[28] S. F. De Medeiros, J. S. Barbosa, and M. M. W. Yamamoto, "Comparison of steroidogenic pathways among normoandrogenic and hyperandrogenic polycystic ovary syndrome patients and normal cycling women," *Journal of Obstetrics and Gynaecology Research*, vol. 41, no. 2, pp. 254–263, 2015.

[29] R. Pasquali, L. Zanotti, F. Fanelli et al., "Defining hyperandrogenism in women with polycystic ovary syndrome: a challenging perspective," *The Journal of Clinical Endocrinology & Metabolism*, vol. 101, no. 5, pp. 2013–2022, 2016.

[30] W. Rosner, R. J. Auchus, R. Azziz, P. M. Sluss, and H. Raff, "Position statement: Utility, limitations, and pitfalls in measuring testosterone: An endocrine society position statement," *The Journal of Clinical Endocrinology & Metabolism*, vol. 92, no. 2, pp. 405–413, 2007.

[31] J. H. Barth, H. P. Field, E. Yasmin, and A. H. Balen, "Defining hyperandrogenism in polycystic ovary syndrome: Measurement of testosterone and androstenedione by liquid chromatography-tandem mass spectrometry and analysis by receiver operator characteristic plots," *European Journal of Endocrinology*, vol. 162, no. 3, pp. 611–615, 2010.

[32] W. A. Salameh, M. M. Redor-Goldman, N. J. Clarke, R. Mathur, R. Azziz, and R. E. Reitz, "Specificity and predictive value of circulating testosterone assessed by tandem mass spectrometry for the diagnosis of polycystic ovary syndrome by the National Institutes of Health 1990 criteria," *Fertility and Sterility*, vol. 101, no. 4, pp. 1135.e2–1141.e2, 2014.

[33] L. M. Bloem, K.-H. Storbeck, P. Swart, T. Du Toit, L. Schloms, and A. C. Swart, "Advances in the analytical methodologies: Profiling steroids in familiar pathways-challenging dogmas," *The Journal of Steroid Biochemistry and Molecular Biology*, vol. 153, article no. 4400, pp. 80–92, 2015.

[34] M. W. O'Reilly, A. E. Taylor, N. J. Crabtree et al., "Hyperandrogenemia predicts metabolic phenotype in polycystic ovary syndrome: the utility of serum androstenedione," *The Journal of Clinical Endocrinology and Metabolism*, vol. 99, no. 3, pp. 1027–1036, 2014.

[35] Y.-A. Sung, J.-Y. Oh, H. Chung, and H. Lee, "Hyperandrogenemia is implicated in both the metabolic and reproductive morbidities of polycystic ovary syndrome," *Fertility and Sterility*, vol. 101, no. 3, pp. 840–845, 2014.

[36] R. A. Birch, V. Padmanabhan, D. L. Foster, W. P. Unsworth, and J. E. Robinson, "Prenatal programming of reproductive neuroendocrine function: Fetal androgen exposure produces progressive disruption of reproductive cycles in sheep," *Endocrinology*, vol. 144, no. 4, pp. 1426–1434, 2003.

[37] R. A. Fordslike, K. Hardy, L. Bull et al., "Disordered follicle decelopment in ovaries of prenatally androgenized ewes," *Journal of Endocrinology*, vol. 192, no. 2, pp. 421–428, 2007.

[38] M. Manikkam, T. L. Steckler, K. B. Welch, E. K. Inskeep, and V. Padmanabhan, "Fetal programming: Prenatal testosterone treatment leads to follicular persistence/luteal defects; partial restoration of ovarian function by cyclic progesterone treatment," *Endocrinology*, vol. 147, no. 4, pp. 1997–2007, 2006.

[39] K. Hogg, J. M. Young, E. M. Oliver, C. J. Souza, A. S. McNeilly, and W. C. Duncan, "Enhanced thecal androgen production is prenatally programmed in an ovine model of polycystic ovary syndrome," *Endocrinology*, vol. 153, no. 1, pp. 450–461, 2012.

[40] D. A. Dumesic, D. H. Abbott, and V. Padmanabhan, "Polycystic ovary syndrome and its developmental origins," *Reviews in Endocrine and Metabolic Disorders*, vol. 8, no. 2, pp. 127–141, 2007.

[41] L. E. Nicol, T. D. O'Brien, D. A. Dumesic, T. Grogan, A. F. Tarantal, and D. H. Abbott, "Abnormal infant islet morphology precedes insulin resistance in PCOS-like monkeys," *PLoS ONE*, vol. 9, no. 9, Article ID 0106527, 2014.

[42] M. Rae, C. Grace, K. Hogg et al., "The pancreas is altered by in utero androgen exposure: implications for clinical conditions such as polycystic ovary syndrome (PCOS)," *PLoS ONE*, vol. 8, no. 2, Article ID e56263, 2013.

[43] D. Abbott, R. Zhou, I. Bird, D. Dumesic, and A. Conley, "Fetal programming of adrenal androgen excess: Lessons from a nonhuman primate model of polycystic ovary syndrome," *Endocrine Development*, vol. 13, pp. 145–158, 2008.

[44] T. Sir-Petermann, C. Hitchsfeld, M. Maliqueo et al., "Birth weight in offspring of mothers with polycystic ovarian syndrome," *Human Reproduction*, vol. 20, no. 8, pp. 2122–2126, 2005.

[45] T. Sir-Petermann, E. Codner, M. Maliqueo et al., "Increased anti-müllerian hormone serum concentrations in prepubertal daughters of women with polycystic ovary syndrome," *The Journal of Clinical Endocrinology & Metabolism*, vol. 91, no. 8, pp. 3105–3109, 2006.

[46] K. Kobaly, P. Vellanki, R. K. Sisk et al., "Parent-of-origin effects on glucose homeostasis in polycystic ovary syndrome," *The Journal of Clinical Endocrinology & Metabolism*, vol. 99, no. 8, pp. 2961–2966, 2014.

[47] N. Xita and A. Tsatsoulis, "Review: fetal programming of polycystic ovary syndrome by androgen excess: Evidence from experimental, clinical, and genetic association studies," *The Journal of Clinical Endocrinology & Metabolism*, vol. 91, no. 5, pp. 1660–1666, 2006.

[48] N. Xita and A. Tsatsoulis, "Fetal origins of the metabolic syndrome," *Annals of the New York Academy of Sciences*, vol. 1205, pp. 148–155, 2010.

[49] D. A. Dumesic, M. O. Goodarzi, G. D. Chazenbalk, and D. H. Abbott, "Intrauterine environment and polycystic ovary syndrome," *Seminars in Reproductive Medicine*, vol. 32, no. 3, pp. 159–165, 2014.

[50] M. Urbanek, R. S. Legro, D. A. Driscoll et al., "Thirty-seven candidate genes for polycystic ovary syndrome: Strongest evidence for linkage is with follistatin," *Proceedings of the National Acadamy of Sciences of the United States of America*, vol. 96, no. 15, pp. 8573–8578, 1999.

[51] N. Shaikh, R. Dadachanji, and S. Mukherjee, "Genetic markers of polycystic ovary syndrome: emphasis on insulin resistance," *International Journal of Medical Genetics*, vol. 2014, Article ID 478972, 10 pages, 2014.

[52] W. L. Miller, "Androgen biosynthesis from cholesterol to DHEA," *Molecular and Cellular Endocrinology*, vol. 198, no. 1-2, pp. 7–14, 2002.

[53] M. Pusalkar, P. Meherji, J. Gokral, S. Chinnaraj, and A. Maitra, "CYP11A1 and CYP17 promoter polymorphisms associate with hyperandrogenemia in polycystic ovary syndrome," *Fertility and Sterility*, vol. 92, no. 2, pp. 653–659, 2009.

[54] V. L. Nelson, R. S. Legro, J. F. Strauss III, and J. M. McAllister, "Augmented androgen production is a stable steroidogenic phenotype of propagated theca cells from polycystic ovaries," *Molecular Endocrinology*, vol. 13, no. 6, pp. 946–957, 1999.

[55] J. K. Wickenheisser, J. M. Biegler, V. L. Nelson-DeGrave, R. S. Legro, J. F. Strauss III, and J. M. McAllister, "Cholesterol side-chain cleavage gene expression in theca cells: augmented transcriptional regulation and mRNA stability in polycystic ovary syndrome," *PLoS ONE*, vol. 7, no. 11, Article ID e48963, 2012.

[56] N. Gharani, D. M. Waterworth, S. Batty et al., "Association of the steroid synthesis gene CYP11a with polycystic ovary syndrome and hyperandrogenism," *Human Molecular Genetics*, vol. 6, no. 3, pp. 397–402, 1997.

[57] S. Daneshmand, S. R. Weitsman, A. Navab, A. J. Jakimiuk, and D. A. Magoffin, "Overexpression of theca-cell messenger RNA in polycystic ovary syndrome does not correlate with polymorphisms in the cholesterol side-chain cleavage and 17α-hydroxylase/C17-20 lyase promoters," *Fertility and Sterility*, vol. 77, no. 2, pp. 274–280, 2002.

[58] K. R. Reddy, M. L. N. Deepika, K. Supriya et al., "CYP11A1 microsatellite (tttta)$_n$ polymorphism in PCOS women from South India," *Journal of Assisted Reproduction and Genetics*, vol. 31, no. 7, pp. 857–863, 2014.

[59] E. Diamanti-Kandarakis, M. I. Bartzis, A. T. Bergiele, T. C. Tsianateli, and C. R. Kouli, "Microsatellite polymorphism (tttta)(n) at —528 base pairs of gene CYP11α influences hyperandrogenemia in patients with polycystic ovary syndrome," *Fertility and Sterility*, vol. 73, no. 4, pp. 735–741, 2000.

[60] M. Gaasenbeek, B. L. Powell, U. Sovio et al., "Large-scale analysis of the relationship between CYP11A Promoter variation, polycystic ovarian syndrome, and serum testosterone," *The Journal of Clinical Endocrinology & Metabolism*, vol. 89, no. 5, pp. 2408–2413, 2004.

[61] Y. Wang, X. Wu, Y. Cao, L. Yi, and J. Chen, "A microsatellite polymorphism (tttta)n in the promoter of the CYP11a gene in Chinese women with polycystic ovary syndrome," *Fertility and Sterility*, vol. 86, no. 1, pp. 223–226, 2006.

[62] J. L. San Millán, J. Sancho, R. M. Calvo, and H. F. Escobar-Morreale, "Role of the pentanucleotide (tttta)n polymorphism in the promoter of the CYP11a gene in the pathogenesis of hirsutism," *Fertility and Sterility*, vol. 75, no. 4, pp. 797–802, 2001.

[63] T. Li and Z. Guijin, "Role of the pentanucleotide (tttta)n polymorphisms of CYP11α gene in the pathogenesis of hyperandrogenism in chinese women with polycystic ovary syndrome," *Journal of Huazhong University of Science and Technology (Medical Sciences)*, vol. 25, no. 2, pp. 212–214, 2005.

[64] C. F. Hao, H. C. Bao, N. Zhang, H. F. Gu, and Z. J. Chen, "Evaluation of association between the CYP11alpha promoter pentannucleotide (TTTTA)n polymorphism and polycystic ovarian syndrome among Han Chinese women," *Neuroendocrinology Letters*, vol. 30, no. 1, pp. 56–60, 2009.

[65] M. S. Perez, G. E. Cerrone, H. Benencia, N. Marquez, E. De Piano, and G. D. Frechtel, "Polymorphism in CYP11α and CYP17 genes and the etiology of hyperandrogenism in patients with polycystic ovary syndrome," *Medicina (B Aires)*, vol. 68, no. 2, pp. 129–134, 2008.

[66] S. Prazakova, M. Vankova, O. Bradnova et al., "(TTTTA)n polymorphism in the promoter of the CYP11A1 gene in the pathogenesis of polycystic ovary syndrome," *Casopis Lekaru Ceskych*, vol. 149, no. 11, pp. 520–525, 2010.

[67] W. Shen, T. Li, Y. Hu, H. Liu, and M. Song, "Common polymorphisms in the CYP1A1 and CYP11A1 genes and polycystic ovary syndrome risk: A meta-analysis and meta-regression," *Archives of Gynecology and Obstetrics*, vol. 289, no. 1, pp. 107–118, 2014.

[68] M. Yu, R. Feng, X. Sun et al., "Polymorphisms of pentanucleotide repeats (tttta)n in the promoter of CYP11A1 and their relationships to polycystic ovary syndrome (PCOS) risk: A meta-analysis," *Molecular Biology Reports*, vol. 41, no. 7, pp. 4435–4445, 2014.

[69] G. H. Gao, Y. X. Cao, L. Yi, Z. L. Wei, Y. P. Xu, and C. Yang, "Polymorphism of CYP11A1 gene in Chinese patients with polycystic ovarian syndrome," *Zhonghua Fu Chan Ke Za Zhi*, vol. 45, no. 3, pp. 191–196, 2010.

[70] C.-W. Zhang, X.-L. Zhang, Y.-J. Xia et al., "Association between polymorphisms of the CYP11A1 gene and polycystic ovary syndrome in Chinese women," *Molecular Biology Reports*, vol. 39, no. 8, pp. 8379–8385, 2012.

[71] L. Sharp, A. H. Cardy, S. C. Cotton, and J. Little, "CYP17 gene polymorphisms: Prevalence and associations with hormone levels and related factors. A HuGE review," *American Journal of Epidemiology*, vol. 160, no. 8, pp. 729–740, 2004.

[72] A. H. Carey, D. Waterworth, K. Patel et al., "Polycystic ovaries and premature male pattern baldness are associated with one allele of the steroid metabolism gene CYP17," *Human Molecular Genetics*, vol. 3, no. 10, pp. 1873–1876, 1994.

[73] N. Gharani, D. M. Waterworth, R. Williamson, and S. Franks, "5′ Polymorphism of the CYP17 gene is not associated with serum testosterone levels in women with polycystic ovaries," *The Journal of Clinical Endocrinology & Metabolism*, vol. 81, no. 11, p. 4174, 1996.

[74] K. Techatraisak, G. S. Conway, and G. Rumsby, "Frequency of a polymorphism in the regulatory region of the 17α-hydroxylase-17,20-lyase (CYP17) gene in hyperandrogenic states," *Clinical Endocrinology*, vol. 46, no. 2, pp. 131–134, 1997.

[75] M. Liović, J. Prezelj, A. Kocijančič, G. Majdič, and R. Komel, "CYP17 gene analysis in hyperandrogenised women with and without exaggerated 17-hydroxyprogesterone response to ovarian stimulation," *Journal of Endocrinological Investigation*, vol. 20, no. 4, pp. 189–193, 1997.

[76] B. Marszalek, M. Laciski, N. Babych et al., "Investigations on the genetic polymorphism in the region of CYP17 gene encoding 5′-UTR in patients with polycystic ovarian syndrome," *Gynecological Endocrinology*, vol. 15, no. 2, pp. 123–128, 2001.

[77] A. K. Chua, R. Azziz, and M. O. Goodarzi, "Association study of CYP17 and HSD11B1 in polycystic ovary syndrome utilizing comprehensive gene coverage," *Molecular Human Reproduction*, vol. 18, no. 6, pp. 320–324, 2012.

[78] M. Kahsar-Miller, L. R. Boots, A. Bartolucci, and R. Azziz, "Role of a CYP17 polymorphism in the regulation of circulating dehydroepiandrosterone sulfate levels in women with polycystic ovary syndrome," *Fertility and Sterility*, vol. 82, no. 4, pp. 973–975, 2004.

[79] J.-M. Park, E.-J. Lee, S. Ramakrishna, D.-H. Cha, and K.-H. Baek, "Association study for single nucleotide polymorphisms in the CYP17A1 gene and polycystic ovary syndrome," *International Journal of Molecular Medicine*, vol. 22, no. 2, pp. 249–254, 2008.

[80] B. Echiburú, F. Pérez-Bravo, M. Maliqueo, F. Sánchez, N. Crisosto, and T. Sir-Petermann, "Polymorphism T → C (-34 base pairs) of gene CYP17 promoter in women with polycystic ovary syndrome is associated with increased body weight and insulin resistance: a preliminary study," *Metabolism*, vol. 57, no. 12, pp. 1765–1771, 2008.

[81] L. Li, Z.-P. Gu, Q.-M. Bo, D. Wang, X.-S. Yang, and G.-H. Cai, "Association of CYP17A1 gene -34T/C polymorphism with polycystic ovary syndrome in Han Chinese population," *Gynecological Endocrinology*, vol. 31, no. 1, pp. 40–43, 2015.

[82] K. Techatraisak, C. Chayachinda, T. Wongwananuruk et al., "No association between CYP17 -34T/C polymorphism and insulin resistance in Thai polycystic ovary syndrome," *Journal of Obstetrics and Gynaecology Research*, vol. 41, no. 9, pp. 1412–1417, 2015.

[83] U. Banerjee, A. Dasgupta, A. Khan et al., "A cross-sectional study to assess any possible linkage of C/T polymorphism in CYP17A1 gene with insulin resistance in non-obese women with polycystic ovarian syndrome," *Indian Journal of Medical Research*, vol. 143, no. 9, pp. 739–747, 2016.

[84] T. Unsal, E. Konac, E. Yesilkaya et al., "Genetic polymorphisms of *FSHR, CYP17, CYP1A1, CAPN10, INSR, SERPINE1* genes in adolescent girls with polycystic ovary syndrome," *Journal of Assisted Reproduction and Genetics*, vol. 26, no. 4, pp. 205–216, 2009.

[85] Y. Li, F. Liu, S. Luo, H. Hu, X.-H. Li, and S.-W. Li, "Polymorphism T→C of gene CYP17 promoter and polycystic ovary syndrome risk: A meta-analysis," *Gene*, vol. 495, no. 1, pp. 16–22, 2012.

[86] S. E. Bulun, K. Takayama, T. Suzuki, H. Sasano, B. Yilmaz, and S. Sebastian, "Organization of the human aromatase P450 (CYP19) gene," *Seminars in Reproductive Medicine*, vol. 22, no. 1, pp. 5–9, 2004.

[87] J. Chen, S. Shen, Y. Tan et al., "The correlation of aromatase activity and obesity in women with or without polycystic ovary syndrome," *Journal of Ovarian Research*, vol. 8, no. 1, article 11, pp. 1–6, 2015.

[88] F. Yang, Y.-C. Ruan, Y.-J. Yang et al., "Follicular hyperandrogenism downregulates aromatase in luteinized granulosa cells in polycystic ovary syndrome women," *Reproduction*, vol. 150, no. 4, pp. 289–296, 2015.

[89] Y.-Y. Yu, C.-X. Sun, Y.-K. Liu, Y. Li, L. Wang, and W. Zhang, "Promoter methylation of CYP19A1 gene in chinese polycystic ovary syndrome patients," *Gynecologic and Obstetric Investigation*, vol. 76, no. 4, pp. 209–213, 2013.

[90] J.-L. Jin, J. Sun, H.-J. Ge et al., "Association between CYP19 gene SNP rs2414096 polymorphism and polycystic ovary syndrome in Chinese women," *BMC Medical Genetics*, vol. 10, p. 139, 2009.

[91] C. J. Petry, K. K. Ong, K. F. Michelmore et al., "Association of aromatase (CYP 19) gene variation with features of hyperandrogenism in two populations of young women," *Human Reproduction*, vol. 20, no. 7, pp. 1837–1843, 2005.

[92] C. J. Petry, K. K. Ong, K. F. Michelmore et al., "Associations between common variation in the aromatase gene promoter region and testosterone concentrations in two young female populations," *The Journal of Steroid Biochemistry and Molecular Biology*, vol. 98, no. 4-5, pp. 199–206, 2006.

[93] P. S. Maier and P. M. Spritzer, "Aromatase gene polymorphism does not influence clinical phenotype and response to oral contraceptive pills in polycystic ovary syndrome women," *Gynecologic and Obstetric Investigation*, vol. 74, no. 2, pp. 136–142, 2012.

[94] N. Xita, L. Lazaros, I. Georgiou, and A. Tsatsoulis, "CYP19 gene: a genetic modifier of polycystic ovary syndrome phenotype," *Fertility and Sterility*, vol. 94, no. 1, pp. 250–254, 2010.

[95] L. Lazaros, N. Xita, E. Hatzi et al., "CYP19 gene variants affect the assisted reproduction outcome of women with polycystic ovary syndrome," *Gynecological Endocrinology*, vol. 29, no. 5, pp. 478–482, 2013.

[96] N. Xita, I. Georgiou, L. Lazaros, V. Psofaki, G. Kolios, and A. Tsatsoulis, "The synergistic effect of sex hormone-binding globulin and aromatase genes on polycystic ovary syndrome phenotype," *European Journal of Endocrinology*, vol. 158, no. 6, pp. 861–865, 2008.

[97] C. F. Hao, N. Zhang, Q. Qu, X. Wang, H. F. Gu, and Z. J. Chen, "Evaluation of the association between the CYP19 tetranucleotide (TTTA)n polymorphism and polycystic ovarian syndrome(PCOS) in Han Chinese women," *Neuroendocrinology Letters*, vol. 31, no. 3, pp. 370–374, 2010.

[98] P. Xu, X. L. Zhang, G. B. Xie et al., "The (TTTA)n polymorphism in intron 4 of CYP19 and the polycystic ovary syndrome risk in a Chinese population," *Molecular Biology Reports*, vol. 40, no. 8, pp. 5041–5047, 2013.

[99] X.-L. Zhang, C.-W. Zhang, P. Xu et al., "SNP rs2470152 in CYP19 is correlated to aromatase activity in Chinese polycystic ovary syndrome patients," *Molecular Medicine Reports*, vol. 5, no. 1, pp. 245–249, 2012.

[100] H. Wang, Q. Li, T. Wang et al., "A common polymorphism in the human aromatase gene alters the risk for polycystic ovary syndrome and modifies aromatase activity in vitro," *Molecular Human Reproduction*, vol. 17, no. 6, pp. 386–391, 2011.

[101] C. Y. Peng, H. J. Xie, Z. F. Guo et al., "The association between androgen receptor gene CAG polymorphism and polycystic ovary syndrome: a case-control study and meta-analysis," *Journal of Assisted Reproduction and Genetics*, vol. 31, no. 9, pp. 1211–1219, 2014.

[102] N. L. Chamberlain, E. D. Driver, and R. L. Miesfeld, "The length and location of CAG trinucleotide repeats in the androgen receptor N-terminal domain affect transactivation function," *Nucleic Acids Research*, vol. 22, no. 15, pp. 3181–3186, 1994.

[103] H. Nam, C. Kim, M. Cha, J. Kim, B. Kang, and H. Yoo, "Polycystic ovary syndrome woman with heterozygous androgen receptor gene mutation who gave birth to a child with androgen insensitivity syndrome," *Obstetrics & Gynecology Science*, vol. 58, no. 2, pp. 179–182, 2015.

[104] K. A. Walters, C. M. Allan, and D. J. Handelsman, "Androgen actions and the ovary," *Biology of Reproduction*, vol. 78, no. 3, pp. 380–389, 2008.

[105] L. H. Lin, M. C. P. Baracat, G. A. R. MacIel, J. M. Soares Jr., and E. C. Baracat, "Androgen receptor gene polymorphism and polycystic ovary syndrome," *International Journal of Gynecology and Obstetrics*, vol. 120, no. 2, pp. 115–118, 2013.

[106] A. N. Schüring, A. Welp, J. Gromoll et al., "Role of the CAG repeat polymorphism of the androgen receptor gene in polycystic ovary syndrome (PCOS)," *Experimental and Clinical Endocrinology & Diabetes*, vol. 120, no. 2, pp. 73–79, 2012.

[107] N. A. Shah, H. J. Antoine, M. Pall, K. D. Taylor, R. Azziz, and M. O. Goodarzi, "Association of androgen receptor CAG repeat polymorphism and polycystic ovary syndrome," *The Journal of Clinical Endocrinology & Metabolism*, vol. 93, no. 5, pp. 1939–1945, 2008.

[108] Y. Xia, Y. Che, X. Zhang et al., "Polymorphic CAG repeat in the androgen receptor gene in polycystic ovary syndrome patients," *Molecular Medicine Reports*, vol. 5, no. 5, pp. 1330–1334, 2012.

[109] F. Van Nieuwerburgh, D. Stoop, P. Cabri, M. Dhont, D. Deforce, and P. De Sutter, "Shorter CAG repeats in the androgen receptor gene may enhance hyperandrogenicity in polycystic ovary syndrome," *Gynecological Endocrinology*, vol. 24, no. 12, pp. 669–673, 2008.

[110] A. Mifsud, S. Ramirez, and E. L. Yong, "Androgen receptor gene CAG trinucleotide repeats in anovulatory infertility and polycystic ovaries," *The Journal of Clinical Endocrinology & Metabolism*, vol. 85, no. 9, pp. 3484–3488, 2000.

[111] S. Rajender, S. J. Carlus, S. K. Bansal et al., "Androgen Receptor CAG Repeats Length Polymorphism and the Risk of Polycystic Ovarian Syndrome (PCOS)," *PLoS ONE*, vol. 8, no. 10, Article ID e75709, 2013.

[112] P. Ferk, M. P. Perme, N. Teran, and K. Gersak, "Androgen receptor gene (CAG)n polymorphism in patients with polycystic ovary syndrome," *Fertility and Sterility*, vol. 90, no. 3, pp. 860–863, 2008.

[113] J. J. Kim, S. H. Choung, Y. M. Choi, S. H. Yoon, S. H. Kim, and S. Y. Moon, "Androgen receptor gene CAG repeat polymorphism in women with polycystic ovary syndrome," *Fertility and Sterility*, vol. 90, no. 6, pp. 2318–2323, 2008.

[114] L. Skrgatic, D. P. Baldani, J. Z. Cerne, P. Ferk, and K. Gersak, "CAG repeat polymorphism in androgen receptor gene is not directly associated with polycystic ovary syndrome but influences serum testosterone levels," *The Journal of Steroid Biochemistry and Molecular Biology*, vol. 128, no. 3-5, pp. 107–112, 2012.

[115] J. Jääskeläinen, S. Korhonen, R. Voutilainen, M. Hippeläinen, and S. Heinonen, "Androgen receptor gene CAG length polymorphism in women with polycystic ovary syndrome," *Fertility and Sterility*, vol. 83, no. 6, pp. 1724–1728, 2005.

[116] P. D. Ramos Cirilo, F. E. Rosa, M. F. Moreira Ferraz, C. A. Rainho, A. Pontes, and S. R. Rogatto, "Genetic polymorphisms associated with steroids metabolism and insulin action in polycystic ovary syndrome," *Gynecological Endocrinology*, vol. 28, no. 3, pp. 190–194, 2012.

[117] M. Möhlig, A. Jürgens, J. Spranger et al., "The androgen receptor CAG repeat modifies the impact of testosterone on insulin resistance in women with polycystic ovary syndrome," *European Journal of Endocrinology*, vol. 155, no. 1, pp. 127–130, 2006.

[118] T. Hickey, A. Chandy, and R. J. Norman, "The androgen receptor CAG repeat polymorphism and X-Chromosome inactivation in australian caucasian women with infertility related to polycystic ovary syndrome," *The Journal of Clinical Endocrinology & Metabolism*, vol. 87, no. 1, pp. 161–165, 2002.

[119] T. Zhang, W. Liang, M. Fang, J. Yu, Y. Ni, and Z. Li, "Association of the CAG repeat polymorphisms in androgen receptor gene with polycystic ovary syndrome: A systemic review and meta-analysis," *Gene*, vol. 524, no. 2, pp. 161–167, 2013.

[120] R. Wang, M. O. Goodarzi, T. Xiong, D. Wang, R. Azziz, and H. Zhang, "Negative association between androgen receptor gene CAG repeat polymorphism and polycystic ovary syndrome? A systematic review and meta-analysis," *Molecular Human Reproduction*, vol. 18, no. 10, pp. 498–509, 2012.

[121] C. Y. Peng, X. Y. Long, and G. X. Lu, "Association of AR rs6152G/A gene polymorphism with susceptibility to polycystic ovary syndrome in Chinese women," *Reproduction, Fertility and Development*, vol. 22, no. 5, pp. 881–885, 2010.

[122] C. Yuan, C. Gao, Y. Qian et al., "Polymorphism of CAG and GGN repeats of androgen receptor gene in women with polycystic ovary syndrome," *Reproductive BioMedicine Online*, vol. 31, no. 6, pp. 790–798, 2015.

[123] G. L. Hammond, "Plasma steroid-binding proteins: Primary gatekeepers of steroid hormone action," *Journal of Endocrinology*, vol. 230, no. 1, pp. R13–R25, 2016.

[124] K. N. Hogeveen, P. Cousin, M. Pugeat, D. Dewailly, B. Soudan, and G. L. Hammond, "Human sex hormone-binding globulin variants associated with hyperandrogenism and ovarian dysfunction," *The Journal of Clinical Investigation*, vol. 109, no. 7, pp. 973–981, 2002.

[125] N. Xita, A. Tsatsoulis, I. Stavrou, and I. Georgiou, "Association of SHBG gene polymorphism with menarche," *Molecular Human Reproduction*, vol. 11, no. 6, pp. 459–462, 2005.

[126] P. Cousin, L. Calemard-Michel, H. Lejeune et al., "Influence of SHBG gene pentanucleotide TAAAA repeat and D327N polymorphism on serum sex hormone-binding globulin concentration in hirsute women," *The Journal of Clinical Endocrinology & Metabolism*, vol. 89, no. 2, pp. 917–924, 2004.

[127] D. P. Baldani, L. Skrgatic, J. Z. Cerne, S. K. Oguic, B. M. Gersak, and K. Gersak, "Association between serum levels and pentanucleotide polymorphism in the sex hormone binding globulin gene and cardiovascular risk factors in females with polycystic ovary syndrome," *Molecular Medicine Reports*, vol. 11, no. 5, pp. 3941–3947, 2015.

[128] P. Ferk, N. Teran, and K. Gersak, "The (TAAAA)n microsatellite polymorphism in the SHBG gene influences serum SHBG levels in women with polycystic ovary syndrome," *Human Reproduction*, vol. 22, no. 4, pp. 1031–1036, 2007.

[129] J. L. Zhao, Z. J. Chen, Y. R. Zhao et al., "Study on the $(TAAAA)_n$ repeat polymorphism in sex hormone-binding globulin gene and the SHBG serum levels in putative association with the glucose metabolic status of Chinese patients suffering from polycystic ovarian syndrome in Shandong province," *Zhonghua Yi Xue Yi Chuan Xue Za Zhi*, vol. 22, no. 6, pp. 644–647, 2005.

[130] N. Xita, A. Tsatsoulis, A. Chatzikyriakidou, and I. Georgiou, "Association of the (TAAAA)n repeat polymorphism in the sex hormone-binding globulin (SHBG) gene with polycystic ovary syndrome and relation to SHBG serum levels," *The Journal of Clinical Endocrinology & Metabolism*, vol. 88, no. 12, pp. 5976–5980, 2003.

[131] W. Fan, S. Li, Q. Chen, and Z. Huang, "Association between the (TAAAA)n SHBG polymorphism and PCOS: A systematic review and meta-analysis," *Gynecological Endocrinology*, vol. 29, no. 7, pp. 645–650, 2013.

[132] B. Hacihanefioğlu, B. Aybey, Y. Hakan Özön, H. Berkil, and K. Karşidağ, "Association of anthropometric, androgenic and insulin-related features with polymorphisms in exon 8 of SHBG gene in women with polycystic ovary syndrome," *Gynecological Endocrinology*, vol. 29, no. 4, pp. 361–364, 2013.

[133] E. P. Wickham III, K. G. Ewens, R. S. Legro, A. Dunaif, J. E. Nestler, and J. F. Strauss III, "Polymorphisms in the SHBG gene influence serum SHBG levels in women with polycystic ovary syndrome," *The Journal of Clinical Endocrinology & Metabolism*, vol. 96, no. 4, pp. E719–E727, 2011.

[134] M. Á. Martínez-García, A. Gambineri, M. Alpañés, R. Sanchón, R. Pasquali, and H. F. Escobar-Morreale, "Common variants in the sex hormone-binding globulin gene (SHBG) and polycystic ovary syndrome (PCOS) in Mediterranean women," *Human Reproduction*, vol. 27, no. 12, pp. 3569–3576, 2012.

[135] T. M. Abu-Hijleh, E. Gammoh, A. S. Al-Busaidi et al., "Common variants in the sex hormone-binding globulin (SHBG) gene influence SHBG levels in women with polycystic ovary syndrome," *Annals of Nutrition and Metabolism*, vol. 68, no. 1, pp. 66–74, 2016.

[136] A.-S. Nazouri, M. Khosravifar, A.-A. Akhlaghi, M. Shiva, and P. Afsharian, "No relationship between most polymorphisms of steroidogenic acute regulatory (StAR) gene with polycystic ovarian syndrome," *Iranian Journal of Reproductive Medicine*, vol. 13, no. 12, pp. 771–778, 2015.

[137] R. Ju, W. Wu, J. Fei et al., "Association analysis between the polymorphisms of HSD17B5 and HSD17B6 and risk of polycystic ovary syndrome in Chinese population," *European Journal of Endocrinology*, vol. 172, no. 3, pp. 227–233, 2015.

[138] K. Qin, D. A. Ehrmann, N. Cox, S. Refetoff, and R. L. Rosenfield, "Identification of a functional polymorphism of the human type 5 17β-hydroxysteroid dehydrogenase gene associated with polycystic ovary syndrome," *The Journal of Clinical Endocrinology & Metabolism*, vol. 91, no. 1, pp. 270–276, 2006.

[139] D. J. Marioli, A. D. Saltamavros, V. Vervita et al., "Association of the 17-hydroxysteroid dehydrogenase type 5 gene polymorphism (-71A/G HSD17B5 SNP) with hyperandrogenemia in polycystic ovary syndrome (PCOS)," *Fertility and Sterility*, vol. 92, no. 2, pp. 648–652, 2009.

[140] M. O. Goodarzi, M. R. Jones, H. J. Antoine, M. Pall, Y.-D. I. Chen, and R. Azziz, "Nonreplication of the type 5 17β-hydroxysteroid dehydrogenase gene association with polycystic ovary syndrome," *The Journal of Clinical Endocrinology & Metabolism*, vol. 93, no. 1, pp. 300–303, 2008.

[141] P. S. Maier, S. S. Mattiello, L. Lages, and P. M. Spritzer, "17-hydroxysteroid dehydrogenase type 5 gene polymorphism (-71A/G HSD17B5 SNP) and treatment with oral contraceptive pills in PCOS women without metabolic comorbidities," *Gynecological Endocrinology*, vol. 28, no. 8, pp. 606–610, 2012.

[142] C. Glister, L. Satchell, R. A. D. Bathgate et al., "Functional link between bone morphogenetic proteins and insulin-like peptide 3 signaling in modulating ovarian androgen production," *Proceedings of the National Acadamy of Sciences of the United States of America*, vol. 110, no. 15, pp. E1426–E1435, 2013.

[143] R. Anand-Ivell, K. Tremellen, Y. Dai et al., "Circulating insulin-like factor 3 (INSL3) in healthy and infertile women," *Human Reproduction*, vol. 28, no. 11, pp. 3093–3102, 2013.

[144] A. Gambineri, L. Patton, O. Prontera et al., "Basal insulin-like factor 3 levels predict functional ovarian hyperandrogenism in the polycystic ovary syndrome," *Journal of Endocrinological Investigation*, vol. 34, no. 9, pp. 685–691, 2011.

[145] D. Szydlarska, W. Grzesiuk, A. Trybuch, A. Kondracka, I. Kowalik, and E. Bar-Andziak, "Insulin-like factor 3—a new hormone related to polycystic ovary syndrome?" *Endokrynologia Polska*, vol. 63, no. 5, pp. 356–361, 2012.

[146] N. Shaikh, R. Dadachanji, P. Meherji, N. Shah, and S. Mukherjee, "Polymorphisms and haplotypes of insulin-like factor 3 gene are associated with risk of polycystic ovary syndrome in Indian women," *Gene*, vol. 577, no. 2, pp. 180–186, 2016.

[147] Z.-J. Chen, H. Zhao, L. He et al., "Genome-wide association study identifies susceptibility loci for polycystic ovary syndrome on chromosome 2p16.3, 2p21 and 9q33.3," *Nature Genetics*, vol. 43, no. 1, pp. 55–59, 2011.

[148] Y. Shi, H. Zhao, Y. Shi et al., "Genome-wide association study identifies eight new risk loci for polycystic ovary syndrome," *Nature Genetics*, vol. 44, no. 9, pp. 1020–1025, 2012.

[149] J. M. McAllister, R. S. Legro, B. P. Modi, and J. F. Strauss, "Functional genomics of PCOS: From GWAS to molecular mechanisms," *Trends in Endocrinology & Metabolism*, vol. 26, no. 3, pp. 118–124, 2015.

[150] M. K. Tee, M. Speek, B. Legeza et al., "Alternative splicing of DENND1A, a PCOS candidate gene, generates variant 2," *Molecular and Cellular Endocrinology*, vol. 434, pp. 25–35, 2016.

[151] M. G. Hayes, M. Urbanek, D. A. Ehrmann et al., "Corrigendum: Genome-wide association of polycystic ovary syndrome implicates alterations in gonadotropin secretion in European ancestry populations," *Nature Communications*, vol. 6, Article ID 7502, 2015.

8

Assessment of Functional EST-SSR Markers (Sugarcane) in Cross-Species Transferability, Genetic Diversity among Poaceae Plants, and Bulk Segregation Analysis

Shamshad Ul Haq,[1,2,3] Pradeep Kumar,[1,4] R. K. Singh,[1] Kumar Sambhav Verma,[5] Ritika Bhatt,[2,3] Meenakshi Sharma,[3] Sumita Kachhwaha,[3] and S. L. Kothari[2,3,5]

[1]Biotechnology Division, UP Council of Sugarcane Research, Shahjahanpur 242001, India
[2]Interdisciplinary Programme of Life Science for Advance Research and Education, University of Rajasthan, Jaipur 302004, India
[3]Department of Botany, University of Rajasthan, Jaipur 302015, India
[4]School of Biotechnology, Yeungnam University, Gyeongsan 712-749, Republic of Korea
[5]Amity Institute of Biotechnology, Amity University Rajasthan, Jaipur 302006, India

Correspondence should be addressed to Shamshad Ul Haq; shamshadbiotech@gmail.com

Academic Editor: Norman A. Doggett

Expressed sequence tags (ESTs) are important resource for gene discovery, gene expression and its regulation, molecular marker development, and comparative genomics. We procured 10000 ESTs and analyzed 267 EST-SSRs markers through computational approach. The average density was one SSR/10.45 kb or 6.4% frequency, wherein trinucleotide repeats (66.74%) were the most abundant followed by di- (26.10%), tetra- (4.67%), penta- (1.5%), and hexanucleotide (1.2%) repeats. Functional annotations were done and after-effect newly developed 63 EST-SSRs were used for cross transferability, genetic diversity, and bulk segregation analysis (BSA). Out of 63 EST-SSRs, 42 markers were identified owing to their expansion genetics across 20 different plants which amplified 519 alleles at 180 loci with an average of 2.88 alleles/locus and the polymorphic information content (PIC) ranged from 0.51 to 0.93 with an average of 0.83. The cross transferability ranged from 25% for wheat to 97.22% for *Schlerostachya*, with an average of 55.86%, and genetic relationships were established based on diversification among them. Moreover, 10 EST-SSRs were recognized as important markers between bulks of pooled DNA of sugarcane cultivars through BSA. This study highlights the employability of the markers in transferability, genetic diversity in grass species, and distinguished sugarcane bulks.

1. Introduction

Sugarcane is a bioenergy crop belonging to the genus *Saccharum* L. of the tribe Andropogoneae (family: Poaceae). This tribe comprises grass species which have high economic value. The noble sugarcane varieties are developed from interspecific hybridization of *Saccharum officinarum* L. ($2n = 80$) which has high sugar content with less disease tolerance and *Saccharum spontaneum* ($2n = 40$ to 120) which provides stress, disease tolerance, and high fiber content for biomass. The taxonomy and genetic constitution of sugarcane are complicated due to complex interspecific aneupolyploid genome which makes chromosome numbers range from 100 to 130 [1]. Moreover, six *Saccharum* spp. (*S. spontaneum, S. officinarum, S. robustum, S. edule, S. barberi,* and *S. sinense*) and four *Saccharum* related genera (*Erianthus, Miscanthus, Sclerostachya,* and *Narenga*) have purportedly undergone interbreeding, forming the "*Saccharum* complex" [2, 3]. The interbreeding has made their genome more complex and added to multigenic and/or multiallelic nature for most agronomic traits that made sugarcane breeding a more difficult task [4].

A vast array of genomic tools has been developed which has opened new ways to define the genetic architecture of sugarcane and helped to explore its functional system [1, 5]. Among the molecular markers, microsatellites are most favored for a variety of genetic applications due to their multiallelic nature, high reproducibility, cross transferability, codominant inheritance, abundance, and extensive genome coverage [6–8]. Microsatellites or simple sequences repeats (SSRs) are monotonous repetitions of very short (one to six) nucleotide motifs, which occur as interspersed repetitive elements in all eukaryotic and prokaryotic genomes. However, transcribed regions of the genome also contain enormous range of microsatellites that correspond to genic microsatellites or EST-SSRs. Therefore, expressed sequence tags (ESTs) are the short transcribed portions and involved in the variety of metabolic functions. The presence of the microsatellites in genes as well as ESTs unveils the biological significance of SSR distribution, expansion, and contraction on the function of the genes themselves [9].

Presently, huge amounts of expressed sequence tags have been deposited in public database (NCBI). In silico approaches to retrieve EST sequences from NCBI and functional annotations provide more constructive EST-SSRs or gene-based SSR (genic SSRs) marker development besides own EST libraries development. This method of the EST-SSR markers development provides the easiest way to reduce cost, time, and labours along with more meaningful marker identifications [10]. The presence of microsatellites in the genic region is found to be more conserved due to which they possess high reproducibility and high interspecific/intraspecific transferability. Hence, EST-SSR could be used for polymorphism, genetic diversity, cross transferability, and comparative mapping in different plant species. Accordingly, several genetic studies were done on sugarcane using microsatellite markers to decipher polymorphism, cross transferability, genetic diversity, informative marker detection through bulk segregation analysis (BSA), and comparative genomics [8, 11–13]. The objective of the present study was to retrieve EST sequences for more informative EST-SSR development and their genetic assessment within and across the taxa through cross transferability, genetic relationships, and bulk segregation analysis.

2. Materials and Methods

2.1. EST Sequences Retrieving, ESTs Assembling, and Microsatellites Identification. Total 10000 EST sequences of the *Saccharum* spp. were downloaded in Fasta format from National Centre for Biotechnology Information (NCBI) for microsatellites deciphering. Further, ESTs assembling was carried out using CAP3 programme (http://mobyle.pasteur .fr/cgi-bin/portal.py#forms::cap3) for minimization of sequences redundancy. Microsatellite identification was carried out using MISA software (http://pgrc.ipk-gatersleben.de/ misa/) and the criteria for SSR detection were 6, 4, 3, 3, and 3 repeat units for di-, tri-, tetra-, penta-, and hexanucleotides, respectively. SSR primer pairs (forward and reverse) were designed for the selected EST sequences having microsatellites using online web tool, batch primer 3 pipeline [14].

2.2. EST-SSR Sequences Annotation. Assessment of EST sequences having SSR was done through blastn/blastx

analysis for homology search and against nonredundant (nr) protein at the NCBI. Furthermore, functional annotation pipeline was also run at online tool for gene ontology (GO) which was intended for different GO functional classes like biological process, cellular component, and molecular function [15].

2.3. PCR Amplification and Electrophoresis. PCR reactions were carried out in a total of 10 μL volume containing 25 ng template DNA, 1.0 μL (10 pmol/μL) of each forward and reverse primer, 100 mM of dNTPs, 0.5 U of *Taq* DNA polymerase, and 1.0 μL of 10x PCR buffer with 2.5 mM of $MgCl_2$. Amplification was performed in a thermal cycler (Bio-Rad) in the following conditions: initial denaturation at 94°C for 5 min followed by 30 amplification cycles of denaturation for 1 min at 94°C followed by annealing temperature (T_a) for 1 min and then extension for 2 min at 72°C; final extension at 72°C for 7 min was allowed. The PCR conditions particularly the annealing temperatures (varying from 52°C to 58°C) for each primer were standardized and amplified products were stored at 4°C. The PCR products were analyzed on a 7% native PAGE in vertical gel electrophoresis unit (Bangalore Genei™) using TBE buffer. The sizes of amplified fragments were estimated using 50 bp DNA ladder (Fermentas). Gels were documented using ethidium bromide (EtBr) stained dye.

2.4. Evaluation of Saccharum EST-SSR across the Taxa through Cross Transferability. The cross transferability of *Saccharum* derived EST-SSR markers was evaluated among the 20 accessions comprising seven cereals (wheat, maize, barley, rice, pearl millet, oat, and *Sorghum*), four *Saccharum* related genera (*Erianthus*, *Miscanthus*, *Narenga*, and *Sclerostachya*), three *Saccharum* species (51NG56 (*S. robustum*), N58 (*S. spontaneum*), and two clones of *S. officinarum* (Bandjermasin Hitam and Gunjera)), and five *Saccharum* commercial cultivars (CoS 88230, CoS 92423, UP 9530, CoS 8436, and CoS 91230). All genotypes were collected from the Sugarcane Research Institute Farm, UPCSR, Shahjahanpur, India. Furthermore, genomic DNA from young juvenile, disease-free, immature leaves was isolated for each genotype using CTAB (cetyl trimethylammonium bromide) method [16]. Isolated DNA samples were treated with RNAase for 1 h at 37°C and purified by phenol extraction (25 phenol : 24 chloroform : 1 isoamyl alcohol, v/v/v) followed by ethanol precipitation [17] and stored at −80°C. DNA was quantified on 0.8% agarose gel and the working concentration of 25 ng/μL was obtained by making final adjustment in 10 mM TE buffer.

2.5. Genetic Diversity Analysis. The assessment of EST-SSRs in genetic diversity analysis was done among 20 plants belonging to distinct groups comprising cereals, *Saccharum* related genera, *Saccharum* species, and *Saccharum* cultivars. The allelic data of 63 EST-SSR primers were used to ascertain the genetic relationships between 20 genotypes by clustering analysis. Amplified bands were scored as binary data in the form of present (1) or absent (0). Dendrogram was constructed by neighbour-joining and Jaccard's algorithm using

FreeTree and TreeView software [18, 19]. The polymorphic information content (PIC) values were calculated for each primer by using the online resource of PIC Calculator (http://www.liv.ac.uk/~kempsj/pic.html).

2.6. Informative Assessment of Functional EST-SSR Markers between Bulks. Plant materials were used as F2 mapping population comprising 209 genotypes of the sugarcane cultivars which were developed from cross between CoS 91230 (Parent; CoS 775 × Co 1148) with CoS 8436 (Parent; MS 68/47 × Co 1148) from September to March (2010-2011). Grouping of genotypes was done according to their stem diameter (contrasting high and low stem diameter genotypes) into two sets. DNA extractions were carried out from both sets and equal quantities of genomic DNA from 10 extreme high stem diameter and 10 extreme low stem diameter genotypes were pooled into two bulks. PCR amplification was done in both bulks with newly developed EST-SSR primers for informative markers identifications through bulk segregation analysis (BSA) [20].

3. Results and Discussion

3.1. Mining of Microsatellites in EST Sequences and SSRs Characterization. Total 10,000 EST sequences related to *Saccharum* spp. were examined from NCBI for the simple sequence repeat (SSR) identification and characterization using computational approach. Prior to the marker deciphering, sequence assembly was performed and 6201 (4201 kb) nonredundant sequences were detected comprising 1752 contigs and 4449 singlets, wherein 406 SSRs were identified with 360 perfect SSRs and 37 sequences containing more than 1 SSR and 30 SSRs in compound formation. Therefore, computational and experimental approach to ascertain microsatellites in EST libraries from public database (NCBI) turned to be very cost effective and reduces time and labour besides expense of own libraries development. EST-SSRs are a more preferable DNA marker in the variety of genetic analysis and found to be more conserved as present in the transcribed region of the genome. These were found to be more transferable across the taxonomic boundaries and could be evaluated as most informative markers for variety of genomics applications [10, 21]. These are more adapted in plants comparative genetic analysis for gene identification, gene mapping, marker-assisted-selection, transferability, and genetic diversity [7, 22–24]. Also, a variety of studies have been reported on sugarcane using EST-SSR markers for desired genetic analysis [8, 13, 25, 26].

The frequency of SSR in EST sequences was 6.4% including all the repeats except mononucleotide repeats. This result is comparatively higher compared to previous studies on sugarcane [8, 27–29]. Contrary to this, Singh et al. [13] reported higher frequency (9.3%) in sugarcane. Kumpatla and Mukhopadhyay [30] also observed high range (2.65% to 10.62%) of SSR frequency in different plant species. In general, about 5% of ESTs contained SSR which has been reported in many plant species [31]. These variations in microsatellite frequency could be attributed to the "search criteria" used, type of SSR motif, size of sequence data, and the mining tools

used [24, 32]. In other words, the density of the microsatellites was one SSR per 10.45 kb which is closely comparable to earlier studies in sugarcane with densities 1 SSR/10.9 kb [8] and 1/9 kb SSR [13].

Analysis revealed that trinucleotide repeats (66.74%) were found to be more frequent followed by di- (26.10%), tetra- (4.67%), penta- (1.5%), and hexanucleotide (1.2%) repeats. Our observation of high frequency of trinucleotide repeats is in agreement with previous reports on sugarcane [8, 13, 27–29, 33]. Several other studies have also represented high frequency of trinucleotide repeats in different plant species [24, 31, 34–36]. A total of 33 different types of motifs were identified of which four belonged to dinucleotide, eight belonged to trinucleotides, twelve belonged to tetranucleotide, five belonged to pentanucleotide, and two belonged to hexanucleotide repeats (Figure 1). We observed that motifs AG/CT and AT/AT were more frequent in dinucleotide repeat followed by motifs CCG/CGG, AGC/CTG, AGG/CCT, and ACG/CGT in trinucleotide repeat, motif AAAG/CTTT in tetranucleotide repeats, motif ACAGG/CCTGT in pentanucleotide repeats, and AACACC/GGTGTT in hexanucleotide repeats. The presence of motif CCG/CGG was also observed in sugarcane by different authors [13, 27]. Kantety et al. [37] also reported CCG/CGG motif as most abundant in wheat and *Sorghum*. Similarly, both Lawson and Zhang [38] and Da Maia et al. [39] also observed abundance of motif CCG/CGG in different member of the grass family. Victoria et al. [35] also decoded motif CCG/CGG in the lower plants (*C. reinhardtii* and *P. patens*). Thus, this predominance of CCG/CGG motif frequency has been related to a high GC-content [5]. Some motifs which are responsible for making unusual DNA folding structure (hairpin formed, bipartite triplex formed, and simple loop folding) also have effect on gene expressions and regulations mechanism, namely, CCT/AGG, CCG/GGC, GGA/TTC, and GAA/TTC motifs [40, 41]. Moreover, the presence of trinucleotide repeats in the coding region formed a distinct group and encoded amino acid tracts within the peptide [42]. We also observed predictable twenty different types of amino acids including stop codon. Alanine, arginine, glycine, proline, and serine were most frequent (Figure 2). This is in agreement with previous studies that reported on different plant species [11, 35, 43].

3.2. Expressed Sequence Tags Annotation and Primers Development. All EST sequences having SSRs were examined by functional annotation (blastn, blastx, and gene ontology). After-effect, sixty-three ESTs having SSRs were successfully identified on the basis of their involvement in the various metabolic processes (Figure 3). After-effect, sixty-three EST-SSRs primer pairs were designed for polymorphic nature, cross transferability, bulk segregation analysis, and genetic diversity in the test plants (Table 1). These selected EST-SSRs comprised all types of repeat motifs (excluding mononucleotide repeat), and among trinucleotide repeats they were highly frequent with GCT/CGA, TCC/AGG, and GGT/CCA repeat motifs. Similarly, Sharma et al. [44] also used functional annotation pipelines for the more prominent molecular markers development related to gene

TABLE 1: Details of selected 63 EST-SSR primer pairs used for cross transferability, genetic diversity, and bulks segregation analysis.

Serial number	Type	Primer sequence	Annealing temperature	SSR motif	PIC value	E-value	Putative identities (blastn/blastx)
SYMS28	F	GCGTCAGAGTGTTAAAACAAG	53	$(GCT)_4$	0.81	$9.39E - 42$	Protein transport protein Sec61 beta
SYMS28	R	GTGTAGAACTGGAGCATTGAG					
SYMS29	F	GGGCAAGCAAGAAACCAC	52	$(TCC)_4$	0.91	$1.62E - 24$	Protein translation factor SUI1
SYMS29	R	GAAGAGGTCAACCAAGAACTC					
SYMS30	F	GCGTCAGAGTGTTAAAACAAG	53	$(GCT)_4$	0.86	$1.00E - 21$	Preprotein translocase Sec
SYMS30	R	GTGTAGAACTGGAGCATTGAG					
SYMS31	F	GAAGCTCCCAAGCTGCTA	53	$(AGCT)_3$	0.76	$2.00E - 12$	*Predicted*: uncharacterized protein
SYMS31	R	CCTACAGGAAAGATTTTAGGG					
SYMS32	F	GTCTCTTCTCCAGTTCTCCTT	55	$(TGCG)_4$	0.84	$2.46E - 63$	*Predicted*: actin-depolymerizing factor
SYMS32	R	GCTCAACAAATGTCTCCCTA					
SYMS33	F	TGCACTAACATGGTTGATGT	54	$(GAAG)_3$	0.86	$2.82E - 90$	Hypothetical protein SORBIDRAFT_03g046450
SYMS33	R	GGTGATTGTAAGGGTCATCTT					
SYMS34	F	GTTAATGGTGGTTCCGTTC	53	$(GGC)_6$	0.88	$4E - 20$	*Predicted*: uncharacterized protein LOC101783547
SYMS34	R	ATTATCAGCGCAGAGACATC					
SYMS35	F	GCGTCAGAGTGTTAAAACAAG	52	$(GCT)_4$	0.75	$1.00E - 21$	Preprotein translocase
SYMS35	R	GTGTAGAACTGGAGCATTGAG					
SYMS36	F	GGACTGTACAAGGACGACAG	53	$(GCT)_4$	0.70	$1.14E - 41$	Protein transport protein Sec61 beta subunit
SYMS36	R	TCTGCTTTCTTGGATATGGTA					
SYMS37	F	AAGAAGGATGCAAAGAAGAAG	54	$(GAT)_4$	0.90	$3.08E - 81$	Hypothetical protein SORBIDRAFT_03g046450
SYMS37	R	AGGCTTAGTAACAGCAGGTTT					
SYMS38	F	AAGAAGGATGCAAAGAAGAAG	56	$(AGA)_4$	0.86	$9.00E - 37$	Hypothetical protein
SYMS38	R	AGGCTTAGTAACAGCAGGTTT					
SYMS39	F	GGACTGTACAAGGACGACAG	—	$(GCT)_4$	—	$1.14E - 41$	Protein transport protein
SYMS39	R	TCTGCTTTCTTGGATATGGTA					
SYMS40	F	GGACTGTACAAGGACGACAG	—	$(GCT)_4$	—	$1.25E - 40$	Preprotein translocase
SYMS40	R	TCTGCTTTCTTGGATATGGTA					
SYMS41	F	GGACTGTACAAGGACGACAG	—	$(GCT)_4$	—	$9.62E - 42$	Protein transport protein Sec61 beta subunit
SYMS41	R	TCTGCTTTCTTGGATATGGTA					
SYMS42	F	CCAAAGAGATCTTGCAGACTA	—	$(ATG)_4$	—	$1.78E - 53$	Jasmonate-induced protein
SYMS42	R	CCCAACACAACAACCAAT					
SYMS43	F	CCACACAAGCAAGAAATAAAC	—	$(GGT)_4$	—	$8.57E - 74$	Dirigent-like protein
SYMS43	R	TCGAACACTATGGTAAAGGTG					
SYMS44	F	GGACTGTACAAGGACGACAG	—	$(GCT)_4$	—	$1.15E - 41$	Homeodomain-like transcription factor
SYMS44	R	TCTGCTTTCTTGGATATGGTA					
SYMS45	F	GCGTCAGAGTGTTAAAACAAG	53	$(GCT)_4$	0.69	$9.86E - 42$	Protein transport protein

TABLE 1: Continued.

Serial number	Type	Primer sequence	Annealing temperature	SSR motif	PIC value	E-value	Putative identities (blastn/blastx)
SYMS45	R	GACTCTGCTTTCTTGGATATG					
SYMS46	F	AGCTATCTTTAGTGGGGACAT	52	$(CGT)_4$	0.90	$1.82E-44$	Hypothetical protein SORBIDRAFT_09g006220
SYMS46	R	GAGGTCTCATCGGAGCTTA					
SYMS47	F	AGGTCGTTTTAATTCCTTCC	53	$(GTTTT)_3$	0.77	$1.00E-21$	Preprotein translocase Sec
SYMS47	R	CGTAAATATGAACGAGGTCAG					
SYMS48	F	AGGTCGTTTTAATTCCTTCC	53	$(TTTA)_6$	0.90	$4.00E-20$	TPA: hypothetical protein
SYMS48	R	CGTAAATATGAACGAGGTCAG					
SYMS49	F	GGACTGTACAAGGACGACAG	—	$(GCT)_4$	—	$1.15E-41$	Zinc finger A20 and AN1 domains-containing protein
SYMS49	R	TCTGCTTTCTTGGATATGGTA					
SYMS50	F	TCCAAGGATTTAGCTATGGAT	—	$(TGT)_{10}$	—	$6.79E-13$	TPA: seed maturation protein
SYMS50	R	TTCAACTACACCCTTCTGTTG					
SYMS51	F	GCGTCAGAGTGTTAAAACAAG	—	$(GCT)_4$	—	$1.22E-41$	Hypothetical protein
SYMS51	R	ATTGTCACTTGCTATCCATTT					
SYMS52	F	CACCTTCTTTCCTTCTCCTC	—	$(CGC)_4$	—	$3.32E-47$	V-type proton ATPase 16 kDa proteolipid subunit
SYMS52	R	GTAGATACCGAGCACACCAG					
SYMS53	F	TCAGTTCAGGGATGACAATAG	56	$(CCGTGG)_3$	0.87	$2.59E-78$	Homeodomain-like transcription factor superfamily protein
SYMS53	R	GGATAGACTGAAATCTGCTCA					
SYMS54	F	CAACTCGACTCTTTTCTCTCA	—	$(CTC)_5$	—	$4.13E-08$	Protein transport protein SEC31
SYMS54	R	GGAGGTGGAACTTCCTGA					
SYMS55	F	GGACTGTACAAGGACGACAG	—	$(GCT)_4$	—	$1.12E-41$	Protein transport protein Sec61 subunit beta-like isoform
SYMS55	R	TCTGCTTTCTTGGATATGGTA					
SYMS56	F	GGACTGTACAAGGACGACAG	—	$(GCT)_4$	—	$8.01E-42$	Protein transport protein Sec61 subunit beta-like isoform
SYMS56	R	TCTGCTTTCTTGGATATGGTA					
SYMS57	F	AAACGATCAGATACCGTTGTA	—	$(CG)_6$	—	$7.84E-27$	Caltractin
SYMS57	R	ATCAAAGAGATCAAAGGCTTC					
SYMS58	F	CATTTCGAAGCTCCTCCT	52	$(CCTCCG)_6$	0.74	$5.97E-66$	Zinc finger A20 and AN1 domains-containing protein
SYMS58	R	TAGGCTGCACAACAATAGTCT					
SYMS59	F	CTCCCCCATTTCTCTTCC	53	$(GCAGCC)_6$	0.80	$4.02E-65$	*Predicted*: reticulon-like protein B1
SYMS59	R	CAAGTACTCCAGCAGAGATGT					
SYMS60	F	CTTTTCCCTCTTCCTCTCTC	—	$(CCG)_5$	—	$1.24E-45$	*Predicted*: uncharacterized tRNA-binding protein
SYMS60	R	TGTCACTAACACGAATCACAA					
SYMS61	F	CCCTCTCCCTGCTCTTTC	54	$(TCC)_5$	0.79	$4.14E-57$	Actin-depolymerizing factor 3

TABLE 1: Continued.

Serial number	Type	Primer sequence	Annealing temperature	SSR motif	PIC value	E-value	Putative identities (blastn/blastx)
SYMS61	R	CAGTCACAAAGTCGAAATCAT					
SYMS62	F	ACAACTCTTCAGTCTTCACGA	54	$(CAAC)_3$	0.85	$4.40E-66$	Truncated alcohol dehydrogenase
SYMS62	R	CCAATCTTGACATCCTTGAC					
SYMS63	F	GCACGGTGAAGTTCTAGTTC	54	$(TCGAT)_4$	0.67	$3.11E-31$	Hypothetical protein SORBIDRAFT_08g002800
SYMS63	R	CAGCTTCACTCATGAATTTTT					
SYMS64	F	GGACTGTACAAGGACGACAG	—	$(GCT)_4$	—	$1.08E-41$	Protein transport protein Sec61 subunit beta-like isoform
SYMS64	R	TCTGCTTTCTTGGATATGGTA					
SYMS65	F	AACACAAGCAAGAAATAAACG	53	$(GGT)_4$	0.51	$3.42E-74$	Dirigent-like protein
SYMS65	R	AACACTATGGTCAAGGTGGTA					
SYMS66	F	GCGTCAGAGTGTTAAAACAAG	52	$(GCT)_4$	0.58	$1.01E-41$	Protein transport protein Sec61 subunit beta-like isoform
SYMS66	R	GAAATCGCTCTATAAGGTTCC					
SYMS67	F	TCTCTCTGAAGATGATGCTTT	52	$(AAG)_5$	0.90	$4.25E-83$	Hypothetical protein SORBIDRAFT_03g005100
SYMS67	R	GTTAAGAGGCTTCCAAAGAAC					
SYMS68	F	CAGCTCGTCGTCTTCTTTT	—	$(GTC)_5$		$2.00E-55$	Putative ubiquitin-conjugating enzyme family
SYMS68	R	GTGGCTTGTTTGGATATTCTT					
SYMS69	F	GGACTGTACAAGGACGACAG	54	$(GCT)_4$	0.79	$9.28E-42$	Protein transport protein Sec61 subunit beta-like isoform
SYMS69	R	CGTCAGACGTACTGAAATGTT					
SYMS70	F	AACACAAGCAAGAAATAAACG	53	$(GGT)_4$	0.77	$1.58E-73$	Putative dirigent protein
SYMS70	R	AACACTATGGTCAAGGTGGTA					
SYMS71	F	GGACTGTACAAGGACGACAG	—	$(GCT)_4$	—	$9.86E-42$	Protein transport protein Sec61 subunit beta-like isoform
SYMS71	R	TCTGCTTTCTTGGATATGGTA					
SYMS72	F	CCCTCTCCCTGCTCTTTC	55	$(TCC)_4$	0.89	$4.36E-57$	Actin-depolymerizing factor 3
SYMS72	R	CAGTCACAAAGTCGAAATCAT					
SYMS73	F	GGACTGTACAAGGACGACAG	55	$(GCT)_4$	0.83	$1.09E-41$	Protein transport protein Sec61 subunit beta-like isoform
SYMS73	R	TCTGCTTTCTTGGATATGGTA					
SYMS74	F	GGACTGTACAAGGACGACAG	52	$(GCT)_4$	0.88	$9.51E-42$	Preprotein translocase Sec
SYMS74	R	TCTGCTTTCTTGGATATGGTA					
SYMS75	F	GCACCCCCAATTCGAACG	52	$(ACG)_3$	0.93	$1.78E-68$	TPA: general regulatory factor 1
SYMS75	R	CGGTAGTCCTTGATGAGTGT					
SYMS76	F	GGACTGTACAAGGACGACAG	52	$(GCT)_4$	0.78	$4.79E-41$	Protein transport protein Sec61 subunit beta-like isoform
SYMS76	R	TCTGCTTTCTTGGATATGGTA					
SYMS77	F	CACGCAACGCAAGCACAG	55	$(CCAT)_3$	0.93	$8.34E-70$	Hypothetical protein SORBIDRAFT_10g030160

TABLE 1: Continued.

Serial number	Type	Primer sequence	Annealing temperature	SSR motif	PIC value	E-value	Putative identities (blastn/blastx)
SYMS77	R	AAGTTGATTCACCCTCATTCT					
SYMS78	F	CACGCAACGCAAGCACAG	53	$(CGATC)_3$	0.92	$1.04E-41$	Translocon-associated protein alpha subunit precursor
SYMS78	R	AAGTTGATTCACCCTCATTCT					
SYMS79	F	GGACTGTACAAGGACGACAG	53	$(GCT)_4$	0.91	$1.04E-41$	Protein transport protein Sec61 subunit beta-like isoform
SYMS79	R	TCTGCTTTCTTGGATATGGTA					
SYMS80	F	CTTGATCCTTGACAAAAGAGA	52	$(AG)_6$	0.87	$2.25E-59$	*Predicted*: ubiquitin-conjugating enzyme E2
SYMS80	R	ATTGCTGTTGATATTTGGATG					
SYMS81	F	GCGTCAGAGTGTTAAAACAAG	53	$(GCT)_4$	0.87	$8.01E-42$	Protein transport protein Sec61 subunit beta-like isoform
SYMS81	R	GTGTAGAACTGGAGCATTGAG					
SYMS82	F	TATCAACAAGCCTTCCATTC	53	$(GTG)_4$	0.90	$1.12E-30$	Glycine-rich RNA-binding protein 2
SYMS82	R	GGCTATAGTCACCACGGTAG					
SYMS83	F	CGACAGGGAGAAGAGTACAG	55	$(GCT)_4$	0.87	$9.39E-42$	Protein transport protein Sec61 subunit beta-like isoform
SYMS83	R	GACTCTGCTTTCTTGGATATG					
SYMS84	F	GCGTCAGAGTGTTAAAACAAG	53	$(GCT)_4$	0.75	$1.14E-41$	Protein transport protein Sec61 subunit beta-like isoform
SYMS84	R	AATCGCTCTATAAGGTTCCTC					
SYMS85	F	CTCTTCTTCACCAATTCCTCT	—	$(CCG)_6$	—	$1.14E-51$	Protein transport protein Sec61 subunit beta-like isoform
SYMS85	R	CAAACCTCATAAAGAGTGCAG					
SYMS86	F	GGGCAAGCAAGAAACCAC	54	$(TCC)_4$	0.93	$1.16E-28$	TPA: translation initiation factor 1
SYMS86	R	CGTACATGAACGTAGTCCTTT					
SYMS87	F	GCGTCAGAGTGTTAAAACAAG	—	$(GCT)_4$	—	$1.19E-41$	Protein transport protein Sec61 beta subunit
SYMS87	R	AATCGCTCTATAAGGTTCCTC					
SYMS88	F	TTATAAGGAAATCCCCCACT	—	$(GCC)_4$	—	$7.71E-55$	Hypothetical protein SORBIDRAFT_09g000970
SYMS88	R	CACCAAGTACTCATCCATCAT					
SYMS89	F	CATCTCCTGCTAACAATTCAC	55	$(TGC)_4$	0.91	$9.64E-60$	*Predicted*: NAC domain-containing protein
SYMS89	R	ATTTATAGGTTGGCACCAGAG					
SYMS90	F	GCGTCAGAGTGTTAAAACAAG	53	$(GCT)_4$	0.85	$1.00E-22$	Protein transport protein Sec61 subunit beta-like isoform
SYMS90	R	GTGTAGAACTGGAGCATTGAG					

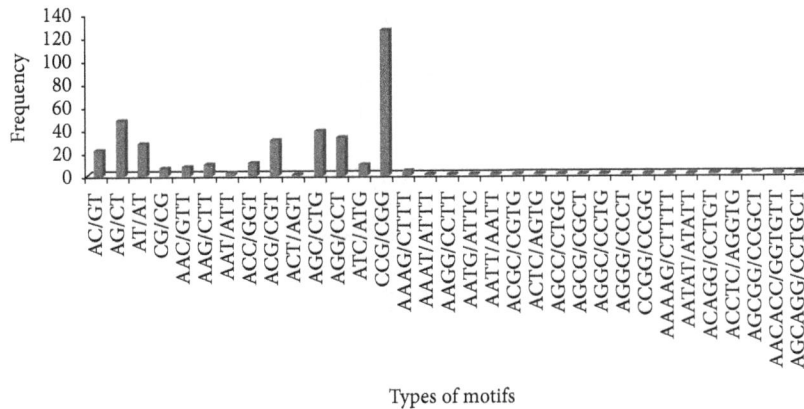

FIGURE 1: Details of 33 different types of nucleotide repeat motifs belonging to di-, tri-, tetra-, penta-, and hexanucleotide repeat motifs with sequence complementarity.

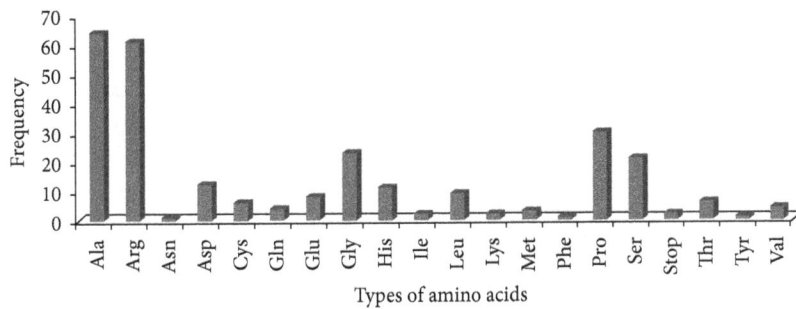

FIGURE 2: Details of different types of predicted amino acids encoded by trinucleotide repeat motifs.

transcripts. Selected EST-SSRs were associated with various pathways of metabolic process, namely, GO:0006281 DNA repair, GO:0006301 postreplication repair, GO:0016070 RNA metabolic process, GO:0016070 RNA metabolic process, GO:0006446 regulation of translational initiation, GO:0015991 ATP hydrolysis coupled proton transport, GO:0006629 lipid metabolic process, GO:0015031 protein transport, GO:0005667 transcription factor complex, GO:0005815 microtubule organizing centre, GO:0003743 translation initiation factor activity, GO:0017005 3′-tyrosyl-DNA phosphodiesterase activity, GO:0030042 actin filament d epolymerization, a nd G O:0015078 h ydrogen ion transmembrane transporter activity, and so forth (see the complete details of the most promising hits of gene ontology of EST-SSRs in the supplementary table).

3.3. Assessment of EST-SSR Marker in Selected Plants.

A set of 63 EST-SSR primers were evaluated for PCR optimization, polymorphism, and cross amplification in twenty genotypes belonging to cereals plants and *Saccharum* related genera and *Saccharum* species and their commercial cultivars, of which 42 EST-SSR primers produced successful amplifications with both expected and unexpected sizes (Figure 4). Among 42 EST-SSRs, twenty-eight belonged to trinucleotide repeats with then seven of tetra-, three of penta-, three of hexa-, and one of dinucleotide repeats. Meanwhile, PCR amplifications

produced 519 alleles (expected size) at 180 loci with an average of 2.88 alleles per locus. This result is comparable with earlier studies that reported on various plant species, namely, 2.79 alleles/locus in rice varieties [45], 2.9 to 6.0 alleles per locus in maize [46], and 3.04 alleles/locus in rye [47]. However, our result of alleles per locus is lower compared to previous studies that reported on sugarcane, that is, 6.04 alleles/locus [28], 7.55 alleles/locus [29], and 6.0 alleles/locus [48]. The polymorphic information content (PIC) was extended from 0.51 to 0.93 with an average of 0.83. It could be encompassed that low and high range of allelic amplifications with EST-SSRs correspond to marker polymorphism and low level of polymorphism from EST-SSRs might be due to possible selection against alterations in the conserved sequences of EST-SSRs [49, 50].

3.4. Cross Transferability.

The potentials of EST-SSR primers were examined for cross transferability among 20 plant species belonging to cereals and *Saccharum* related genera and *Saccharum* species and their cultivars under the same PCR conditions. However, 42 EST-SSRs showed successful amplifications among all the selected plants. The cross transferability was estimated to be 27.22% in wheat, 27.22% in maize, 47.22% in barely, 46.66% in rice, 36.11% in pearl millet, 55.55% in oat, 26.11% in *Sorghum*, 88.33% in *Narenga*, 98.88% in *Sclerostachya*, 71.11% in *Erianthus*, 60.0% in *Miscanthus*, 73.33% in *Bandjermasin Hitam*, 55.55% in *Gunjera*, 75.55% in

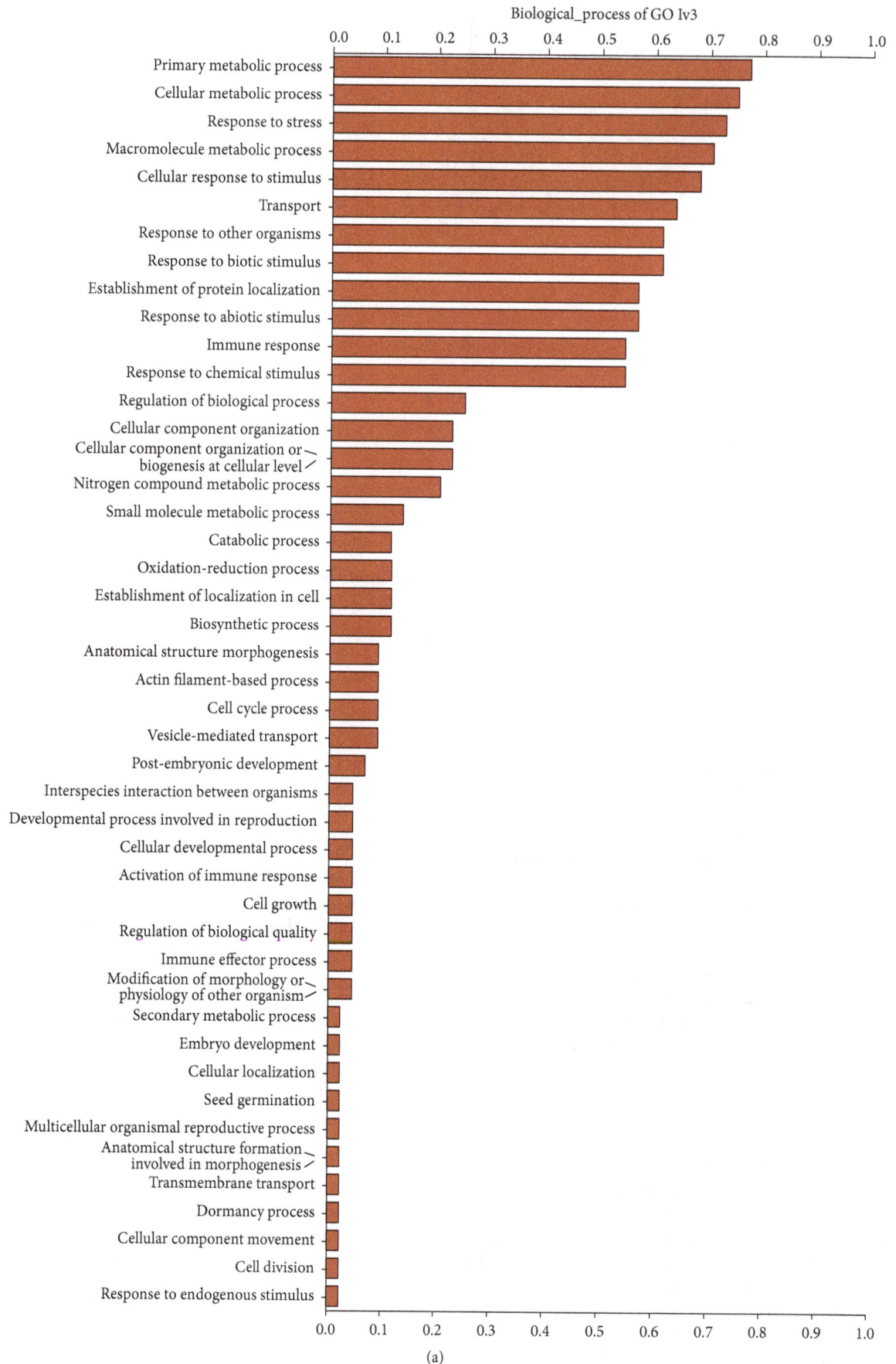

Figure 3: Continued.

Cellular_component of GO Iv3

(b)

Molecular_function of GO Iv3

(c)

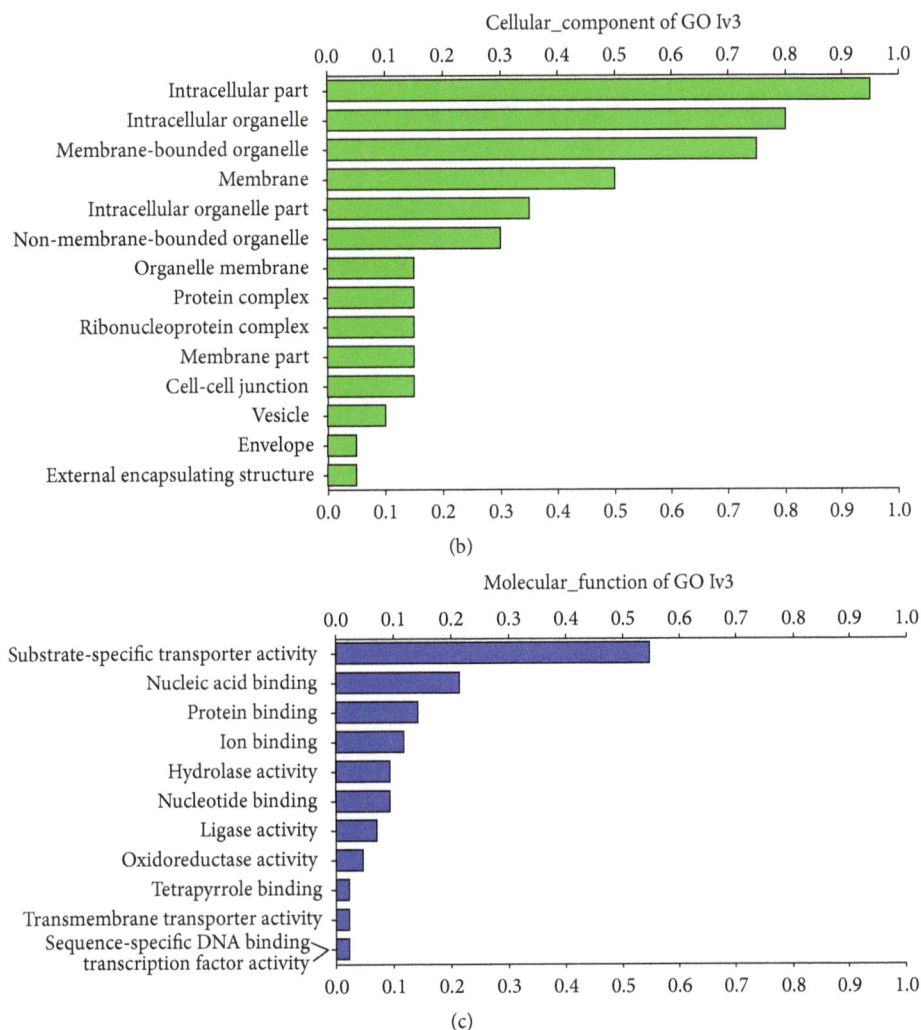

FIGURE 3: Most promising results of gene ontology (GO) as horizontal bar graphs. These graphs represent the distribution of GO terms categorized as a biological process (a), cellular component (b), and molecular function (c).

51NG56, 55.0% in N58, 50.56% in CoS 92423, 58.88% in CoS 88230, 51.11% in UP 9530, 52.78% in CoS 91230, and 60.0% in CoS 8436. Meanwhile, the frequency distributions of cross transferability of EST-SSRs ranged from 26.11% for *Sorghum* to 98.88% for *Sclerostachya*, with an average of 55.86% (Table 2). *Saccharum* related genera (79.58%) and *Saccharum* species (64.86%) showed high rate of cross transferability compared to other groups. This is in agreement with previous studies reported on *Saccharum* species and *Saccharum* related genera [12, 13, 51]. Several earlier studies related to cross transferability have been reported on distinct plant groups from different families using EST-SSRs markers [7, 52, 53]. This suggests that transferring ability of genic markers makes it compatible to determine genetic studies across the taxa for utilization in mapping of genes from related species along with genera and identification of suspended hybridization. This can also aid vigilance of the introgression of genetic entity from wild relatives to cultivated, comparative mapping and establishing evolutionary relationship between them. Thus, microsatellites derived from expressed region of the

genome are expected to be more conserved and more transferable across taxa.

3.5. Genetic Diversity Analysis by EST-SSRs. In order to evaluate the potential of EST-SSRs, the genetic analysis was done among 20 genotypes belonging to 7 cereals (wheat, maize, barley, rice, pearl millet, oat, and *Sorghum*), 4 *Saccharum* related genera (*Erianthus*, *Miscanthus*, *Narenga*, and *Sclerostachya*), 3 *Saccharum* species (51NG56 (*S. robustum*), N58 (*S. spontaneum*), and two of *S. officinarum* clones (*Gunjera* and *Bandjermasin Hitam*)), and 5 sugarcane commercial cultivars (CoS 8436, CoS 91230, CoS 88230, UP 9530, and CoS 92423). The generated allelic data were used for genetic relationships analysis by making dendrogram based on Jaccard's and neighbour-joining algorithm using FreeTree and TreeView software. The dendrogram fell into three major clusters with several edges, cluster I with eight genotypes comprising most of *Saccharum* species and their commercial cultivars, cluster II encompassing six genotypes of most of cereals species, and cluster III with six species comprising

TABLE 2: Details of cross transferability of 42 EST-SSR markers in twenty genotypes belonging to cereals and *Saccharum* related genera and *Saccharum* species and their cultivars. Lanes 1 to 20 represent number of bands produced in wheat, maize, barley, rice, pearl millet, oat, *Sorghum*, *Narenga*, *Sclerostachya*, *Erianthus*, *Miscanthus*, *Bandjermasin Hitam*, *Gunjera*, 51NG56, N58, CoS 92423, CoS 88230, UP 9530, CoS 91230, and CoS 8436, respectively.

S. number/lane	1	2	3	4	5	6	7	8	9	10	11	12	13	14	15	16	17	18	19	20	Primer polymorphism (%)
SY 28	4	5	6	1	2	3	0	2	5	5	3	2	3	2	3	3	1	3	2	3	95
SY 29	6	3	5	9	3	12	3	5	12	6	7	7	8	7	7	6	7	6	6	8	100
SY 30	1	0	4	0	0	7	0	5	6	6	3	3	4	3	2	3	2	3	2	3	80
SY 31	0	0	0	0	0	0	0	2	3	2	1	1	2	2	2	0	2	2	0	0	50
SY 32	1	0	0	1	0	2	0	2	4	3	1	3	0	0	1	0	0	0	0	0	55
SY 33	0	0	2	2	3	5	3	7	6	4	6	5	5	5	2	3	0	2	3	3	85
SY 34	0	0	0	0	0	0	0	0	0	0	0	3	2	5	4	5	5	5	9	9	85
SY 35	0	0	0	1	0	4	0	5	2	1	0	0	1	1	1	2	2	1	2	1	65
SY 36	0	0	0	0	0	0	0	0	3	0	0	5	2	2	2	0	2	0	0	2	35
SY 37	9	7	5	0	6	5	5	9	12	10	11	8	8	8	8	6	7	6	5	6	100
SY 38	0	0	0	3	0	2	0	7	8	1	0	0	2	2	1	2	0	0	3	0	50
SY 45	0	0	0	0	0	2	0	5	1	0	1	2	1	1	0	0	0	0	0	0	35
SY 46	0	1	4	4	1	0	0	4	4	0	0	0	0	0	0	0	0	0	0	0	35
SY 47	0	1	3	3	2	0	0	1	0	0	0	0	0	0	0	0	0	0	0	0	25
SY 48	0	0	0	0	5	3	0	8	12	0	5	3	3	2	1	3	2	2	0	0	60
SY 53	0	0	0	1	1	3	3	12	9	2	1	1	2	2	1	3	2	1	3	1	85
SY 58	0	0	0	1	1	0	3	2	1	0	0	0	0	0	0	0	0	0	0	0	25
SY 59	0	0	0	0	0	0	0	4	0	3	3	0	0	0	0	0	0	0	0	0	20
SY 61	0	3	7	5	4	0	3	4	5	6	3	4	1	5	0	5	5	5	5	5	90
SY 62	0	0	0	1	0	2	0	4	5	3	2	2	0	1	0	0	0	0	0	0	35
SY 63	0	0	0	0	0	0	0	1	1	4	1	6	1	1	1	0	1	0	2	1	55
SY 65	0	0	0	0	0	0	0	0	0	1	1	1	0	1	2	3	2	2	2	1	50
SY 66	0	0	0	0	0	0	0	1	0	0	1	3	1	2	0	0	3	2	2	1	40
SY 67	1	2	5	6	3	3	0	2	5	6	3	2	0	0	2	0	0	0	0	0	60
SY 69	1	5	2	0	1	3	0	3	1	2	1	1	1	1	1	2	1	1	3	2	80
SY 70	1	0	0	1	0	1	0	5	2	0	0	1	0	0	2	0	3	3	0	0	45
SY 72	1	0	2	0	0	0	0	1	4	5	8	8	5	8	1	3	6	4	6	8	70
SY 73	0	0	2	0	1	2	0	2	2	0	3	3	1	4	1	0	3	0	0	3	60
SY 74	2	1	1	0	1	1	3	1	3	2	1	4	1	5	1	0	3	0	0	2	80
SY 75	1	0	2	3	3	0	0	0	6	3	9	7	6	5	6	0	4	6	0	2	65
SY 76	0	0	0	0	0	0	0	1	2	1	4	0	0	0	1	0	0	6	0	0	30
SY 77	2	5	6	6	4	4	3	9	6	9	3	0	4	4	3	4	3	0	2	0	85
SY 78	2	2	5	6	6	6	4	5	7	6	4	10	9	10	3	2	8	7	5	9	100
SY 79	0	0	0	0	1	0	0	3	5	2	0	0	1	1	7	3	1	0	0	6	55
SY 80	1	1	0	4	0	1	0	4	5	6	2	5	4	5	5	3	5	4	5	5	85
SY 81	0	0	0	1	2	6	0	6	6	8	7	3	3	4	3	3	3	3	3	2	80
SY 82	0	0	10	3	1	4	3	9	1	1	6	5	4	3	5	3	4	4	3	2	90
SY 83	1	2	0	6	2	3	4	2	7	2	1	3	2	3	1	3	3	2	3	3	95
SY 84	0	0	0	1	2	3	0	0	2	0	0	3	3	1	0	3	3	3	3	3	60

TABLE 2: Continued.

S. number/lane	1	2	3	4	5	6	7	8	9	10	11	12	13	14	15	16	17	18	19	20	Primer polymorphism (%)
SY 86	1	5	6	3	2	5	4	5	8	11	6	5	3	11	5	12	7	6	7	10	100
SY 89	9	6	5	4	1	7	3	7	4	3	10	4	4	5	7	3	4	7	3	5	100
SY 90	5	0	3	2	5	2	3	4	3	4	0	4	3	6	4	4	2	3	6	4	90
Average of transferability	27.22	27.22	47.22	46.67	36.11	55.56	26.11	88.33	98.89	71.11	60.0	73.33	55.56	75.56	55.0	50.56	58.89	51.11	52.78	60.0	

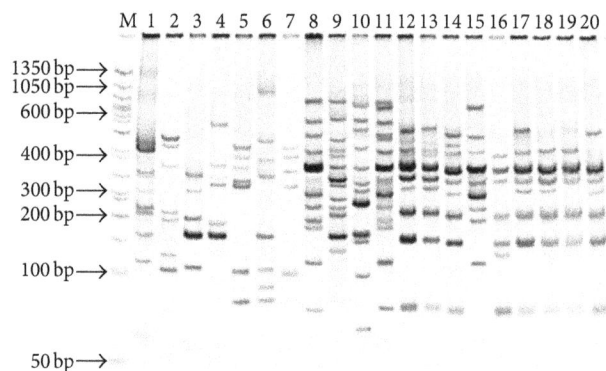

FIGURE 4: The gel represents PCR amplification profile with SYMS37 primer among twenty different plant species. Lanes: 1 *wheat*, 2 *maize*, 3 *barley*, 4 *rice*, 5 *pearl millet*, 6 *oat*, 7 *Sorghum*, 8 *Narenga*, 9 *Schlerostachya*, 10 *Erianthus*, 11 *Miscanthus*, 12 *Bandjermasin Hitam*, 13 *Gunjera*, 14 *51NG56*, 15 *N58*, 16 *CoS 92423*, 17 *CoS 88230*, 18 *UP 9530*, 19 *CoS 91230*, and 20 *CoS 8436*.

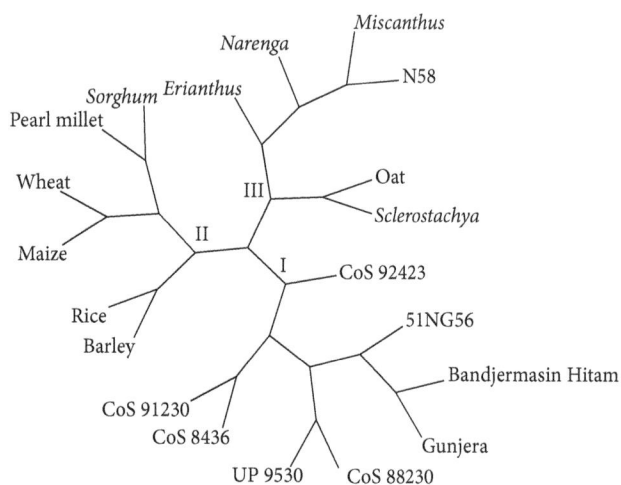

FIGURE 5: Dendrogram is constructed based on allelic data produced from 42 EST-SSR markers using FreeTree and TreeView software.

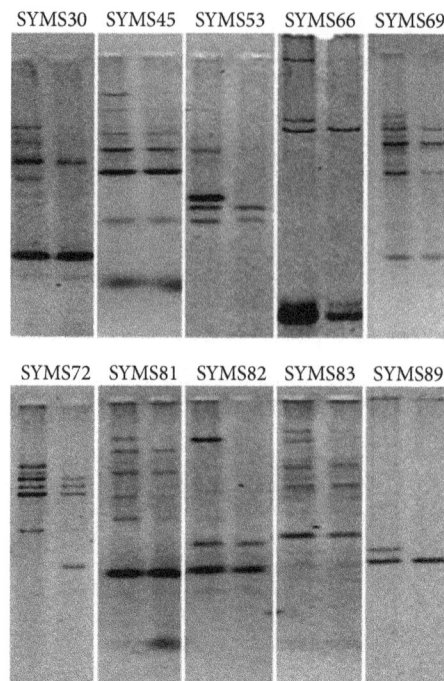

FIGURE 6: The gel represents polymorphism and discrimination between bulks of pooled DNA with contrasting high and low plant diameter through bulk segregation analysis.

most of the *Saccharum* related genera along with some interventions (Figure 5). This relationship is in agreement with previous studies reported by other authors [12, 13, 51, 54]. Our EST-SSRs markers showed close syntenic relationship and their evolutionary nature among the 20 genotypes into three major clusters with some genotypes divergence. These relationships have resulted from the expansion and contraction of SSRs in conserved EST sequences within the same group of plant species along with some variation having resulted from higher evolutionary divergence among them. Several earlier studies also reported on genetic diversity analysis within and across the plant taxa using molecular marker [7, 8, 24, 48, 52, 55–57]. Thus, microsatellite markers distinguished all the genotypes to certain extent and also provided the realistic estimate of genetic diversity among them.

3.6. Bulk Segregation Analysis (BSA) in Sugarcane.
All the 42 EST-SSR markers were evaluated in pooled DNA bulks of contrasting trait of sugarcane cultivars (CoS 91230 (CoS 775 × Co 1148) cross with CoS 8436 (MS 68/47 × Co 1148)) for the identification of reporter EST-SSR markers based on their allelic differences between them. Interestingly, 10 markers showed polymorphic nature and apparently discriminating potential between bulks through bulk segregation analysis (Figure 6). Among these, markers SYMS30, SYMS53, SYMS82, and SYMS89 showed a better response to discriminating the bulks. BSA is the strategy that involves the identification of genetic markers associated with character or trait which are based on their allelic differences between bulks [20]. Earlier studies have been established in sugarcane for the most prominent molecular markers detection linked to desirable traits through BSA. For example, molecular markers apparently linked to high fiber content in *Saccharum* species [58–60] and molecular markers used for QTL analysis and utilized for generating genetic maps around resistance genes in sugarcane against diseases and pests through BSA [12, 61, 62]. Several other studies also reported on selection of different agronomic traits in sugarcane for breeding programme with the development of molecular markers through BSA [1, 63–65]. Alternatively, BSA approach has been recently used for various purposes against the identification of differential expressed gene associated with both qualitative and quantitative using of the cDNA-AFLP approach [66–69]. Thus, BSA approach provides the easiest way in the direction of trait linked marker identification and also makes it possible to select informative markers beside evaluations of each marker in the whole progeny.

4. Conclusion

The present study was intended for identification and characterization of SSR in *Saccharum* spp. expressed sequence tag which is retrieved from public database (NCBI). Further, functional annotation was feasible to identify the most eminent EST-SSR markers selection. Therefore, this is the bypass way for EST-SSR markers development which reduces cost and time and provides an efficient way to analyze the transcribed portion of genome besides expense of own libraries development. A total of 63 EST-SSR markers were developed and experimentally validated for cross transferability along with their genetic relationships and also used for differentiation between pooled DNA bulks of *Saccharum* cultivars. These markers showed successful transferability rate among the twenty genotypes and established genetic diversity among cereals, *Saccharum* species/cultivars, and *Saccharum* related genera with some inconsistency. Further, some prominent marker also distinguished pooled DNA bulks of sugarcane cultivars based on stem diameter. Consequently, these EST-SSR markers were found to be more convenient which made it easy for us to use them as informative markers in further genetic studies in sugarcane breeding programme.

Competing Interests

The authors declare that there is no conflict of interests regarding the publication of this paper.

Acknowledgments

Authors are highly grateful to the Division of Biotechnology, UP Council of Sugarcane Research, for providing an opportunity and facilities for research works. Authors are also grateful to Director, UP Council of Sugarcane Research, Shahjahanpur, UP, India for their moral support. Authors also acknowledge University of Rajasthan for providing DBT-IPLS and DBT-BIF facilities.

References

[1] J. Y. Hoarau, G. Souza, A. D'Hont et al., *Sugarcane, A Tropical Crop with a Highly Complex Genome*, Plant Genomics, Sciences Publishers, Enfield, NH, USA, 2006.

[2] B. Roach and J. Daniels, "A review of the origin and improvement of sugarcane," *Copersucar International Sugarcane Breeding Workshop*, vol. 1, pp. 1–31, 1987.

[3] A. D'Hont, D. Ison, K. Alix, C. Roux, and J. C. Glaszmann, "Determination of basic chromosome numbers in the genus *Saccharum* by physical mapping of ribosomal RNA genes," *Genome*, vol. 41, no. 2, pp. 221–225, 1998.

[4] R. E. Casu, J. M. Manners, G. D. Bonnett et al., "Genomics approaches for the identification of genes determining important traits in sugarcane," *Field Crops Research*, vol. 92, no. 2-3, pp. 137–147, 2005.

[5] M. Morgante and A. M. Olivieri, "PCR-amplified microsatellites as markers in plant genetics," *Plant Journal*, vol. 3, no. 1, pp. 175–182, 1993.

[6] M. Agarwal, N. Shrivastava, and H. Padh, "Advances in molecular marker techniques and their applications in plant sciences," *Plant Cell Reports*, vol. 27, no. 4, pp. 617–631, 2008.

[7] S. U. Haq, R. Jain, M. Sharma, S. Kachhwaha, and S. L. Kothari, "Identification and characterization of microsatellites in expressed sequence tags and their cross transferability in different plants," *International Journal of Genomics*, vol. 2014, Article ID 863948, 12 pages, 2014.

[8] S. K. Parida, A. Pandit, K. Gaikwad et al., "Functionally relevant microsatellites in sugarcane unigenes," *BMC Plant Biology*, vol. 10, no. 1, article 251, 2010.

[9] X. Li, W. Gao, H. Guo, X. Zhang, D. D. Fang, and Z. Lin, "Development of EST-based SNP and InDel markers and their utilization in tetraploid cotton genetic mapping," *BMC Genomics*, vol. 15, no. 1, article 1046, 2014.

[10] Q. Kong, C. Xiang, Z. Yu et al., "Mining and charactering microsatellites in *Cucumis melo* expressed sequence tags from sequence database," *Molecular Ecology Notes*, vol. 7, no. 2, pp. 281–283, 2007.

[11] S. Gupta, K. P. Tripathi, S. Roy, and A. Sharma, "Analysis of unigene derived microsatellite markers in family solanaceae," *Bioinformation*, vol. 5, no. 3, pp. 113–121, 2010.

[12] R. K. Singh, R. B. Singh, S. P. Singh, and M. L. Sharma, "Identification of sugarcane microsatellites associated to sugar content in sugarcane and transferability to other cereal genomes," *Euphytica*, vol. 182, no. 3, pp. 335–354, 2011.

[13] R. K. Singh, S. N. Jena, S. Khan et al., "Development, cross-species/genera transferability of novel EST-SSR markers and their utility in revealing population structure and genetic diversity in sugarcane," *Gene*, vol. 524, no. 2, pp. 309–329, 2013.

[14] F. M. You, N. Huo, Y. Q. Gu et al., "BatchPrimer3: a high throughput web application for PCR and sequencing primer design," *BMC Bioinformatics*, vol. 9, article 253, 2008.

[15] T.-W. Chen, R.-C. R. Gan, T. H. Wu et al., "FastAnnotator—an efficient transcript annotation web tool," *BMC Genomics*, vol. 13, no. 7, article S9, 2012.

[16] D. Hoisington, M. Khairallah, and D. González-de-León, *Laboratory Protocols: CIMMYT Applied Molecular Genetics Laboratory*, CIMMYT, Mexico DF, Mexico, 2nd edition, 1994.

[17] J. Sambrook and D. W. Russell, *Molecular Cloning: A Laboratory Manual*, Cold Spring Harbor Laboratory Press, Cold Spring Harbor, New York, NY, USA, 2001.

[18] A. Pavlíček, Š. Hrdá, and J. Flegr, "Free-tree—freeware program for construction of phylogenetic trees on the basis of distance data and bootstrap/jackknife analysis of the tree robustness. Application in the RAPD analysis of genus Frenkelia," *Folia Biologica*, vol. 45, no. 3, pp. 97–99, 1999.

[19] R. D. Page, "TreeView: an application to display phylogenetic trees on personal computers," *Computer Applications in the Biosciences*, vol. 12, no. 4, pp. 357–358, 1996.

[20] R. W. Mlchelmore, I. Paran, and R. V. Kesseli, "Identification of markers linked to disease-resistance genes by bulked segregant analysis: a rapid method to detect markers in specific genomic regions by using segregating populations," *Proceedings of the National Academy of Sciences of the United States of America*, vol. 88, no. 21, pp. 9828–9832, 1991.

[21] J. R. Ellis and J. M. Burke, "EST-SSRs as a resource for population genetic analyses," *Heredity*, vol. 99, no. 2, pp. 125–132, 2007.

[22] J. Squirrell, P. M. Hollingsworth, M. Woodhead et al., "How much effort is required to isolate nuclear microsatellites from plants?" *Molecular Ecology*, vol. 12, no. 6, pp. 1339–1348, 2003.

[23] C. Duran, N. Appleby, D. Edwards, and J. Batley, "Molecular genetic markers: discovery, applications, data storage and visualisation," *Current Bioinformatics*, vol. 4, no. 1, pp. 16–27, 2009.

[24] K. Kumari, M. Muthamilarasan, G. Misra et al., "Development of eSSR-markers in setaria italica and their applicability in studying genetic diversity, cross-transferability and comparative mapping in millet and non-millet species," *PLoS ONE*, vol. 8, no. 6, article e67742, 2013.

[25] M. Rossi, P. G. Araujo, F. Paulet et al., "Genomic distribution and characterization of EST-derived resistance gene analogs (RGAs) in sugarcane," *Molecular Genetics and Genomics*, vol. 269, no. 3, pp. 406–419, 2003.

[26] K. S. Aitken, P. A. Jackson, and C. L. McIntyre, "A combination of AFLP and SSR markers provides extensive map coverage and identification of homo(eo)logous linkage groups in a sugarcane cultivar," *Theoretical and Applied Genetics*, vol. 110, no. 5, pp. 789–801, 2005.

[27] G. M. Cordeiro, R. Casu, C. L. McIntyre, J. M. Manners, and R. J. Henry, "Microsatellite markers from sugarcane (*Saccharum* spp.) ESTs cross transferable to erianthus and sorghum," *Plant Science*, vol. 160, no. 6, pp. 1115–1123, 2001.

[28] L. R. Pinto, K. M. Oliveira, E. C. Ulian, A. A. F. Garcia, and A. P. De Souza, "Survey in the sugarcane expressed sequence tag database (SUCEST) for simple sequence repeats," *Genome*, vol. 47, no. 5, pp. 795–804, 2004.

[29] K. M. Oliveira, L. R. Pinto, T. G. Marconi et al., "Characterization of new polymorphic functional markers for sugarcane," *Genome*, vol. 52, no. 2, pp. 191–209, 2009.

[30] S. P. Kumpatla and S. Mukhopadhyay, "Mining and survey of simple sequence repeats in expressed sequence tags of dicotyledonous species," *Genome*, vol. 48, no. 6, pp. 985–998, 2005.

[31] R. K. Varshney, A. Graner, and M. E. Sorrells, "Genic microsatellite markers in plants: features and applications," *Trends in Biotechnology*, vol. 23, no. 1, pp. 48–55, 2005.

[32] E. Portis, I. Nagy, Z. Sasvári, A. Stágel, L. Barchi, and S. Lanteri, "The design of *Capsicum* spp. SSR assays via analysis of in silico DNA sequence, and their potential utility for genetic mapping," *Plant Science*, vol. 172, no. 3, pp. 640–648, 2007.

[33] B. T. James, C. Chen, A. Rudolph et al., "Development of microsatellite markers in autopolyploid sugarcane and comparative analysis of conserved microsatellites in sorghum and sugarcane," *Molecular Breeding*, vol. 30, no. 2, pp. 661–669, 2012.

[34] L. Qiu, C. Yang, B. Tian, J.-B. Yang, and A. Liu, "Exploiting EST databases for the development and characterization of EST-SSR markers in castor bean (*Ricinus communis* L.)," *BMC Plant Biology*, vol. 10, no. 1, article 278, 2010.

[35] F. C. Victoria, L. C. da Maia, and A. C. de Oliveira, "In silico comparative analysis of SSR markers in plants," *BMC Plant Biology*, vol. 11, article 15, 2011.

[36] M. Muthamilarasan, B. V. Suresh, G. Pandey, K. Kumari, S. K. Parida, and M. Prasad, "Development of 5123 intron-length polymorphic markers for large-scale genotyping applications in foxtail millet," *DNA Research*, vol. 21, no. 1, pp. 41–52, 2014.

[37] R. V. Kantety, M. La Rota, D. E. Matthews, and M. E. Sorrells, "Data mining for simple sequence repeats in expressed sequence tags from barley, maize, rice, sorghum and wheat," *Plant Molecular Biology*, vol. 48, no. 5-6, pp. 501–510, 2002.

[38] M. J. Lawson and L. Zhang, "Distinct patterns of SSR distribution in the *Arabidopsis thaliana* and rice genomes," *Genome Biology*, vol. 7, no. 2, article R14, 2006.

[39] L. C. Da Maia, V. Q. De Souza, M. M. Kopp, F. I. F. De Carvalho, and A. C. De Oliveira, "Tandem repeat distribution of gene transcripts in three plant families," *Genetics and Molecular Biology*, vol. 32, no. 4, pp. 822–833, 2009.

[40] I. Fabregat, K. S. Koch, T. Aoki et al., "Functional pleiotropy of an intramolecular triplex-forming fragment from the 3′-UTR of the rat Pigr gene," *Physiological Genomics*, vol. 5, no. 2, pp. 53–65, 2001.

[41] Y.-C. Li, A. B. Korol, T. Fahima, A. Beiles, and E. Nevo, "Microsatellites: genomic distribution, putative functions and mutational mechanisms: a review," *Molecular Ecology*, vol. 11, no. 12, pp. 2453–2465, 2002.

[42] J. R. Gatchel and H. Y. Zoghbi, "Diseases of unstable repeat expansion: mechanisms and common principles," *Nature Reviews Genetics*, vol. 6, no. 10, pp. 743–755, 2005.

[43] R. K. Mishra, B. H. Gangadhar, J. W. Yu, D. H. Kim, and S. W. Park, "Development and characterization of EST based SSR markers in Madagascar periwinkle (*Catharanthus roseus*) and their transferability in other medicinal plants," *Plant Omics Journal*, vol. 4, no. 3, pp. 154–162, 2011.

[44] R. K. Sharma, P. Bhardwaj, R. Negi, T. Mohapatra, and P. S. Ahuja, "Identification, characterization and utilization of unigene derived microsatellite markers in tea (*Camellia sinensis* L.)," *BMC Plant Biology*, vol. 9, no. 1, article 53, 2009.

[45] V. Pachauri, N. Taneja, P. Vikram, N. K. Singh, and S. Singh, "Molecular and morphological characterization of Indian farmers rice varieties (*Oryza sativa* L.)," *Australian Journal of Crop Science*, vol. 7, no. 7, pp. 923–932, 2013.

[46] B. Stich, A. E. Melchinger, M. Frisch, H. P. Maurer, M. Heckenberger, and J. C. Reif, "Linkage disequilibrium in European elite maize germplasm investigated with SSRs," *Theoretical and Applied Genetics*, vol. 111, no. 4, pp. 723–730, 2005.

[47] B. Hackauf and P. Wehling, "Identification of microsatellite polymorphisms in an expressed portion of the rye genome," *Plant Breeding*, vol. 121, no. 1, pp. 17–25, 2002.

[48] T. G. Marconi, E. A. Costa, H. R. Miranda et al., "Functional markers for gene mapping and genetic diversity studies in sugarcane," *BMC Research Notes*, vol. 4, no. 1, article 264, 2011.

[49] M. C. Saha, M. A. R. Mian, I. Eujayl, J. C. Zwonitzer, L. Wang, and G. D. May, "Tall fescue EST-SSR markers with transferability across several grass species," *Theoretical and Applied Genetics*, vol. 109, no. 4, pp. 783–791, 2004.

[50] S. Feingold, J. Lloyd, N. Norero, M. Bonierbale, and J. Lorenzen, "Mapping and characterization of new EST-derived microsatellites for potato (*Solanum tuberosum* L.)," *Theoretical and Applied Genetics*, vol. 111, no. 3, pp. 456–466, 2005.

[51] S. K. Parida, S. K. Kalia, S. Kaul et al., "Informative genomic microsatellite markers for efficient genotyping applications in sugarcane," *Theoretical and Applied Genetics*, vol. 118, no. 2, pp. 327–338, 2009.

[52] S. Gupta and M. Prasad, "Development and characterization of genic SSR markers in *Medicago truncatula* and their transferability in leguminous and non-leguminous species," *Genome*, vol. 52, no. 9, pp. 761–771, 2009.

[53] S. K. Gupta, R. Bansal, U. J. Vaidya, and T. Gopalakrishna, "Development of EST-derived microsatellite markers in mungbean [*Vigna radiata* (L.) Wilczek] and their transferability to other Vigna species," *Indian Journal of Genetics and Plant Breeding*, vol. 72, no. 4, pp. 468–471, 2012.

[54] M. S. Khan, S. Yadav, S. Srivastava, M. Swapna, A. Chandra, and R. K. Singh, "Development and utilisation of conserved-intron

scanning marker in sugarcane," *Australian Journal of Botany*, vol. 59, no. 1, pp. 38–45, 2011.

[55] L. Y. Zhang, M. Bernard, P. Leroy, C. Feuillet, and P. Sourdille, "High transferability of bread wheat EST-derived SSRs to other cereals," *Theoretical and Applied Genetics*, vol. 111, no. 4, pp. 677–687, 2005.

[56] A. Selvi, N. V. Nair, J. L. Noyer et al., "AFLP analysis of the phenetic organization and genetic diversity in the sugarcane complex, *Saccharum* and *Erianthus*," *Genetic Resources and Crop Evolution*, vol. 53, no. 4, pp. 831–842, 2006.

[57] M. Raghami, A. I. López-Sesé, M. R. Hasandokht, Z. Zamani, M. R. F. Moghadam, and A. Kashi, "Genetic diversity among melon accessions from Iran and their relationships with melon germplasm of diverse origins using microsatellite markers," *Plant Systematics and Evolution*, vol. 300, no. 1, pp. 139–151, 2014.

[58] B. W. S. Sobral and R. J. Honeycutt, "High output genetic mapping of polyploids using PCR-generated markers," *Theoretical and Applied Genetics*, vol. 86, no. 1, pp. 105–112, 1993.

[59] N. Msomi and F. Botha, "Identification of molecular markers linked to fibre using bulk segregant analysis," in *Proceedings of the South African Sugar Technologists' Association (SASTA '94)*, vol. 68, pp. 41–45, 1994.

[60] L. Nivetha, N. Nair, P. Prathima, and A. Selvi, "Identification of microsatellite markers for high fibre content in sugarcane," *Journal of Sugarcane Research*, vol. 3, no. 1, pp. 14–19, 2013.

[61] C. M. Dussle, M. Quint, A. E. Melchinger, M. L. Xu, and T. Lübberstedt, "Saturation of two chromosome regions conferring resistance to SCMV with SSR and AFLP markers by targeted BSA," *Theoretical and Applied Genetics*, vol. 106, no. 3, pp. 485–493, 2003.

[62] C. Asnaghi, D. Roques, S. Ruffel et al., "Targeted mapping of a sugarcane rust resistance gene (*Bru1*) using bulked segregant analysis and AFLP markers," *Theoretical and Applied Genetics*, vol. 108, no. 4, pp. 759–764, 2004.

[63] J. H. Daugrois, L. Grivet, D. Roques et al., "A putative major gene for rust resistance linked with a RFLP marker in sugarcane cultivar 'R570'," *Theoretical and Applied Genetics*, vol. 92, no. 8, pp. 1059–1064, 1996.

[64] C. T. Guimarães, G. R. Sills, and B. W. S. Sobral, "Comparative mapping of Andropogoneae: *Saccharum* L. (sugarcane) and its relation to sorghum and maize," *Proceedings of the National Academy of Sciences of the United States of America*, vol. 94, no. 26, pp. 14261–14266, 1997.

[65] R. Ming, S.-C. Liu, P. H. Moore, J. E. Irvine, and A. H. Paterson, "QTL analysis in a complex autopolyploid: genetic control of sugar content in sugarcane," *Genome Research*, vol. 11, no. 12, pp. 2075–2084, 2001.

[66] J. Guo, R. H. Y. Jiang, L. G. Kamphuis, and F. Govers, "A cDNA-AFLP based strategy to identify transcripts associated with avirulence in *Phytophthora infestans*," *Fungal Genetics and Biology*, vol. 43, no. 2, pp. 111–123, 2006.

[67] A. Fernández-del-Carmen, C. Celis-Gamboa, R. G. F. Visser, and C. W. B. Bachem, "Targeted transcript mapping for agronomic traits in potato," *Journal of Experimental Botany*, vol. 58, no. 11, pp. 2761–2774, 2007.

[68] B. Kloosterman, M. Oortwijn, J. uitdeWilligen et al., "From QTL to candidate gene: genetical genomics of simple and complex traits in potato using a pooling strategy," *BMC Genomics*, vol. 11, no. 1, article 158, 2010.

[69] X. Chen, P. E. Hedley, J. Morris, H. Liu, R. E. Niks, and N. R. Waugh, "Combining genetical genomics and bulked segregant analysis-based differential expression: an approach to gene localization," *Theoretical and Applied Genetics*, vol. 122, no. 7, pp. 1375–1383, 2011.

9

Analysis of Mutation Rate of 17 Y-Chromosome Short Tandem Repeats Loci using Tanzanian Father-Son Paired Samples

Fidelis Charles Bugoye ⓘ,[1] Elias Mulima,[1] and Gerald Misinzo[2]

[1]*Department of Forensic Science and DNA Services, Government Chemist Laboratory Authority, Dar es Salaam, Tanzania*
[2]*Department of Veterinary Microbiology, Parasitology and Biotechnology, Sokoine University of Agriculture, Morogoro, Tanzania*

Correspondence should be addressed to Fidelis Charles Bugoye; bugoye81@yahoo.co.uk

Academic Editor: Giuseppe Novelli

Hundred unrelated father-son buccal swab sample pairs collected from consented Tanzanian population were examined to establish mutation rates using 17 Y-STRs loci DYS19, DYS389I, DYS389II, DYS390, DYS391, DYS392, DYS393, DYS385a, DYS385b, DYS437, DYS438, DYS439, DYS448, DYS456, DYS458, DYS635, and Y-GATA-H4 of the AmpFlSTRYfiler kit used in forensics and paternity testing. Prior to 17 Y-STRs analysis, father-son pair biological relationships were confirmed using 15 autosomal STRs markers and found to be paternally related. A total of four single repeat mutational events were observed between father and sons. Two mutations resulted in the gain of a repeat and the other two resulted in a loss of a repeat in the son. All observed mutations occurred at tetranucleotide loci DYS389II, DYS385a, and DYS385b. The locus specific mutation rate varied between 0 and 1.176 x10^{-3} and the average mutation rate of 17Y-STRs loci in the present study was 2.353x10^{-3} (6.41x10^{-4} - 6.013x10^{-3}) at 95% CI. Furthermore the mean fathers' age with at least one mutation at son's birth was 32 years with standard error of 2.387 while the average age of all fathers without mutation in a sampled population at son's birth was 26.781 years with standard error of 0.609. The results shows that fathers' age at son's birth may have an effect on Y-STRs mutation rate analysis, though this age difference was statistically not significant using unpaired samples t-test (p = 0.05). As a consequence of observed mutation rates in this study, the precise and reliable understanding of mutation rate at Y-chromosome STR loci is necessary for a correct evaluation and interpretation of DNA typing results in forensics and paternity testing involving males. The criterion for exclusion in paternity testing should be defined, so that an exclusion from paternity has to be based on exclusion constellations at a minimum of two 17 Y-STRs loci.

1. Introduction

Research and application of Y-chromosome short tandem repeats (Y-STRs) have proven beneficial in a number of fields including paternity, anthropology, and genealogical studies [1]. The very useful application of Y-STR systems is due to their potential in detecting and discriminating male DNA. Human Y-STR polymorphisms or microsatellites are useful in resolving and relating male lineages in forensics especially in sexual assault cases where there is a large proportion of mixed male/female stains [2], genealogical [3], evolutionary studies [4], and anthropological applications [5].

The interpretation of DNA evidence in forensic analysis and paternity testing is based on the similarities or differences at a genetic loci used. In parenthood testing, the difference at inheritable genetic marker loci between the putative father and the offspring is attributed to nonbiological paternity and therefore leads to exclusion of biological paternity. On the other hand, the spontaneous mutations in the germline of the putative father at any genetic marker locus used in the analysis can lead to an erroneous exclusion because such mutation results in differences between the parent and offspring. Since new alleles occur due to the mutation events, there is natural correlation between the degree of polymorphism and the underlined mutations rate of any given locus; i.e., the higher the mutation rate is, the more variable the locus is [6].

In forensic DNA typing applications, highly polymorphic loci are usually preferred due to their high power of discrimination. Therefore, short tandem repeat (STR) loci or microsatellites are considered to be the markers of choice in forensics because of their high power of discrimination

TABLE 1: Mutation count and Y-STRs loci mutation characteristics events as revealed by direct observation on father-son paired samples of previously confirmed biological relationship.

Sample ID's	Loci	Repeat sequence[a]	Father's profile	Son's profile	Mutation characteristics	Mutation count
F/C003	DYS385b	GAAA	16	17	Gain	1
F/C012	DYS385a	AAGG	15	16	Gain	1
F/C074	DYS385a	AAGG	16	15	Loss	1
F/C082	DYS389II	CTGT/CTAT	31	30	Loss	1

[a]Repetitive sequence structure previously reported by Gusmao and Carracedo (2003).

and ease of analysis. For criminal and paternity testing investigations, which involve males with deceased alleged father, Y-STRs are used as the marker of choice [7]. Y-STRs are preferred because they are transmitted without recombination from fathers to sons and therefore are able to characterise paternal pedigree. In addition, Y-STRs are suitable for sexual assault investigations as they provide male specific DNA profiles which avoid problems of mixed stain interpretation [1]. However, since highly polymorphic Y-STR loci applied in forensic investigations constantly evolve through mutations, the evaluation and interpretation of the genetic profiles requires precise knowledge on mutation rates at each loci used. Reliable estimations of mutation rates for these loci are valuable in assisting the interpretation of Y-STRs test results. Most of investigations have reported mutation rates for the minimal haplotype loci in different populations, but very few articles have reported results with the 17 Y-STRs loci mutation rates using African populations.

2. Materials and Methods

During this study, buccal swab samples were collected from consented father-son paired samples whose biological relationship was confirmed by autosomal STRs using AmpFlSTR Identifiler kit [8]. A total of 100 father-son pairs from Tanzanian population were collected in Dar es Salaam after obtaining informed consent for participation in the study. DNA extraction was done using Chelex method and the extracted DNA were amplified using 17Y-STRs of AmpFlSTRYfiler™ kit (DYS456, DYS389I, DYS390, DYS389II, DYS458, DYS19, DYS385a, DYS385b, DYS393, DYS391, DYS439, DYS635, DYS392, Y-GATA-H4, DYS437, DYS438, and DYS448) [8] using the following conditions: PCR amplification of the Y filer loci was performed using 0.5–1 ng of DNA template and the total $25\,\mu L$ reaction volumes of PCR amplification were used, as recommended by the manufacturer. The PCR amplicons were analyzed using capillary electrophoresis in an ABI Prism 3130xl genetic analyser [8]. Analysis of DNA fragments was performed using a Gene Mapper IDv.3.2 [8]. Data analysis was carried out using the Excel statistical Software [9], and the confidence interval (CI) was estimated from the binomial standard deviation [10].

3. Results and Discussion

3.1. 17 Y-STRs Locus Specific Mutation Characteristics. Analysis of locus specific mutation characteristics using 17 Y-STRs loci in Tanzanian father-son pairs of DNA confirmed

biological paternity revealed four mutations events which were identified on DYS385a, DYS385b, and DYS389II among 17Y-STRs loci analyzed [9] (Table 1).

However, no mutation event was observed for DYS19, DYS389I, DYS390, DYS391, DYS392, DYS393, DYS437, DYS348, DYS439, DYS448, DYS456, DYS458, DYS635, and Y-GATA-H4 loci analyzed. The observed locus specific mutation rate ranged between 0 for DYS19, DYS389I, DYS390, DYS391, DYS392, DYS393, DYS437, DYS348, DYS439, DYS448, DYS456, DYS458, DYS635, and Y-GATA-H4 loci and 1.765×10^{-3} ($1.43 \times 10^{-4} - 4.243 \times 10^{-3}$) for DYS385a locus at 95% CI [9]. Among 100 father-son pairs analyzed at the same 17 Y-STRs loci, there was no observation of multiple Y-chromosome microsatellite mutation within the same germline transmission or nonuniform alleles such as microvariants, duplication, and triplication that have been previously reported by Laouina [11] in Moroccan population.

The highly polymorphic Y-STR locus DYS385 was observed to have a higher mutation rate compared to all other Y-STRs loci analyzed (Table 2). In this study, the observed higher specific locus mutation rate for Y-STR locus DYS385a/b (if treated as single locus) was 1.765×10^{-3} followed by mutation rate of 5.88×10^{-4} for locus 389II [9].

All observed mutation events were characterised by single-step mutations (Table 1), in accordance with the generally accepted mutation model for microsatellites, in which the alleles are known to mutate primarily through the gain and loss of single repeat units [12, 13].

In addition, two tetranucleotide microsatellites loci DYS385 and 389II appeared to consist of higher average mutation rates among all 17 Y-STRs analyzed compared to all other trinucleotide and dinucleotide microsatellite loci [9]. Similar locus specific mutation characteristics were found in Moroccan's population in a sample of 252 father-son pairs using 17 Y-STRs loci by Laouina [11] in which average locus specific mutation rate was higher at tetra nucleotide microsatellites loci. This higher mutation rate on tetranucleotide microsatellites was also observed by Kayser et al. [1] using 15 Y-STRs loci for a total of 4999 male germline transmission from father-son pairs of previously confirmed paternity. Furthermore, the single loss mutation characteristics event observed in this study was in agreement with research results found by Farfán and Prieto [14] where three single-step loss mutations were observed at DYS389II loci, during mutations analysis at 17 Y-STR loci in father-son pairs from southern Spain.

TABLE 2: Mutation count, mutation rate and 95% confidence interval (CI) for the 17 Y-STRs loci studied using Tanzanian father-son paired samples.

Loci	Repetitive DNA sequence[a]	Mutation count	Allele transmission	Mutation rate	95% CI
DYS19	CTAT/CTAC	0	1700	0.000	$0.000 - 2.168 \times 10^{-3}$
DYS389I	AAGG/GAAA	0	1700	0.000	$0.000 - 2.168 \times 10^{-3}$
DYS389II	CTGT/CTAT	1	1700	5.88×10^{-4}	$1.5 \times 10^{-5} - 3.273 \times 10^{-3}$
DYS390	CTGT/CTAT	0	1700	0.000	$0.000 - 2.168 \times 10^{-3}$
DYS391	CTGT/CTAT	0	1700	0.000	$0.000 - 2.168 \times 10^{-3}$
DYS392	ATT	0	1700	0.000	$0.000 - 2.168 \times 10^{-3}$
DYS393	GATA	0	1700	0.000	$0.000 - 2.168 \times 10^{-3}$
DYS385a	AAGG	2	1700	1.176×10^{-3}	$1.43 \times 10^{-4} - 4.243 \times 10^{-3}$
DYS385b	GAAA	1	1700	5.88×10^{-4}	$1.5 \times 10^{-5} - 3.273 \times 10^{-3}$
DYS438	TTTTC/TTTTA	0	1700	0.000	$0.000 - 2.168 \times 10^{-3}$
DYS439	GATA	0	1700	0.000	$0.000 - 2.168 \times 10^{-3}$
DYS437	TCTA/TCTG	0	1700	0.000	$0.000 - 2.168 \times 10^{-3}$
DYS448	AGAGAT	0	1700	0.000	$0.000 - 2.168 \times 10^{-3}$
DYS458	GAAA	0	1700	0.000	$0.000 - 2.168 \times 10^{-3}$
DYS456	AGAT	0	1700	0.000	$0.000 - 2.168 \times 10^{-3}$
DYS635	TCTA/TGTA	0	1700	0.000	$0.000 - 2.168 \times 10^{-3}$
Y GATA H4	TAGA	0	1700	0.000	$0.000 - 2.168 \times 10^{-3}$
Average		**4**	**1 700**	**2.353×10^{-3}**	**$6.41 \times 10^{-4} - 6.013 \times 10^{-3}$**

3.2. Analysis of 17 Y-STRs Locus Specific Mutation Rate. Using 100 Tanzanian father-son paired samples with confirmed paternity covering 1700 meioses were used to estimate 17Y-STRs locus specific mutation rate. The observed average estimates of 17 Y-STRs locus specific mutation rate ranged from 0 to 1.765×10^{-3} (1.43×10^{-4} - 4.243×10^{-3}) at 95% CI. The higher average locus specific mutation rate was found at DYS385a locus while mutation rates of 5.88×10^{-4} (1.5×10^{-5} - 3.273×10^{-3}) at 95% CI were observed for both DYS385b and DYS389II loci. There was no mutation observed for DYS19, DYS389I, DYS390, DYS391, DYS392, DYS393, DYS437, DYS348, DYS439, DYS448, DYS456, DYS458, DYS635, and Y-GATA-H4 loci analyzed [9].

The average mutation rate across all markers in this study was 2.353×10^{-3} (6.41×10^{-4} - 6.013×10^{-3}) at 95% CI [9] (Table 2). This overall average mutation rate is nearly similar to those reported by Sanchez-Diz [15] in which a mutation rate of 2.2×10^{-3} in 701 father-son pairs in Iberian and Latin America groups was found. Dupuy et al. [16] used 1766 father-son pairs to analyze 9Y-STRs alone in Norway's population and found the average mutation rate of 2.3×10^{-3}. A similar mutation rate of 2.3×10^{-3} was found by Lee et al. [17] in collection of Y-STRs mutation events for Korean population using a high number of loci, 22 Y-STRs in a sample size equal to 369 father-son pairs. Furthermore, the average mutation rate estimated in this study is not significantly different from the average mutation rate of autosomal STR loci commonly used in forensics as previously reported from family analysis. In their findings, the autosomal STRs mutation rate of 2.1×10^{-3} was reported by Brinkman et al., 1998; mutation rate of 2.7×10^{-3} was reported by Henke and Henke [18] whereas mutation rate of 0.6×10^{-3} was reported by Sajantila [19].

The average mutation rate for 17Y-STRs loci found in this research study is greater than those calculated by Viera-Silva [20] in a sample of 95 father-son pairs from Portugal (1.85×10^{-3}) and by Farfán and Prieto [14] using 17 Y-STRs loci from southern Spain population was 1.563×10^{-3} (0.322×10^{-3} - 4.559×10^{-3}) at 95% CI, but the average mutation rate found in this study is less than those calculated by Decker [21] for Caucasian and Asians populations cited up; in US admixed sample of 399 father-son pairs using 17 Y-STRs, the mutation rate found was 3.13×10^{-3}.

The results of present study are in general agreement with the fore mentioned research findings in which all the same 17 Y-STRs set or other number Y-STRs loci used the average mutation rate observed were in the order of 10^{-3} though number of father-son pairs varied between the mentioned studies above. Since there were no significant differences in mutation rate observed, therefore the mutation rates analysis does not depend on the sample size or number of Y-STRs loci used but population diversity [9].

3.3. Effects of Father's Age on 17 Y-STR Mutation Rate Analysis. The present study shows that the average fathers' age with at least one mutation at son's birth was 32 years with standard error of 2.387 while the average age of all fathers without mutation in a sampled population at son's birth was 26.781 years with standard error of 0.609 (Table 3). Results shows undoubtedly the age of the mutated father from our study which is marginally older than that without mutations. The results clearly shows that fathers' age at son's birth may have an effect on Y-STRs mutation rate analysis, though this age difference is statistically not significant using unpaired samples t-test (p = 0.05) [9]. The results of present study are in agreement with results of research findings by Sanchez-Diz

TABLE 3: Fathers's age at the time of Sons' birth with at least one mutation and without any mutation on 17 Y-STRs loci.

Fathers' age (Years) with at least one mutation	Fathers' age (Years) without any mutation
25,40,32,31	19,20,26,25,30,21,25,26,24,30,32,30,21,28,30,30,24,25,34,20
	35,18,19,18,19,20,32,36,32,33,34,32,21,20,25,34,29,28,35,25
	24,26,25,24,26,25,34,23,25,24,28,29,25,26,23,24,27,28,36,26
	37,25,34,20,34,18,19,18,19,39,32,33,32,33,36,32,42,34,20,18
	18,19,26,19,20,32,33,32,33,36,25,20,24,18,29,19
Average = 32.000	**Average = 26.781**
Standard error = 2.387	**Standard error = 0.609**
P Value = 0.060	

[15], Lee [17], and Goebloed [22] who also reported relatively older average age of mutated fathers in their studies but the age difference between fathers' age with at least one mutation and fathers' age without mutation was found statistically not significant.

4. Conclusion

The results of 17-Y-STRs mutation observed from this study revealed that the precise and reliable understanding of mutation rate at Y-chromosome short tandem repeats loci is necessary for a correct evaluation and interpretation of DNA typing results in forensics and paternity testing involving males. Based on the findings, the criterion for exclusion in paternity testing should be defined in any DNA testing laboratory using 17-Yfiler Amplification kits, so that an exclusion from paternity has to be based on exclusion constellations at the minimum of two 17 Y-STRs loci.

Acknowledgments

This study was funded by the Government Chemist Laboratory Authority (GCLA). Fidelis Charles Bugoye is a recipient of a scholarship from GCLA through technical capacity training program.

References

[1] M. Kayser, P. De Knijff, P. Dieltjes et al., "Application of microsatellite-based Y chromosome haplotyping," Electrophoresis, vol. 18, no. 9, pp. 1602–1607, 1997.

[2] A. J. Redd, A. B. Agellon, V. A. Kearney et al., "Forensic value of 14 novel STRs on the human Y chromosome," Forensic Science International, vol. 130, no. 2-3, pp. 97–111, 2002.

[3] M. Kayser, M. Vermeulen, H. Knoblauch, H. Schuster, M. Krawczak, and L. Roewer, "Relating two deep-rooted pedigrees from Central Germany by high-resolution Y-STR haplotyping," Forensic Science International: Genetics, vol. 1, no. 2, pp. 125–128, 2007.

[4] M. A. Jobling and C. Tyler-Smith, "The human Y chromosome: An evolutionary marker comes of age," Nature Reviews Genetics, vol. 4, no. 8, pp. 598–612, 2003.

[5] P. De Knijff, "Messages through bottlenecks: On the combined use of slow and fast evolving polymorphic markers on the human Y chromosome," American Journal of Human Genetics, vol. 67, no. 5, pp. 1055–1061, 2000.

[6] M. Kayser and A. Sajantila, "Mutations at Y-STR loci: Implications for paternity testing and forensic analysis," Forensic Science International, vol. 118, no. 2-3, pp. 116–121, 2001.

[7] L. Roewer, M. Kayser, P. De Knijff et al., "A new method for the evaluation of matches in non-recombining genomes: Application to Y-chromosomal short tandem repeat (STR) haplotypes in European males," Forensic Science International, vol. 114, no. 1, pp. 31–43, 2000.

[8] AmpFlSTR _Yfiler _PCR Amplification Kit User Guide, Applied Biosystems, Foster City, CA, USA, 2006.

[9] F. Charles, Analysis of Mutation Rate of 17 Y-Chromosome Short Tandem Repeats Loci Using Tanzanian Father-Son Paired Samples, 2015, suanet.ac.tz: 8080/xmlui/handle/123456789/1290., http://www.suaire.

[10] http://statpages.org/confint.html.

[11] A. Laouina, S. Nadifi, R. Boulouiz et al., "Mutation rate at 17 Y-STR loci in "Father/Son" pairs from moroccan population," Legal Medicine, vol. 15, no. 5, pp. 269–271, 2013.

[12] J. L. Weber and C. Wong, "Mutation of human short tandem repeats," Human Molecular Genetics, vol. 2, no. 8, pp. 1123–1128, 1993.

[13] L. A. Zhivotovsky and M. W. Feldman, "Microsatellite variability and genetic distances," Proceedings of the National Acadamy of Sciences of the United States of America, vol. 92, no. 25, pp. 11549–11552, 1995.

[14] M. J. Farfán and V. Prieto, "Mutations at 17 Y-STR loci in father-son pairs from Southern Spain," Forensic Science International: Genetics Supplement Series, vol. 2, no. 1, pp. 425-426, 2009.

[15] P. Sánchez-Diz, C. Alves, E. Carvalho et al., "Population and segregation data on 17 Y-STRs: Results of a GEP-ISFG collaborative study," International Journal of Legal Medicine, vol. 122, no. 6, pp. 529–533, 2008.

[16] B. M. Dupuy, M. Stenersen, T. Egeland, and B. Olaisen, "Y-Chromosomal Microsatellite Mutation Rates: Differences in Mutation Rate between and Within Loci," Human Mutation, vol. 23, no. 2, pp. 117–124, 2004.

[17] H. Y. Lee, M. J. Park, U. Chung et al., "Haplotypes and mutation analysis of 22 Y-chromosomal STRs in Korean father-son pairs," International Journal of Legal Medicine, vol. 121, no. 2, pp. 128–135, 2007.

[18] J. Henke and L. Henke, "Mutation rate in human microsatellites," American Journal of Human Genetics, vol. 64, no. 5, pp. 1473-1474, 1999.

[19] A. Sajantila, M. Lukka, and A.-C. Syvänen, "Experimentally observed germline mutations at human micro- and minisatellite loci," European Journal of Human Genetics, vol. 7, no. 2, pp. 263–266, 1999.

[20] C. Vieira-Silva, P. Dario, T. Ribeiro, I. Lucas, H. Geada, and R. Espinheira, "Y-STR mutational rates determination in South Portugal Caucasian population," *Forensic Science International: Genetics Supplement Series*, vol. 2, no. 1, pp. 60-61, 2009.

[21] A. E. Decker, M. C. Kline, J. W. Redman, T. M. Reid, and J. M. Butler, "Analysis of mutations in father-son pairs with 17 Y-STR loci," *Forensic Science International: Genetics*, vol. 2, no. 3, pp. e31–e35, 2008.

[22] M. Goedbloed, M. Vermeulen, R. N. Fang et al., "Comprehensive mutation analysis of 17 Y-chromosomal short tandem repeat polymorphisms included in the AmpFSTR® Yfiler® PCR amplification kit," *International Journal of Legal Medicine*, vol. 123, no. 6, pp. 471–482, 2009.

10

Association of RBP4 Genotype with Phenotypic Reproductive Traits of Sows

A. Marantidis, G. P. Laliotis, and M. Avdi

Laboratory of Physiology of Reproduction of Farm Animals, Department of Animal Production, School of Agriculture, Aristotle University of Thessaloniki, 54124 Thessaloniki, Greece

Correspondence should be addressed to G. P. Laliotis; georgelaliotis@hotmail.com

Academic Editor: Norman A. Doggett

PCR-RFLP was applied to a commercial crossbred pig population in order to investigate the association between polymorphism (SNP) of Retinol-binding protein 4 (RBP4) gene and reproductive performance. 400 sows were genotyped and 2000 records of reproductive traits were used in order to retrieve information about the allele frequencies and the association of the RBP4 gene with main reproductive characteristics of the population. A deviation from the Hardy-Weinberg equilibrium was observed as a result of the AB genotype excess. In addition, the AA genotype saw statistically significant higher values of (i) the total number of born piglets ($p < 0.05$), (ii) the number of piglets born alive ($p < 0.01$), and (iii) the number of weaned piglets ($p < 0.01$). The number of the mummified piglets and the number of the piglets born dead did not differ between the various RBP4 genotypes. Interestingly, the AA genotype had a negative impact ($p < 0.05$) on the number of piglets born dead, resulting indirectly in a larger litter size. In conclusion, the AA genotype and in extension the A allele of RBP4 gene are in favor of producing larger litter size, suggesting that the RBP4 gene may be used in Marker-Assisted Selection (MAS) programs for a rapid improvement of the reproductive characteristics in pigs.

1. Introduction

The implementation of reliable genetic markers on the Marker-Assisted Selection (MAS) programs applied to the pig industry may result in small to moderate increases in litter size that would improve farm economic performance. In addition, the early selection of the sows, before they reach reproductive maturity, would stop breeding low-performance sows [1]. An effective way to detect genetic markers is the candidate gene approach, where the identification of gene polymorphisms that cause variation in a trait is based on physiological, immunological, or endocrine evidence [2].

The RBP4 gene is expressed during the period of fast elongation of the pig blastocyst [3]. This is a critical period for the survival of the embryos. Harney et al. [4] reported an increased expression of the RBP4 gene in the gravid endometrium, between the 10th to 12th day of pregnancy of sows, suggesting that the respective coding protein (RBP4) plays an important role in uterine and conceptus physiology

during the establishment of pregnancy. In addition, Retinol-binding protein 4 (RBP4) has been found to be a major secretory product of the pig conceptus prior to implantation [5]. Therefore, RBP4 is reported as a candidate gene for litter size owing to its possible role at the time of embryonic development [1].

An initial study in hyperprolific and control sows yielded an estimated additive effect (0.4 piglets/birth) of RBP4 gene [6]. In a follow-up study, Rothschild et al. [1] checked whether this effect was also detectable in commercial lines of pigs. They found an approximate additive effect of 0.23 pigs per litter ($p < 0.05$) for TNB (total number of born piglets) and 0.15 pigs per litter for NBA (number of piglets born alive) suggesting that RBP4 probably affects the litter size in some commercial lines. However, later studies in various crossbreds and pure breeds have not always confirmed these results. Even though the same trend was obtained in most cases, a statistically significant difference was not always observed [7, 8] or it was restricted to some parities [9, 10].

One of the major problems of the Greek pig industry is the low prolificacy performance reflecting a narrow economic income. To the best of our knowledge there is no previous reported study investigating any interaction between RBP4 gene polymorphism and reproductive traits in Greek pig farming, as a tool of identification of the more productive animals. Therefore, the aim of this study was to determine any possible associations of the RBP4 genotypes with main reproductive traits of sows in a commercial population reared in Greece, so as to enable such information to be used in the future for breeding selection schemes.

2. Materials and Methods

2.1. Animals and Data Collection. The pig population was derived from a Greek commercial farm (North Western of Greece). 400 sows in total (crossbreed from Large White × Landrace) were genotyped. The sows were randomly selected from the whole population of farm among those who gave their first birth at the age of 12 months and had at least 5 continuous litters. Sows were artificially inseminated with fresh semen derived from Duroc × Pietrain boars. All sows were kept under the same feeding and housing conditions and their reproductive performance was permanently recorded by the staff of the farm. Reproductive traits that were taken into account were (i) the total number of born piglets (TNB), (ii) the number of piglets born alive (NBA), (iii) the number of piglets born dead (NBD), (iv) the number of mummified piglets (NBM), (v) the number of aborted piglets (ABRT), and (vi) the number of weaned piglets (NW).

2.2. DNA Isolation and Genotyping. Polymerase chain reaction-restriction fragment length polymorphism (PCR-RFLP) for the RBP4 gene genotyping was performed. DNA was extracted from hair roots or blood of the sows, using the Nucleospin blood or tissue kits (Macherey-Nagel, Germany). An electrophoresis was performed to ensure the integrity of the DNA samples. Concerning the PCR procedure, the primer sequences used were 5′-GAGCAAGATGGAATG-GGTT-3′ and 5′-CTCGGTGTCTGTAAAGGTG-3′ for the forward and the reverse primer, respectively [1]. The PCR amplification was performed as follows: approximately 150 ng of genomic DNA was used as template and amplified in a final volume of 25 μL containing 200 nM of each primer, 1 mM dNTPs, and 1 unit MyTaq DNA Polymerase (Bioline). PCR amplification was performed using the following conditions: initial denaturation at 95°C for 5 min, 30 amplification cycles including denaturation at 95°C for 30 s, annealing at 56°C for 30 s, and extension at 72°C for 45 s and a final extension step at 72°C for 10 min. Finally, 15 μL of PCR product (550 bp) was digested in a total volume of 20 μL, containing 10 U of enzyme MspI (Takara), 2 μL of restriction buffer, and 2 μL of BSA for 3 hours. Restriction fragments were examined by electrophoresis on 2.5% agarose gel with 1x TBE buffer.

2.3. Statistical Analysis. Genotype frequencies, allele frequencies, and Hardy-Weinberg equilibrium estimations were calculated using PopGene Software v. 1.32 [11]. The statistical

TABLE 1: Genotype distribution and allele frequencies of the RBP4 gene in the analyzed population.

Genotype	Observed	Expected	Frequencies (P)
AA	113	125	0.28
AB	221	197	0.55
BB	66	78	0.17

$p = 0.56$; $q = 0.44$; $\chi^2 = 5.81$; df = 1.

procedures were performed using the SPSS program (version 19.0). A mixed statistical model was used for the analysis of associations between the RBP4 genotypes and the total number of born piglets (TNB), the number of piglets born alive (NBA), the number of piglets born dead (NBD), the number of piglets born mummified (BMUM), the number of aborted piglets (ABRT), and the number of weaned piglets (NW). Due to the uneven distribution of the first litters, their parameters were analyzed separately. The statistical models used in the analysis were as follows:

$$Y_{ijk} = \mu + G_i + L_j + \left(G_i * L_j\right) + T_k + e_{ijk}, \qquad (1)$$

where Y_{ijk} is trait value, μ is the general mean, G_i is the fixed effect of RBP4 genotype ($i = 1, 2, 3$), L_j is the fixed effect of the litter parity ($j = 2, 3, 4, 5$), $G_i * L_j$ is the effect of the interaction between the i genotype and the j litter parity, T_k is the random effect of the sow ($k = 1, 2, \ldots, 400$), and e_{ijk} is the random error.

When the data of the examined traits were analyzed for each parity separately, the factors regarding L_j (the fixed effect of litter parity) and $G_i * L_j$ (the effect of interaction between the i genotype and the j litter parity) were excluded from the model.

3. Results

3.1. Genotype and Allele Distribution of RBP4 Gene. Two RBP4 alleles (A, B) and three genotypes, namely, AA, AB, and BB, were identified in the examined population. The allelic and genotypic frequencies are presented in Table 1. Allele frequencies were 0.56 and 0.44 for allele A and allele B, respectively. A heterozygosity excess was observed, while the population was found to deviate ($p < 0.05$) from the Hardy-Weinberg equilibrium.

3.2. Association of Sows' Reproductive Traits and RBP4 Genotypes. The results of mean TNB, NBA, NBD, NBM, ABRT, and NW values (piglets/birth) in regard to the observed RBP4 genotypes are presented in Table 2. Statistically significant differences were detected between genotypes AA and AB as well as between AA and BB genotypes for almost all analyzed reproductive traits. Specifically, in regard to the TNB value, the AA genotype showed a higher ($p < 0.05$) number of piglets/litter (13.82 ± 0.10) compared to the AB (13.51 ± 0.07) and the BB (13.44 ± 0.13) genotype. The same significant ($p < 0.01$) trend was also observed for the number of born piglets (NBA) in reference to the respective RBP4 genotypes. Nonsignificant differences were observed

TABLE 2: Association of major reproductive traits with RBP4 genotypes among four parities (N = 400, 2–5 parities). The values presented as mean ± SEM[†].

Genotype	TNB	NBA	NBD	NBM	ABRT	NW
AA	13.82 ± 0.10[a,*]	13.02 ± 0.11[a,**]	0.54 ± 0.04[a,*]	0.25 ± 0.03[NS]	0.04 ± 0.10[NS]	12.30 ± 0.11[a,**]
AB	13.51 ± 0.07[b,*]	12.61 ± 0.08[b,**]	0.64 ± 0.03[b,*]	0.27 ± 0.02[NS]	0.05 ± 0.07[NS]	11.91 ± 0.08[b,**]
BB	13.44 ± 0.13[b,*]	12.55 ± 0.14[b,**]	0.60 ± 0.05[b,*]	0.29 ± 0.04[NS]	0.03 ± 0.01[NS]	11.78 ± 0.14[b,**]

[†]TNB: total number of born piglets; NBA: number of piglets born alive; NBD: number of piglets born dead; NBM: number of mummified piglets; ABRT: number of aborted piglets; NW: number of weaned piglets. [a,b]Different superscripts in the same column indicate significant difference ([*] $p < 0.05$; [**] $p < 0.01$; [NS]not significant).

TABLE 3: Genotype performance among the parities of 400 sows.

Parity	Genotype	n	TNB (mean + SD)	NBA (mean + SD)	NBD (mean + SD)	NBM (mean + SD)	ABRT (mean + SD)	NW (mean + SD)
1st	AA	113	11.73 ± 2.13	10.90 ± 2.18	0.56 ± 0.72	0.27 ± 0.67	0.04 ± 0.18	10.47 ± 2.15[a,**]
	AB	221	11.52 ± 1.98	10.59 ± 2.10	0.66 ± 0.78	0.28 ± 0.62	0.05 ± 0.23	10.03 ± 2.05[b]
	BB	66	11.56 ± 2.21	10.48 ± 2.50	0.65 ± 1.30	0.42 ± 1.42	0.03 ± 0.17	10.15 ± 2.66[b]
2nd	AA	113	13.25 ± 2.00	12.58 ± 2.14	0.44 ± 0.61	0.22 ± 0.56	0.04 ± 0.18	11.85 ± 2.19[a,*]
	AB	221	12.96 ± 2.04[a,**]	12.15 ± 2.34[a,**]	0.53 ± 0.70	0.27 ± 0.78	0.05 ± 0.23	11.39 ± 2.419[b,*]
	BB	66	12.71 ± 2.38[b,**]	11.97 ± 2.46[b,**]	0.53 ± 0.98	0.21 ± 0.45	0.03 ± 0.17	11.29 ± 2.58[a,**]
3rd	AA	113	14.04 ± 2.15[a,*]	13.11 ± 2.22[a,*]	0.69 ± 0.73	0.24 ± 0.51	0.04 ± 0.18	12.31 ± 2.29[a,*]
	AB	221	13.51 ± 2.17[b,*]	12.56 ± 2.26[b,*]	0.69 ± 0.83	0.26 ± 0.52	0.05 ± 0.23	11.78 ± 2.29[b,*]
	BB	66	13.53 ± 2.32[a,*]	12.58 ± 2.42[a,*]	0.55 ± 0.86	0.41 ± 0.66	0.03 ± 0.17	11.67 ± 2.46[a,*]
4th	AA	113	14.12 ± 2.39	13.30 ± 2.59[a,*]	0.54 ± 0.67	0.27 ± 0.52	0.04 ± 0.19	12.34 ± 2.55[a,*]
	AB	221	13.8 ± 2.31	12.80 ± 2.40[b,*]	0.72 ± 0.79	0.28 ± 0.53	0.05 ± 0.23	11.98 ± 2.44[b,*]
	BB	66	13.98 ± 2.50	13.11 ± 2.61[a,*]	0.62 ± 0.67	0.26 ± 0.62	0.03 ± 0.17	11.97 ± 2.74[a,*]
5th	AA	113	13.87 ± 2.16[a,*]	13.08 ± 2.09[a,*]	0.50 ± 0.63	0.29 ± 0.66	0.04 ± 0.19	12.71 ± 1.95[a,*]
	AB	221	13.76 ± 1.95[b,*]	12.93 ± 1.76[b,*]	0.60 ± 0.67	0.28 ± 0.52	0.05 ± 0.23	12.51 ± 1.73[b,*]
	BB	66	13.52 ± 2.15	12.54 ± 2.12[a,*]	0.71 ± 0.82	0.28 ± 0.55	0.03 ± .0.17	12.18 ± 1.97[a,*]

[†]TNB: total number of born piglets; NBA: number of piglets born alive; NBD: number of piglets born dead; NBM: number of mummified piglets; ABRT: number of aborted piglets; NW: number of weaned piglets. [a,b]Different superscripts in the same column and parity indicate significant difference ([*] $p < 0.05$; [**] $p < 0.01$).

for the number of piglets born mummified (BMUM) and the aborted piglets (ABRT) among the observed RBP4 genotypes. Interestingly, the AA genotype gave a statistically significant ($p < 0.05$) lower number of piglets born dead (NBD) per litter (0.54 ± 0.04) with respect to the two other analyzed genotypes. Moreover, the sows with the AA genotype had also larger number ($p < 0.01$) of piglets than the sows carrying the AB and the BB genotypes. Specifically, the AA genotype produced 0.52 weaned piglets (NW) more than the BB genotypic sows. This difference also remained statistically significant ($p < 0.01$) in the first parity (Table 3), with the AA genotype still having the largest litter size (10.47 ± 2.15) compared to the AB (10.03 ± 2.05) and the BB (10.15 ± 2.66) genotype. As far as it concerns TNB and NBA traits, it revealed that the AA and/or the AB genotypes are in favor of producing larger litter size in regard to the BB genotype among all analyzed parities (Table 3), except the first parity.

4. Discussion

Herein we reported the influence of the RBP4 gene on the prolificacy of a crossbreed population reared in Greece.

400 sows were genotyped and 2000 records were used in order to retrieve information about allele frequencies and the association of the RBP4 gene on main reproductive characteristics of the population.

According to our data, an excess of the AB genotype was observed and a higher frequency of allele A (0.56) with respect to allele B was noted. Similar allelic frequencies have been reported for Large White and Landrace × Large White populations [1, 7, 12–14] and in Black Slavonian sows [15]. Higher allelic values for A allele have been previously reported in Duroc [16] and in Polish Landrace [17] sows populations, while lower values have been reported by Kapelański et al. [18] in Polish Landrace and in Police Large White populations. The analyzed population was found to deviate from the Hardy-Weinberg equilibrium due to the higher value of heterozygous genotype, as also reported by Omelka et al. [19].

As far as it concerns the effect of RBP4 genotype on main reproductive traits of the studied population, it was revealed that both the AA and the AB genotypes were favored in producing statistically significant higher values of TNB, NBA, and NW traits (piglets/birth), rendering the A allele as an allele with an additive effect. Our results are in agreement with previous studies [2, 6, 20], which reported that the AA genotypes were associated with higher TNB and NBA piglets/birth. Moreover, previous authors [1, 9] concluded that the presence of B allele had a negative effect on the litter size, suggesting that A allele was in favor of prolificacy. In addition, Sun et al. [21] reported that crossbreed pigs with the AA genotype produced 0.72 TNB, 0.64 NBA more than the BB genotypic sows, while Gonçalves et al. [22] reported that A allele of RBP4 gene produced more piglets (TNB) and more live piglets per litter (NBA).

Contrary to our results, data obtained in other sows' populations (crossbreeds or pure breeds) failed to reach statistically significant difference in regard to the RBP4 genotypes and prolificacy [16, 23]. Furthermore, other researchers noted that BB genotypes displayed higher litter sizes than the AA and the AB genotypes [7, 24]. The B allele originates from a Chinese pig [25] and is associated also with high fertility performance [26]. The fact that in the studied population the B allele had a negative effect on the examined reproductive traits may reflect the absence of Chinese ancestors in our population.

It is worth noting that the AA genotype had also a negative impact ($p < 0.05$) on the number of piglets that were born dead, with respect to the other two genotypes (AB and BB) reflecting indirectly a greater litter size. To our knowledge this is the first time that a certain genotype is associated with the number of piglets that may be born dead in a litter.

Recent developments in the porcine genome maps set the basis for the identification of individual genes that affect reproduction. Therefore, the application of MAS on swine production may become more efficient as more associations between markers and traits are identified. This seems to be promising especially for litter size due to the low heritability and the sex limited nature of these traits [27]. Allele effects may differ between lines or populations due to the genetic background, rendering the establishment of a certain genotype expressing an improved reproductive trait not an easy task [15]. The A allele of RBP4 seems to impart an additive effect on the litter size rendering itself as a potential molecular marker in pig breeding schemes.

5. Conclusion

Our results on the RBP4 gene polymorphism studied in a commercial pig population showed that polymorphism of the RBP4 gene can be related to litter size. Statistical analysis revealed that sows with AA genotype had statistically higher litter sizes than those with BB genotypes, which displayed lower TNB, NBA, and NW values and higher NBD value. In addition, according to our results, A allele of the RBP4 gene seems to render an additive effect to the desired phenotypic reproductive traits (litter size), suggesting that this allele can be included in future Marker-Assisted Selection programs in sows' populations.

Acknowledgments

The authors would like to thank the owner and staff of the pig farm for their support to the data and samples collection. The authors would also like to acknowledge M.-A. Driancourt for his useful comments on the final paper. A. Marantidis was funded by the Greek State Scholarships Foundation, I.K.Y.

References

[1] M. F. Rothschild, L. Messer, A. Day et al., "Investigation of the retinol-binding protein 4 (RBP4) gene as a candidate gene for increased litter size in pigs," *Mammalian Genome*, vol. 11, no. 1, pp. 75–77, 2000.

[2] A. Spötter, S. Müller, H. Hamann, and O. Distl, "Effect of polymorphisms in the genes for LIF and RBP4 on Litter Size in Two German Pig Lines," *Reproduction in Domestic Animals*, vol. 44, no. 1, pp. 100–105, 2009.

[3] J. V. Yelich, D. Pomp, and R. D. Geisert, "Detection of transcripts for retinoic acid receptors, retinol-binding protein, and transforming growth factors during rapid trophoblastic elongation in the porcine conceptus," *Biology of Reproduction*, vol. 57, no. 2, pp. 286–294, 1997.

[4] J. P. Harney, T. L. Ott, R. D. Geisert, and F. W. Bazer, "Retinol-binding protein gene expression in cyclic and pregnant endometrium of pigs, sheep, and cattle," *Biology of Reproduction*, vol. 49, no. 5, pp. 1066–1073, 1993.

[5] W. E. Trout, J. J. McDonnell, K. K. Kramer, G. A. Baumbach, and R. M. Roberts, "The retinol-binding protein of the expanding pig blastocyst: molecular cloning and expression in trophectoderm and embryonic disc," *Molecular Endocrinology*, vol. 5, no. 10, pp. 1533–1540, 1991.

[6] L. Ollivier, L. A. Messer, M. F. Rothschild, and C. Legault, "The use of selection experiments for detecting quantitative trait loci with an application to the INRA hyperprolific pig," *Genetical Research*, vol. 69, no. 3, pp. 227–232, 1997.

[7] X. Wang, A. Wang, J. Fu, and H. Lin, "Effects of ESR1, FSHB and RBP4 genes on litter size in a large white and a landrace herd," *Archiv für Tierzucht*, vol. 49, no. 1, pp. 64–70, 2006.

[8] M. Muñoz, A. I. Fernández, C. Ovilo et al., "Non-additive effects of RBP4, ESR1 and IGF2 polymorphisms on litter size at different parities in a Chinese-European porcine line," *Genetics Selection Evolution*, vol. 42, article 23, 2010.

[9] A. Terman, M. Kmiec, D. Polasik, and K. Pradziadowicz, "Retinol binding protein 4 gene and reproductive traits in pigs," *Archiv Tierzucht, Dummerstorf*, vol. 50, pp. 181–185, 2007.

[10] A. Terman, M. Kmiec, D. Polasik, and A. Rybarczyk, "Association between RBP4 gene polymorphism and reproductive traits in polish sows," *Journal of Animal and Veterinary Advances*, vol. 10, no. 20, pp. 2639–2641, 2011.

[11] F. C. Yeh, R. C. Yang, and I. Boyle, *POPGENE Version 1.32. Microsoft Windows-Based Freeware for Population Genetic Analysis*, Center for International Forestry Research, University of Alberta, Edmonton, Canada, 1999.

[12] T. H. Short, M. F. Rothschild, O. I. Southwood et al., "Effect of the estrogen receptor locus on reproduction and production traits in four commercial pig lines," *Journal of Animal Science*, vol. 75, no. 12, pp. 3138–3142, 1997.

[13] J. P. Gibson, Z. H. Jiang, J. A. B. Robinson, A. L. Archibald, and C. S. Haley, "No detectable association of the ESR PvuII mutation with sow productivity in a Meishan x Large White F2 population," *Animal Genetics*, vol. 33, no. 6, pp. 448–450, 2002.

[14] S. Dall'Olio, L. Fontanesi, L. Tognazzi, and V. Russo, "Genetic structure of candidate genes for litter size in Italian Large White pigs," *Veterinary Research Communications*, vol. 34, supplement 1, pp. 203–206, 2010.

[15] E. A. Kabalin, K. Starčević, S. Menčik, M. Maurić, V. Sušić, and I. Štoković, "Analysis of ESR and RBP polymorphisms in black Slavonian sows: prel results," *Acta argiculturae Slovenica Supplement*, vol. 4, pp. 45–48, 2013.

[16] G. Drogemuller, H. Hamann, and O. Distl, "Candidate gene markers for litter size in different German pig lines," *Journal of Animal Science*, vol. 79, no. 10, pp. 2565–2570, 2001.

[17] M. Kmieć, J. Dvořák, and I. Vrtková, "Study on a relation between estrogen receptor (ESR) gene polymorphism and some pig reproduction performance characters in Polish Landrace breed," *Czech Journal of Animal Science*, vol. 47, no. 5, pp. 189–193, 2002.

[18] W. Kapelański, R. Eckert, H. Jankowiak, A. Mucha, M. Bocian, and S. Grajewska, "Polymorphism of ESR, FSHß, RBP4, PRL, OPN genes and their influence on morphometric traits of gilt reproductive tract before sexual maturity," *Acta Veterinaria Brno*, vol. 82, no. 4, pp. 369–374, 2013.

[19] R. Omelka, M. Martiniaková, D. Peškovičová, and M. Bauerová, "Associations between RBP4/MspI polymorphism and reproductive traits in pigs: an application of animal model," *Journal of Agrobiology*, vol. 25, pp. 77–80, 2008.

[20] L. A. Messer, L. Wang, J. Yelich, D. Pomp, R. D. Geisert, and M. F. Rothschild, "Linkage mapping of the retinol-binding protein 4 (RBP4) gene to porcine Chromosome 14," *Mammalian Genome*, vol. 7, article 396, 1996.

[21] Y.-X. Sun, Y.-Q. Zeng, H. Tang et al., "Relationship of genetic polymorphism of PRLR and RBP4 genes with litter size traits in pig," *Zhongguo yi Chuan Xue Hui Bian Ji*, vol. 31, no. 1, pp. 63–68, 2009.

[22] I. D. V. Gonçalves, P. B. D. Gonçalves, J. C. da Silva et al., "Interaction between estrogen receptor and retinol-binding protein-4 polymorphisms as a tool for the selection of prolific pigs," *Genetics and Molecular Biology*, vol. 31, no. 2, pp. 481–486, 2008.

[23] C. D. Blowe, K. E. Boyette, M. S. Ashwell, E. J. Eisen, O. W. Robison, and J. P. Cassady, "Characterization of a line of pigs previously selected for increased litter size for RBP4 and follistatin," *Journal of Animal Breeding and Genetics*, vol. 123, no. 6, pp. 389–395, 2006.

[24] A. Korwin-Kossakowska, W. Kapelański, M. Bocian, M. Kamyczek, and G. Sender, "Preliminary study of the RBP4, EGF and PTGS2 genes polymorphism in pigs and its association with reproduction traits of sows," *Animal Science Papers and Reports*, vol. 23, no. 2, pp. 95–105, 2005.

[25] B. J. Isler, K. M. Irvin, S. M. Neal, S. J. Moeller, and M. E. Davis, "Examination of relationship between estrogen receptor gene and reproductive traits in pig," *Journal of Animal Science*, vol. 80, no. 9, pp. 2334–2339, 2002.

[26] L. Alfonso, "Use of meta-analysis to combine candidate gene association studies: application to study the relationship between the ESR PvuII polymorphism and sow litter size," *Genetics Selection Evolution*, vol. 37, no. 4, pp. 417–435, 2005.

[27] M. F. Rothschild, "Genetics and reproduction in the pig," *Animal Reproduction Science*, vol. 42, no. 1–4, pp. 143–151, 1996.

11

Electronic Northern Analysis of Genes and Modeling of Gene Networks Underlying Bovine Milk Fat Production

Bhaskar Ganguly, Tanuj Kumar Ambwani, and Sunil Kumar Rastogi

Animal Biotechnology Center, Department of Veterinary Physiology and Biochemistry, College of Veterinary and Animal Sciences, G. B. Pant University of Agriculture and Technology, Pantnagar 263145, India

Correspondence should be addressed to Bhaskar Ganguly; vetbhaskar@gmail.com

Academic Editor: Norman A. Doggett

Milk fat is one of the most important economic traits in dairy animals. Yet, the biological machinery involved in milk fat synthesis remains poorly understood. In the present study, expression profiling of 45 genes involved in lipid biosynthesis and secretion was performed using a computational approach to identify those genes that are differentially expressed in mammary tissue. Transcript abundance was observed for genes associated with nine bioprocesses, namely, fatty acid import into cells, xenobiotic and cholesterol transport, acetate and fatty acid activation and intracellular transport, fatty acid synthesis and desaturation, triacylglycerol synthesis, sphingolipid synthesis, lipid droplet formation, ketone body utilization, and regulation of transcription in mammary, skin, and muscle tissue. Relative expression coefficient of the genes was derived based on the transcript abundance across the three tissue types to determine the genes that were preferentially expressed during lactation. 13 genes (*ACSS1, ACSS2, ADFP, CD36, FABP3, FASN, GPAM, INSIG1, LPL, SCD5, SPTLC1, SREBF1,* and *XDH*) showed higher expression in the mammary tissue of which 6 (*ADFP, FASN, GPAM, LPL, SREBF1,* and *XDH*) showed higher expression during adulthood. Further, interaction networks were mapped for these genes to determine the nature of interactions and to identify the major genes in the milk fat biosynthesis and secretion pathways.

1. Introduction

Milk fat content is regarded as one of the most important economic traits of milch animals; identification of gene networks that regulate lipid biosynthesis and secretion in the mammary gland is essential to our understanding of lactation physiology. Finding candidate genes for improved fat content represents a constant research goal [1] that may further provide opportunities for genetic manipulations to derive more or better milk fat. Comparing biomolecular composition of mammary tissue with other tissues can allow insights into the molecular responses that govern milk fat production. Transcriptional regulation is a major long-term mechanism for the control of metabolism, and switching on and off gene expression essentially drives a cell's biological function and activity [2]. In the present study, an attempt has been made to identify the genes, which are differentially expressed during milk fat production in bovines, and determine their interaction networks using a computational approach.

2. Materials and Methods

2.1. Identification of Differentially Expressed Genes. The reference bovine gene sequences for 45 genes, previously known to be involved in lipid synthesis (Table 1) [3], were obtained from Ensembl [4]. Electronic Northern (*e*-Northern) was performed using dbEST and UniGene; briefly, the dbEST [5] was queried for these sequences by BLASTN *v*2.2.27 [6] using default parameters and the significant hits were looked up in UniGene ESTProfile [7] for transcript abundance based on normalized "transcripts per million" (TPM) values in mammary tissue (TPM_{ma}), skin (TPM_s), and muscles (TPM_{mu}). Where information was available, transcript

TABLE 1: Genes involved in milk fat synthesis and secretion. 45 genes previously reported to be involved in nine different bioprocesses (in bold) of milk fat biosynthesis and secretion [3] were studied.

Gene name	Gene product
(1) FA import into cells	
CD36	CD36 molecule (thrombospondin receptor)
LPL	Lipoprotein lipase
VLDLR	Very-Low-Density Lipoprotein Receptor
(2) Xenobiotic and Cholesterol transport	
ABCA1	ATP-binding cassette, subfamily A (ABC1), member 1
ABCG2	ATP-binding cassette, subfamily G (WHITE), member 2
(3) Acetate and FA activation and intracellular transport	
ACBP	Acyl-CoA binding protein (diazepam binding inhibitor)
ACSL1	Acyl-CoA synthetase long-chain family member 1
ACSS1	Acyl-CoA synthetase short-chain family member 1
ACSS2	Acyl-CoA synthetase short-chain family member 2
FABP3	Fatty acid-binding protein, heart
(4) Fatty acid synthesis and desaturation	
ACACA	Acetyl-coenzyme A carboxylase alpha
FADS1	Fatty acid desaturase 1 (delta-5 desaturase)
FADS2	Fatty acid desaturase 2 (delta-6 desaturase)
FASN	Fatty acid synthase
SCD5	Stearoyl-CoA desaturase (delta-9-desaturase)
(5) Triacylglycerol synthesis	
AGPAT6	1-Acylglycerol-3-phosphate O-acyltransferase 6
DGAT1	Diacylglycerol acyltransferase 1
DGAT2	Diacylglycerol acyltransferase 2
GPAM	Glycerol-3-phosphate acyltransferase, mitochondrial
LPIN1	Lipin 1
(6) Sphingolipid synthesis	
ASAHL	N-Acylsphingosine amidohydrolase-like
LASS2	LAG1 homolog, ceramide synthase 2
OSBP	Oxysterol-binding protein 1
OSBPL10	Oxysterol-binding protein-like 10
OSBPL2	Oxysterol-binding protein-like 2
SGPL1	Sphingosine-1-phosphate lyase
SPHK2	Sphingosine kinase 2
SPTLC1	Serine palmitoyltransferase, long-chain base subunit 1
SPTLC2	Serine palmitoyltransferase, long-chain base subunit 2
UGCG	Ceramide glucosyltransferase
(7) Lipid droplet formation	
ADFP	Adipose differentiation related protein (adipophilin, PLIN2)
BTN1A1	Butyrophilin, subfamily 1, member A1
PLIN	Perilipin
XDH	Xanthine dehydrogenase
(8) Ketone body Utilization	
BDH1	3-Hydroxybutyrate dehydrogenase, type 1
OXCT1	3-Oxoacid CoA transferase 1
(9) Regulation of transcription	
INSIG1	Insulin-induced gene 1
INSIG2	Insulin-induced gene 2
PPARG	Peroxisome proliferator-activated receptor gamma
PPARGC1A	PPAR gamma, coactivator 1 alpha
PPARGC1B	PPAR gamma, coactivator 1 beta
SCAP	SREBP cleavage activating protein

TABLE 1: Continued.

Gene name	Gene product
SREBF1	Sterol regulatory element-binding transcription factor 1
SREBF2	Sterol regulatory element-binding transcription factor 2
THRSP	Thyroid hormone responsive SPOT14

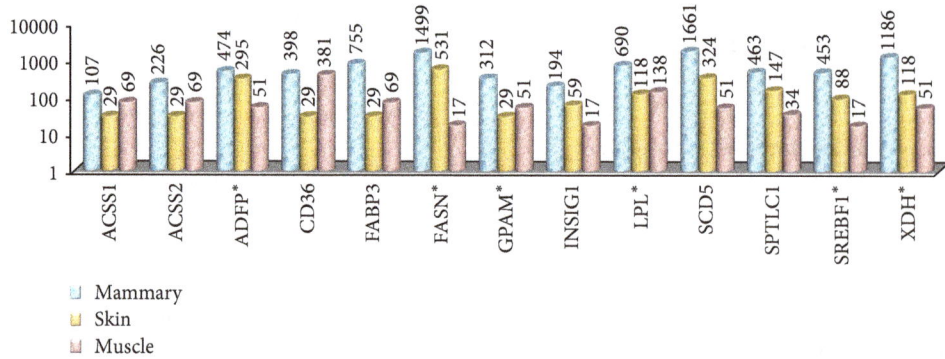

FIGURE 1: Transcript abundance of overexpressed genes. Based on TPM_{ma}: TPM_s and E_r values, 13 genes appeared to be overexpressed in mammary tissue. Transcript abundance values for these genes in mammary tissue, skin, and muscle have been shown for comparison. Genes marked with "∗" showed preferential expression in adult-derived tissues.

abundance (value not shown) was also compared between adult and young stages.

Percent mammary transcript abundance for a gene x was calculated using the formula:

$$\% \text{ Transcript abundance} = \left[\frac{TPM_{ma}x}{\sum TPM_{ma} \text{ all genes}} \right] \times 100. \qquad (1)$$

To confirm preferential mammary expression, relative expression coefficient (E_r) was calculated as the ratio of mammary transcript abundance to the geometric mean of cutaneous and muscle transcript abundance; that is,

$$E_r = \frac{TPM_{ma}}{\sqrt{(TPM_s \cdot TPM_{mu})}}. \qquad (2)$$

A twofold change in E_r was, arbitrarily, assumed to be significant; that is, upregulation of expression was inferred when $TPM_{ma} > TPM_s$ and $E_r \geq 2$. Similarly, downregulation was inferred when $TPM_{ma} < TPM_s$ and $E_r \leq 0.5$.

2.2. Gene Network Analysis.
Interaction networks and coexpression profiles for the genes were derived using STRING v9.1 with default settings [8]. STRING is a web-based application for network generation and visualization that uses a database of physical and functional protein interactions derived from four separate sources, namely, genomic context, high-throughput experimental data, coexpression, and existing literature. It quantitatively combines the information from these four sources to generate a weighted interaction network.

3. Results and Discussion

3.1. Transcript Abundance.
Transcript abundance was inferred from UniGene ESTProfile on the basis of normalized TPM values (Table 2) for the 45 genes involved in nine bioprocesses including fatty acid import into cells (CD36, LPL, and VLDLR); xenobiotic and cholesterol transport (ABCA1, ABCG2); acetate and fatty acid activation and intracellular transport (ACSS1, ACSS2, ACSL1, ACBP, and FABP3); fatty acid synthesis and desaturation (ACACA, FADS1, FADS2, FASN, and SCD5); triacylglycerol synthesis (AGPAT6, DGAT1, DGAT2, GPAM, and LPIN1); sphingolipid synthesis (ASAHL, LASS2, OSBP, OSBPL10, OSBPL2, SGPL1, SPHK2, SPTLC1, SPTLC2, and UGCG); lipid droplet formation (ADFP, BTN1A1, PLIN, and XDH); ketone body utilization (BDH1, OXCT1); and transcriptional regulation (INSIG1, INSIG2, PPARG, PPARGC1A, PPARGC1B, SCAP, SREBF1, SREBF2, and THRSP). Of the 45 genes included in the study, 23 genes did not have complete ESTProfiles and hence could not be included for further analysis. Notably, the absence of ESTProfiles of these 23 genes does not depress the robustness of the methodology that has been employed in the present study. Clearly, as more and more ESTProfiles get submitted to UniGene, it would become possible to use the same approach for analyzing the expression patterns of different genes including those of these 23 genes. Further, though ESTProfile TPM values lack exactitude as a measure of gene expression, the differences in TPM values tend to correlate with overall expression patterns.

3.2. TPM_{ma}/TPM_s and Percent Transcript Abundance: Functional Inferences.
Based on E_r and TPM_{ma}: TPM_s values, 13 genes (ACSS1, ACSS2, ADFP, CD36, FABP3, FASN, GPAM, INSIG1, LPL, SCD5, SPTLC1, SREBF1, and XDH; Table 2; Figure 1) were found to exhibit higher mammary expression

TABLE 2: Summary of results of transcript abundance studies. Of the 45 genes involved, 23 genes (S. numbers "23–45") did not have complete UniGene ESTProfile and were precluded from further studies. Of the 22 genes studied (S. numbers "1–22"), 13 genes (in bold) appeared to be overexpressed in mammary tissue. Of these, six genes (marked with an asterisk) further showed preferential expression in adult-derived tissues. TPM: transcripts per million; ma: mammary; s: skin; mu: muscle.

S. number	Gene	TPM_{ma}	TPM_{s}	TPM_{mu}	% transcript abundance	TPM_{ma}/TPM_{s}	E_{r}
(1)	ACBP	64	118	86	0.601	0.542	0.635
(2)	ACSL1	172	295	363	1.616	0.583	0.526
(3)	**ACSS1**	107	29	69	1.005	3.690	2.392
(4)	**ACSS2**	226	29	69	2.124	7.793	5.052
(5)	**ADFP***	474	295	51	4.454	1.607	3.864
(6)	**CD36**	398	29	381	3.740	13.724	3.786
(7)	DGAT1	32	88	17	0.301	0.364	0.827
(8)	**FABP3**	755	29	69	7.095	26.034	16.878
(9)	**FASN***	1499	531	17	14.086	2.823	15.777
(10)	**GPAM***	312	29	51	2.932	10.759	8.113
(11)	**INSIG1**	194	59	17	1.823	3.288	6.126
(12)	LASS2	194	324	17	1.823	0.599	2.614
(13)	**LPL***	690	118	138	6.484	5.847	5.407
(14)	OSBP	21	59	17	0.197	0.356	0.663
(15)	PLIN	32	29	34	0.301	1.103	1.019
(16)	PPARG	75	177	17	0.705	0.424	1.367
(17)	SCAP	21	51	17	0.197	0.412	0.713
(18)	**SCD5**	1661	324	51	15.608	5.127	12.921
(19)	SGPL1	10	177	34	0.094	0.056	0.129
(20)	**SPTLC1**	463	147	34	4.351	3.150	6.549
(21)	**SREBF1***	453	88	17	4.257	5.148	11.712
(22)	**XDH***	1186	118	51	11.145	10.051	15.288
(23)	ABCA1	0	0	0	0	—	—
(24)	ABCG2	0	0	0	0	—	—
(25)	ACACA	10	0	17	0.094	—	—
(26)	AGPAT6	593	29	0	5.572	20.448	—
(27)	ASAHL	0	0	17	0	—	—
(28)	BDH1	0	118	51	0	—	0
(29)	BTN1A1	744	0	0	6.991	—	—
(30)	DGAT2	0	88	103	0	—	0
(31)	FADS1	75	0	51	0.705	—	—
(32)	FADS2	0	0	0	0	—	—
(33)	INSIG2	0	0	0	0	—	—
(34)	LPIN1	0	0	138	0	—	—
(35)	OSBPL10	140	0	17	1.316	—	—
(36)	OSBPL2	0	88	0	0	—	—
(37)	OXCT1	10	0	0	0.094	—	—
(38)	PPARGC1A	21	0	51	0.197	—	—
(39)	PPARGC1B	0	0	0	0	—	—
(40)	SPHK2	0	0	0	0	—	—
(41)	SPTLC2	10	0	17	0.094	—	—
(42)	SREBF2	0	295	51	0	—	0
(43)	THRSP	0	29	69	0	—	0
(44)	UGCG	0	29	0	0	—	—
(45)	VLDLR	0	0	34	0	—	—

over skin or muscle; 6 of these 13 genes (*ADFP, FASN, GPAM, LPL, SREBF1,* and *XDH*) further showed preferential expression during adulthood.

The skin has been included for comparison because the mammary tissue is known to be modified cutaneous tissue [9] and differences in the expression pattern of genes between mammary and cutaneous tissue are likely to signify functional differences; muscle tissue has been included as a control. *ACSS1, ACSS2, ADFP, CD36, FABP3, FASN, GPAM, INSIG1, LPL, SCD5, SPTLC1, SREBF1,* and *XDH* had higher mammary expression over skin or muscle; *ADFP, FASN, GPAM, LPL, SREBF1,* and *XDH* showed preferential expression during adulthood and, hence, was considered most likely to be differentially expressed during milk fat synthesis.

Among genes responsible for fatty acid import into cells, both *LPL* and *CD36* appeared to have greater expression in mammary tissue. *LPL* primarily functions in the hydrolysis of triglycerides of circulating chylomicrons and very low-density lipoproteins (VLDL). *CD36* binds long-chain fatty acids and functions in their transport and also as a regulator of fatty acid transport. *LPL* showed more than 5-fold increase in TPM values in mammary tissue over cutaneous tissue whereas *CD36* showed a more than 13-fold increase. Further, the expression of *LPL* was greater in adult-derived tissues than in tissues derived from young ones. Our findings support the predication that *LPL* has higher mammary activity by virtue of high transcript abundance [10]. *LPL* was the fifth most abundant transcript. Also, more than 8-fold increase in transcript abundance of *CD36* has been previously reported during *in vivo* studies [3].

Among the five genes for acetate/fatty acid activation and intracellular transport, three showed relatively higher expression in mammary tissue. *ACSS1* showed a >3-fold increase, and *ACSS2* showed more than 7-fold increase in transcript abundance. These findings are comparable to previous findings; Bionaz and Loor have reported a higher (~13-fold) increase in *ACSS2* over *ACSS1* (~4-fold) [3]. *ACSS1* and *ACSS2* are responsible for activation of short-chain fatty acids; while *ACSS1*, primarily mitochondrial enzyme, activates acetate for energy production, *ACSS2*, the cytosolic enzyme, activates acetate for fatty acid synthesis [11]. With acetate being the chief substrate for energy production and fatty acid synthesis in the mammary tissue [9], overexpression of *ACSS1* and *ACSS2* during lactation is teleologically expected. In the same study [3], *FABP3* was the second most abundant transcript with a nearly 80-fold change in transcript abundance at 60 days of lactation. However, the relative change in transcript abundance at onset and 15, 30, 120, and 240 days of lactation ranged about 20–40. In our study, *FABP3* showed a >26-fold increase and was also the fourth most abundant transcript among all ones considered in the study. *FABP3* is involved in the intracellular trafficking long-chain fatty acids and their acyl coesters.

Fatty acid synthesis and desaturation per se are the most important step in milk fat synthesis. However, of the five genes studied, only two appear to be involved during the milk fat synthesis response in the mammary tissue. *FASN* that catalyzes the formation of long-chain fatty acids from acetyl-CoA, malonyl-CoA, and NADPH was the second most abundant transcript and showed ~3-fold increase in expression. *SCD5*, responsible for introducing a double bond in fatty acyl-coenzyme A at the delta 9 position, was the most abundant transcript (~15.6%) with more than 5-fold increase in mRNA expression. Bionaz and Loor have also reported *SCD5* to be the most abundant (~23%) among transcripts of genes involved in milk fat synthesis. However, in their study, the relative increase in expression has been reported to be much higher (~10–40-fold increase) [3].

GPAM, with more than 2% of all transcripts studied, was the only one of five genes involved in triacylglycerol synthesis found to be overexpressed (>10-fold increase). Bionaz and Loor have reported identical values of transcript abundance and relative expression of this gene [3]. Among the genes involved in sphingolipid synthesis, *SPTLC1* appeared to be overexpressed (>3-fold) whereas the expression of *SGPL1* appeared to be downregulated at about 1/20th of cutaneous expression.

Among the genes involved in lipid droplet formation, *ADFP* and *XDH* were overexpressed with 1.6- and a 10-fold increase in relative expression, respectively, over the cutaneous tissue. Both of these genes also showed preferential expression in adult-derived tissues. *XDH* includes xanthine dehydrogenase and xanthine oxidase; the enzyme can be converted from the dehydrogenase form (D) into the oxidase form (O) irreversibly by proteolysis or reversibly through the oxidation of sulfhydryl groups. *XDH* was the third most abundant of all transcripts (>11%). Bionaz and Loor have similarly reported >7% abundance of *XDH* transcripts and about 8-fold increase in its relative expression in the lactating mammary tissue [3].

Among transcriptional regulators that drive or sustain milk fat synthesis, *INSIG1* and *SREBF1* appeared to be overexpressed. Percent transcript abundance and relative increase in expression for the genes were about 1.8%, ~3-fold, and 4.2%, ~5-fold, respectively; 2.4- and 2.5-fold increases in the expression of these two genes have been reported previously [12]. Increase in *INSIG1* [3] and *SREBF1* [13] activities during lactation to much greater extents than being reported in the present study have also been reported earlier. A greater function of *SREBF2* than *SREBF1* in milk fat synthesis has been hypothesized [3]. Our study could not include *SREBF2* due to insufficient information on this gene in the UniGene ESTProfile. However, based on our results, *SREBF1* is expected to play a role at least equivalent to, if not greater than, *SREBF2* in regulating the transcriptional response during milk fat production in the mammary tissue. None of the genes involved in xenobiotic and cholesterol transport and ketone body utilization appeared to be differentially expressed as part of the lactational milk fat synthesis response.

3.3. Gene Interaction Networks. Interaction network for all the 45 genes, obtained using STRING, has been shown in Figure 2. The interactions were further purged to map only those 13 genes that showed preferential expression in mammary tissue in UniGene ESTProfile (Figure 3); in Figure 3(a), the weight of the edges shows the strength of

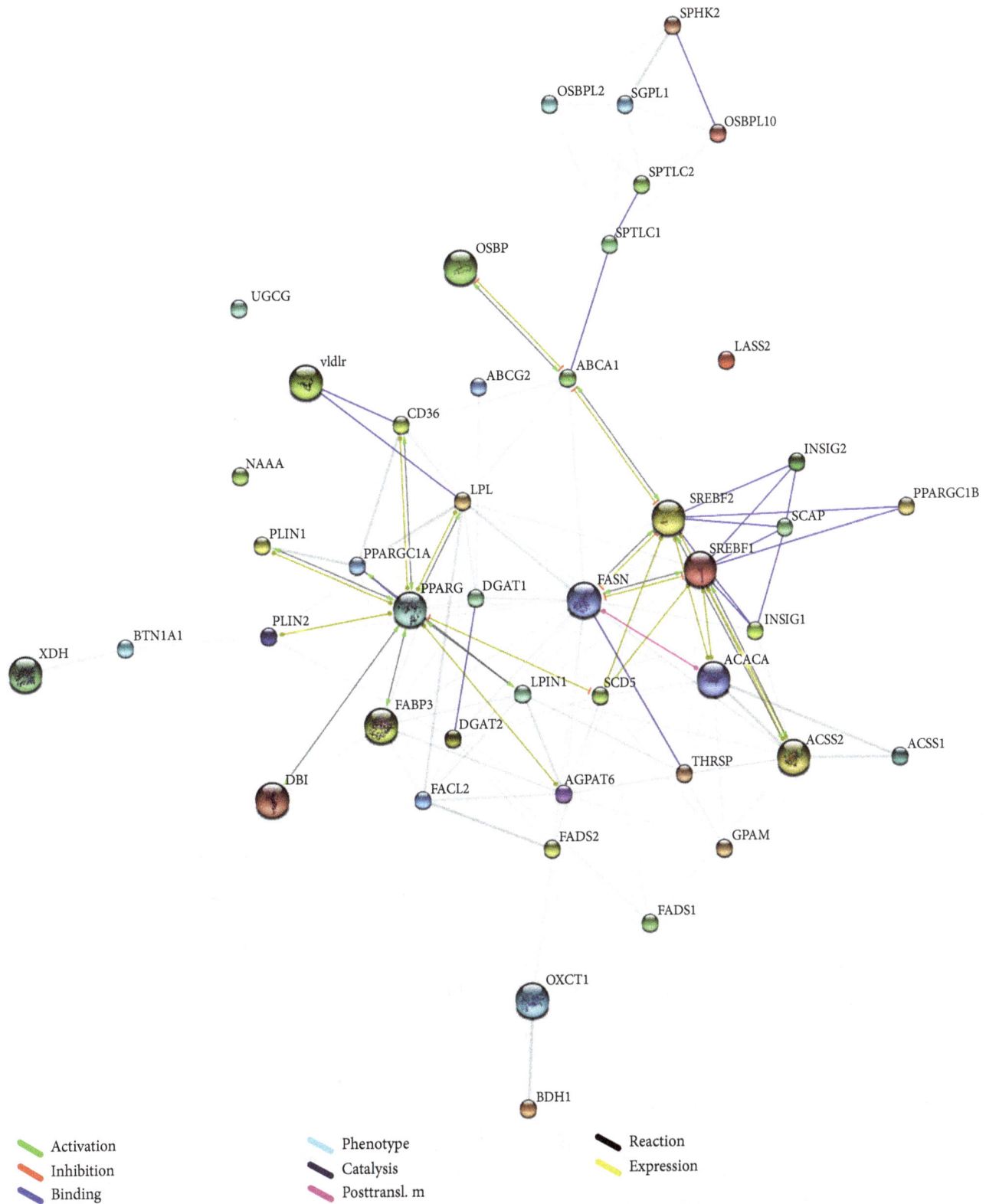

FIGURE 2: Interaction network of the genes involved in milk fat biosynthesis and secretion. STRING *v*9.1 was used to derive the network among genes involved in milk fat synthesis and secretion. *FASN* appears to be the central component in milk fat synthesis. The entire network appears to operate under two different control systems: one under *PPARG* and another under the joint control of *SREBF1* and *SREBF2*. Three genes (*ASAHL/NAAA*, *LASS2*, and *UGCG*) involved in sphingolipid synthesis did not interact with any other gene/gene product in the network.

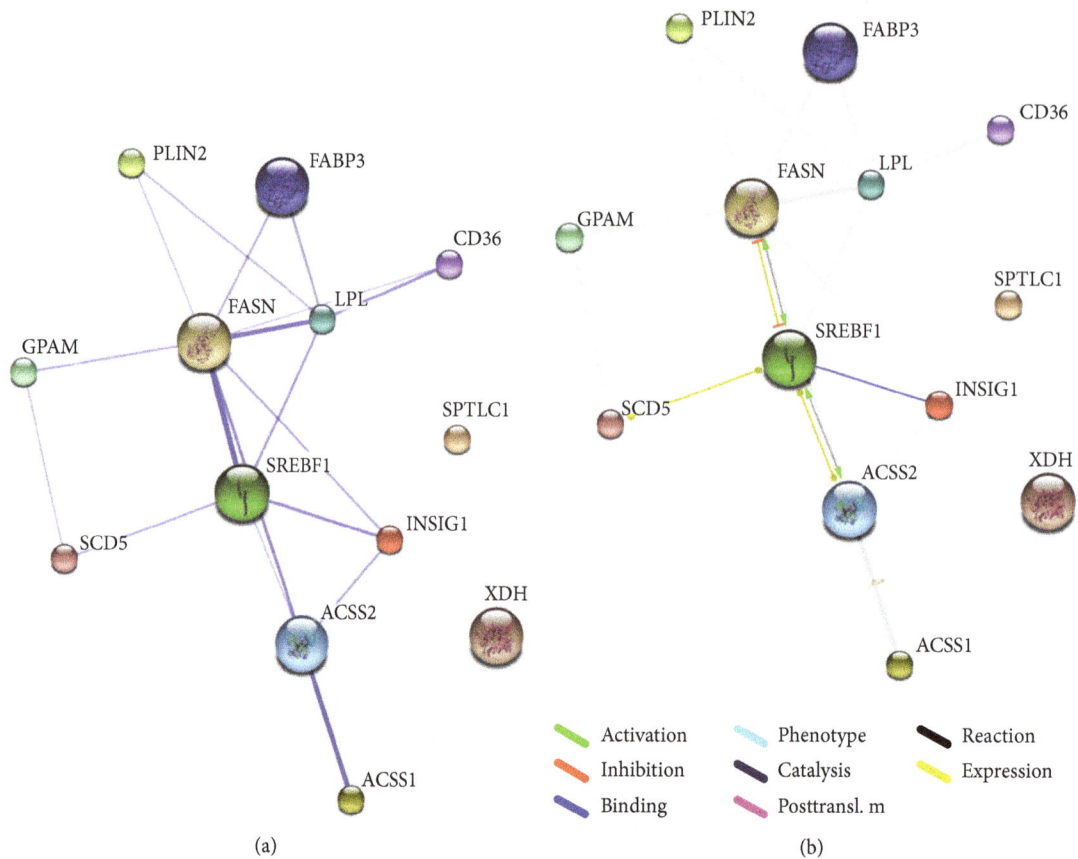

FIGURE 3: Interaction network of overexpressed genes. STRING *v*9.1 was also used to map the interaction networks between the genes that appeared to be overexpressed based on transcript abundance studies. In (a), the weight of the edges represents the confidence of the interaction; the nature of these interactions has been shown in (b). *SPTLC1* and *XDH* did not interact with any other gene of the 11 genes.

the interactions. The nature of these interactions has been depicted in Figure 3(b).

Network analysis shows that *FASN*, *SREBF1*, *SREBF2*, *PPARG*, and *ACSS2* are the major components of the milk fat synthesis pathway. Two subnetworks are evident: one under the predominant control of *PPARG* and the other one majorly under the joint control of *SREBF1* and *SREBF2*; both these subnetworks appear to converge at *FASN*. *SCD5*, the most abundant transcript, was the only gene under the direct control of *PPARG*, *SREBF1*, and *SREBF2*. Also, three of the four genes showing the maximum relative change in expression, namely, *FABP3*, *CD36*, and *XDH*, were chiefly under the control of *PPARG*. Thus, *PPARG*, though not found to be overexpressed based on TPM values, appears to play a major role in the transcriptional regulation of milk fat synthesis. Bionaz and Loor [3] have also advocated a role of *PPARG* in regulating the entire bovine milk fat synthesis machinery notwithstanding its downregulation and low mRNA abundance in mammary tissue. The genes involved in sphingolipid synthesis and ketone body utilization appeared to form two nearly independent clusters with sparse interaction with the rest of the network; *ASAHL (NAAA)*, *LASS2*, and *UGCG* did not interact with any other gene at all. *THRSP* did not form part of the cluster of genes involved in transcriptional

regulation. While all other gene products were involved directly or indirectly in interactions with each other, *SPTLC1* and *XDH* did not interact with any of these gene products.

STRING was also used to determine coexpression patterns between these genes; a functional association of the gene products can be assumed if a group of genes exhibits strong coexpression. Only a low level of association could be inferred between some of the genes based on the coexpression pattern (Figure 4). Again, *FASN* appeared to be the central component of the milk fat synthesis pathway.

To conclude, in this study we have put forward a simplistic approach for determining the relative expression of genes based on their transcript abundance values in UniGene ESTProfile. Further, we used this approach for the expression profiling of genes involved in milk fat biosynthesis and secretion in bovines. Based on our findings, an updated model of the transcriptional profile of the genes involved in milk fat production by the mammary gland has been presented. For the genes studied, the results were in good agreement with the previously reported results from wetlab studies, indicating the satisfactory performance of our computational approach. Our study included cutaneous tissue as a control assuming its ontogenetic equivalence to the quiescent, nonlactating mammary gland; the congruity of our findings

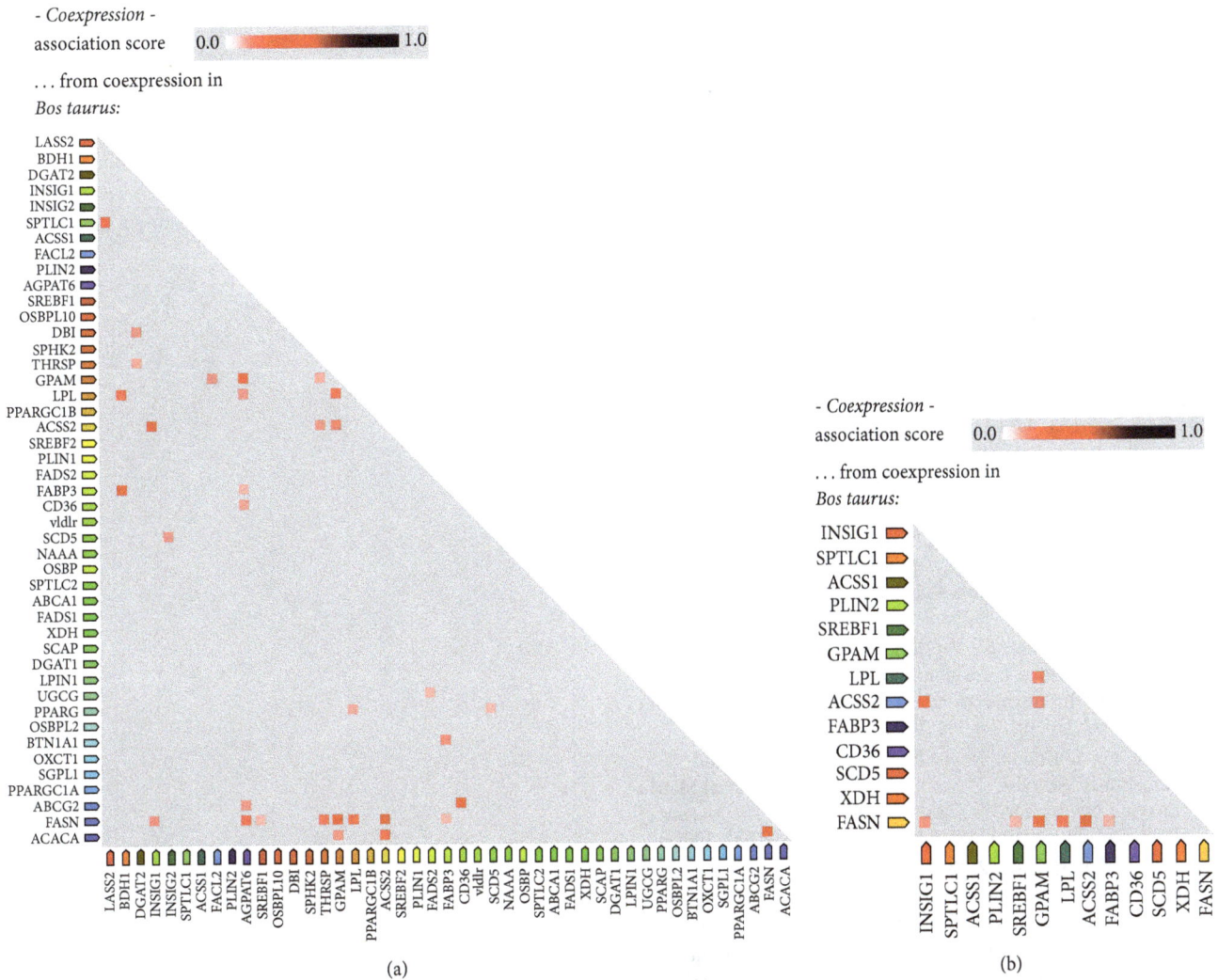

FIGURE 4: Coexpression of genes involved in milk fat biosynthesis and secretion. Coexpression pattern of the genes involved in milk fat synthesis and secretion in bovines (a) was derived from STRING. Analysis of the coexpression pattern of the thirteen genes that appeared to be overexpressed based on transcript abundance (b) showed a weak association between these genes. *FASN* shows coexpression based functional association with the maximum number of genes in both (a) and (b).

with those from previous studies projects this equivalence beyond the histological landscape to a biomolecular level. Previously, *SREBF2* has been upheld as the major regulator of transcription during milk fat biosynthesis, refuting the role of *SREBF1*. Our results reinstate *SREBF1* as a major transcriptional regulator, along with *INSIG1*, during the process. Using interaction network analysis of the genes, we could also show two separate transcriptional controls under *PPARG* and *SREBFs*. *FASN*, *SREBF1*, *SREBF2*, *PPARG*, and *ACSS2* were the major components of the milk fat synthesis pathway. However, expression profiles could not be studied for nearly half of the genes due to incomplete UniGene ESTProfile. Also, the inferences would have been more conclusive if UniGene ESTProfile also included information on the stage of lactation during which the mammary glands had been sampled. Thus, further studies are warranted to verify the proposed model and to fill in the research gaps in the present study.

Acknowledgments

Bhaskar Ganguly acknowledges the financial assistance received from Council of Scientific and Industrial Research (CSIR), India, in the form of a Senior Research Fellowship during this study.

References

[1] A. Tăbăran, V. A. Balteanu, E. Gal et al., "Influence of DGAT1 K232A Polymorphism on Milk Fat Percentage and Fatty Acid Profiles in Romanian Holstein Cattle," *Animal Biotechnology*, vol. 26, no. 2, pp. 105–111, 2015.

[2] D. Murray, P. Doran, P. MacMathuna, and A. C. Moss, "In silico gene expression analysis - An overview," *Molecular Cancer*, vol. 6, article no. 50, 2007.

[3] M. Bionaz and J. J. Loor, "Gene networks driving bovine milk fat synthesis during the lactation cycle," *BMC Genomics*, vol. 9, no. 1, article 366, 2008.

[4] J. Stalker, B. Gibbins, P. Meidl et al., "The Ensembl web site: Mechanics of a genome browser," *Genome Research*, vol. 14, no. 5, pp. 951–955, 2004.

[5] M. S. Boguski, T. M. J. Lowe, and C. M. Tolstoshev, "dbEST — database for "expressed sequence tags"," *Nature Genetics*, vol. 4, no. 4, pp. 332-333, 1993.

[6] Z. Zhang, S. Schwartz, L. Wagner, and W. Miller, "A greedy algorithm for aligning DNA sequences," *Journal of Computational Biology*, vol. 7, no. 1-2, pp. 203–214, 2000.

[7] U. J. Pontius, L. Wagner, and G. D. Schuler, "UniGene: A Unified View of the Transcriptome," in *The NCBI Handbook. Bethesda, MD: National Library of Medicine (US)*, NCBI, 2003.

[8] A. Franceschini, D. Szklarczyk, S. Frankild et al., "STRING v9.1: protein-protein interaction networks, with increased coverage and integration," *Nucleic Acids Research*, vol. 41, no. 1, pp. D808–D815, 2013.

[9] S. C. Park and G. L. Lindberg, "The mammary gland and lactation," in *Duke's Physiology of Domestic Animals*, W. O. Reece, Ed., pp. 720–741, Panima Publishing Corp, New Delhi, 12th edition, 2005.

[10] L. Bernard, C. Richard, V. Gelin, C. Leroux, and Y. Heyman, "Milk fatty acid composition and mammary lipogenic genes expression in bovine cloned and control cattle," *Livestock Science*, vol. 176, pp. 188–195, 2015.

[11] T. Fujino, J. Kondo, M. Ishikawa, K. Morikawa, and T. T. Yamamoto, "Acetyl-CoA Synthetase 2, a Mitochondrial Matrix Enzyme Involved in the Oxidation of Acetate," *The Journal of Biological Chemistry*, vol. 276, no. 14, pp. 11420–11426, 2001.

[12] Y. Gao, X. Lin, K. Shi, Z. Yan, and Z. Wang, "Bovine mammary gene expression profiling during the onset of lactation," *PLoS ONE*, vol. 8, no. 8, Article ID e70393, 2013.

[13] K. J. Harvatine and D. E. Bauman, "SREBP1 and thyroid hormone responsive spot 14 (S14) are involved in the regulation of bovine mammary lipid synthesis during diet-induced milk fat depression and treatment with CLA," *Journal of Nutrition*, vol. 136, no. 10, pp. 2468–2474, 2006.

12

Unique AGG Interruption in the CGG Repeats of the *FMR1* Gene Exclusively Found in Asians Linked to a Specific SNP Haplotype

**Pornprot Limprasert,[1] Janpen Thanakitgosate,[2]
Kanoot Jaruthamsophon,[1] and Thanya Sripo[1]**

[1]*Department of Pathology, Faculty of Medicine, Prince of Songkla University, Songkhla 90110, Thailand*
[2]*Department of Pathology, Faculty of Medicine, Ramathibodi Hospital, Mahidol University, Bangkok 10400, Thailand*

Correspondence should be addressed to Pornprot Limprasert; lpornpro@yahoo.com

Academic Editor: Norman A. Doggett

Fragile X syndrome (FXS) is the most common inherited intellectual disability. It is caused by the occurrence of more than 200 pure CGG repeats in the *FMR1* gene. Normal individuals have 6–54 CGG repeats with two or more stabilizing AGG interruptions occurring once every 9- or 10-CGG-repeat blocks in various populations. However, the unique (CGG)6AGG pattern, designated as 6A, has been exclusively reported in Asians. To examine the genetic background of AGG interruptions in the CGG repeats of the *FMR1* gene, we studied 8 SNPs near the CGG repeats in 176 unrelated Thai males with 19–56 CGG repeats. Of these 176 samples, we identified AGG interruption patterns from 95 samples using direct DNA sequencing. We found that the common CGG repeat groups (29, 30, and 36) were associated with 3 common haplotypes, GCGGATAA (Hap A), TTCATCGC (Hap C), and GCCGTTAA (Hap B), respectively. The configurations of 9A9A9, 10A9A9, and 9A9A6A9 were commonly found in chromosomes with 29, 30, and 36 CGG repeats, respectively. Almost all chromosomes with Hap B (22/23) carried at least one 6A pattern, suggesting that the 6A pattern is linked to Hap B and may have originally occurred in the ancestors of Asian populations.

1. Introduction

The cause of fragile X syndrome (FXS) is the expansion of CGG repeats in the 5′UTR of the *FMR1* gene and subsequent hypermethylation at the CpG island in the promoter region of this gene, leading to transcriptional silence of the mRNA and absence of FMRP translation [1, 2]. Affected full mutation individuals have >200 pure CGG repeats. Premutation carriers have 55–200 CGG repeats with one AAG interruption or absent AGG interruption resulting in increasing length of pure CGG repeats at the 3′ end of the CGG repeat tracts. Normal individuals have 6–54 CGG repeats with two or more stabilizing AGG interruptions occurring once every 9 or 10 CGG repeat blocks [3, 4]. The common patterns are (CGG)9AGG and (CGG)10AGG,

found in various populations. However, the (CGG)6AGG pattern (designated as 6A) has been reported exclusively in Asian populations [5–11], leading to the possibility that this 6A pattern may have originated in Asia.

To explore the evolution of the 6A pattern, we studied 176 unrelated Thai males with 19–56 CGG repeats using 8 SNPs near the CGG repeats of the *FMR1* gene. Of these 176 samples, we identified AGG interruption patterns from 95 samples with different CGG repeats using direct DNA sequencing. We found a specific SNP haplotype linked to the 6A pattern, and we also found something new that the SNP haplotypes showed strong associations between the common CGG repeat groups (29, 30, and 36) and AGG interruption patterns, suggesting different evolutionary lineages in the common CGG repeats of the *FMR1* gene.

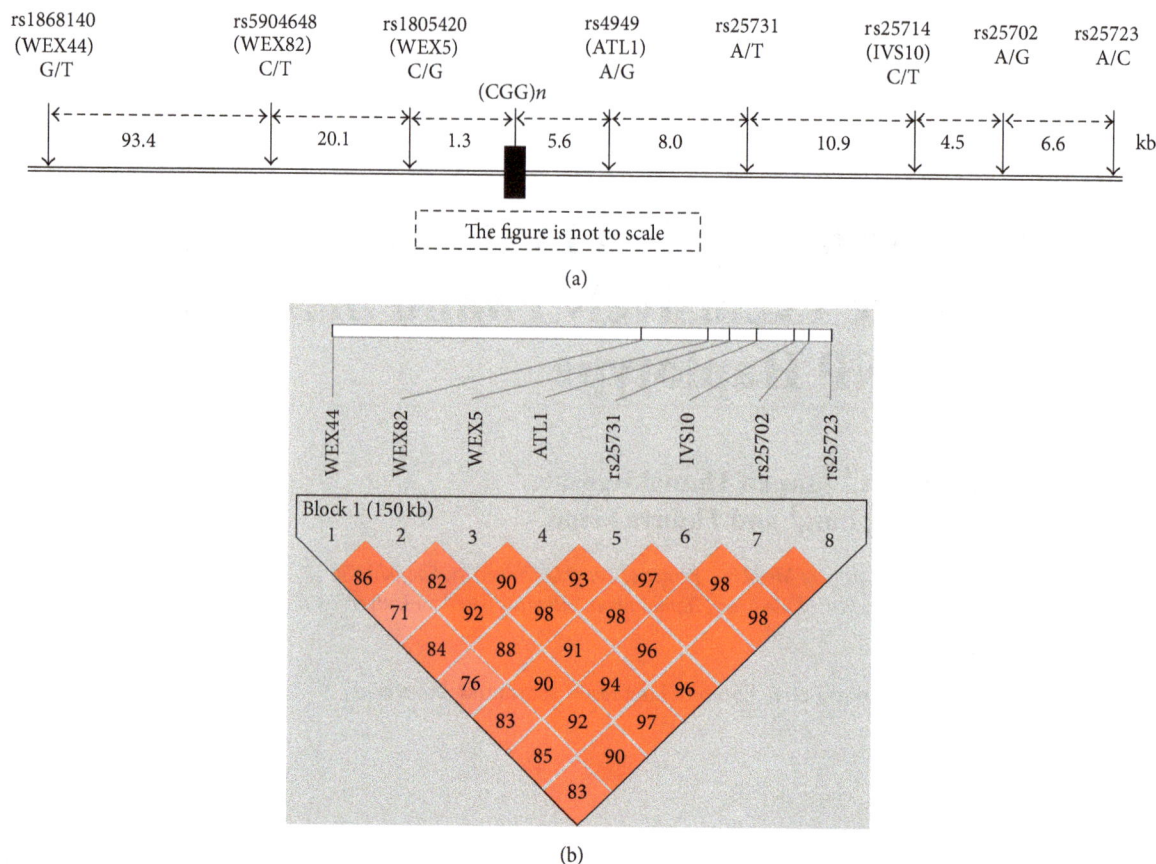

FIGURE 1: (a) The locations of the 8 SNPs. SNP-alleles of each locus are indicated under the SNPs. The distance between each SNP in Kb is shown below. The figure is not to scale. (b) Linkage disequilibrium (D') plot of the 8 SNPs within 150 kb of the CGG-*FMR1* gene. All SNPs pairs have high D' values, more than 80 or close to 80.

2. Materials and Methods

2.1. DNA Samples.

DNA was extracted from whole blood using the standard phenol/chloroform method. The PCR for the *CGG-FMR1* gene and methylation specific PCR were used with minor modification as previous reports [15, 16]. We selected 176 unrelated Thai males in this study, ranging from 19 to 56 CGG repeats. At this time the Thai population is known to have three common alleles, 29, 30, and 36 CGG repeats [15]. In the analysis, samples were divided into 6 groups corresponding to common and uncommon CGG repeats: 19–28, 29, 30, 31–35, 36, and 37–56. The study protocol was approved by the Institutional Ethics Committee.

2.2. SNP Study.

We selected 2 prior investigated SNPs, ATL1 or rs4949, IVS10 or rs25714 [17]. Six additional SNPs, WEX44 (rs1868140), WEX82 (rs5904648), WEX5 (rs1805420), rs25731, rs25702, and rs25723, were obtained from the previous reports [12, 13, 18]. The *FMR1* genomic and SNP position references were according to GenBank reference sequences L29074 and NC_000023.11. These SNPs are located both proximally and distally to the CGG repeats region of the *FMR1* gene (Figure 1(a)). Primer sequences and PCR conditions of all SNPs are shown in Table 1. A single-tube multiplex PCR was performed in a 10 μL reaction

containing 50 ng of genomic DNA, 1x PCR buffer, 200 μM dNTPs, and 0.5 U Taq DNA polymerase (Invitrogen). The MgCl$_2$ concentration and the presence or absence of an adjuvant in the PCR reactions were optimized to obtain the maximum yield of multiplex PCR products. In order to enhance the efficiency of allele-specific amplification, the concentration ratios of tetraprimer for each SNP assay were adjusted to produce a similar band intensity of each PCR product after gel electrophoresis. For the rs25731 SNP locus, PCR reactions were performed in a 20 μL PCR reaction consisting of 100 ng of genomic DNA, 1x PCR buffer, 200 μM dNTPs, 1.5 mM MgCl$_2$, 0.25 μM of each primer, and 1 U Taq DNA polymerase. The reactions were initially denatured for 5 min at 95°C, followed by 35 cycles of 30 sec at 95°C, 30 sec at appropriate annealing temperature, and 30 sec at 72°C and a final extension at 72°C for 10 min. Then 4 μL of the rs25731 PCR reaction was digested with 4 units of *DraI*. Direct PCR products or digested PCR products were electrophoresed on 2.5% agarose gel and stained with ethidium bromide.

2.3. Sequencing Analysis of AGG Interruption Patterns.

For accurate AGG interruption patterns, direct sequencing across the CGG repeats region was performed with primer A [1] and primer 571R [19] in a 50 μL reaction volume comprised of 250 ng of genomic DNA, 50.25 mM Tris-HCl pH 8.8,

TABLE 1: The oligonucleotide sequences of primers and the PCR conditions.

Locus	Name	Sequence (5′ to 3′)	References	Conc. (µM)	MgCl$_2$ Conc. (mM)	Adjuvant (Conc.)	Annealing temperature (°C)	Product size (bp.)
WEX44 (rs1868140)	WEX44F	CTATCTGGGGGCAAATGAACCATAG	This study	0.1	1.5	Q-solution (1x)	57	Control (333)
	WEX44R	CTCTGAGTTTACGCTCCCAG	This study	0.1				T allele (134)
	WEX44TF	CCTAACAGTATAGACCATGATGGAAACATAT	Ennis et al. (2007) [12]	1.5				G allele (262)
	WEX44GR	CAGAGAATAGTTTCAGTTTCTCAGTTTAAACTC	Ennis et al. (2007) [12]	1.5				
WEX82 (rs5904648)	WEX82F	GACAACCCATAATCTGTCATTGG	This study	0.08	2.5	Betaine (0.6 M)	62	Control (398)
	WEX82R	CCACCAGTACTTCCTAATGATA	This study	0.08				C allele (152)
	WEX82CF	TATAACCATGTAAAAAGATCTTCAATC	This study	3				T allele (295)
	WEX82TR	CCTCTGATTATTAATTTATTAATTGCA	This study	7.84				
WEX5 (rs1805420)	WEX5F	GAATGTGGCCCTAGATCCAC	This study	0.1	1.25	BSA (0.13 mg/mL)	60	Control (361)
	WEX5R	GTGCTAACGAGAAATCGGTG	This study	0.1				C allele (143)
	WEX5CF	CTTATCACAGCTGCAACTACAC	This study	1				G allele (261)
	WEX5GR	CAAATTGTCAGACAAGTAAACC	This study	3				
ATL1 (rs4949)	ATL1F	ACCCTGATGAAGAACTTGTATCTCT	Brightwell et al. (2002) [13]	0.1	1.5	BSA (0.13 mg/mL)	53	Control (302)
	ATL1R	GAAATTACACACATAGGTGGCACT		0.1				A allele (107)
	ATL1AF	TGTACATTTTCCAAATGCAAAGA	*Modified	3				G allele (239)
	ATL1GR2*	AGAGACACAGAATCATAAATGC		0.1				
rs25731	731F	AGATTCCCACCTCCTGTAGG	This study	0.25	1.5	—	60	Product (269) Cut by *DraI* A allele (125, 119, 25) T allele (150, 119)
	731R	CATGCTCTGAGTACTGCTC	This study	0.25				
IVS10 (rs25714)	IVS10F	AAAGCTGATTCAGGAGATTGTG	This study	0.1	1.5	BSA (0.13 mg/mL)	53	Control (268)
	IVS10R	ACTGCATTAGAGGACAGAGA	Xu et al. (1999) [14]	0.1				T allele (189)
	IVS10TF	CAAGAAGAGGTATGTTACAGTAT	This study	3				C allele (127)
	IVS10CR	TTATTATATGTGCCACAAAATATTGG	This study	3				
rs25702	702F	ACTCAGTTTAGGCAATCCTG	This study	0.15	1.5	BSA (0.13 mg/mL)	55	Control (379)
	702R	CACAGCTAGTTCATTTGCTG	This study	0.15				A allele (148)
	702AF	TCAGTTTAGTTAGTGTGATGTA	This study	3				G allele (274)
	702GR	GAAATTTTAAGGAGGCATAATC	This study	2.8				
rs25723	723F	GAGCGAGACTGTCTGGGAA	This study	0.05	1.5	BSA (0.13 mg/mL)	55	Control (330)
	723R	TGGAAGGACTGGAATCCTAG	This study	0.05				A allele (263)
	723AF	ACATTTAAAACAATGCACATATA	This study	0.6				C allele (114)
	723CR	TTTCAAAGTATGTTTAAGTAGTAG	This study	4				

12.45 mM $(NH_4)_2SO_4$, 1 mM $MgCl_2$, 200 μM dATP, 200 μM dCTP, 200 μM dTTP, 100 μM dGTP, 100 μM 7-deaza dGTP, 0.25 μM of each primer, 10% DMSO, 128 μg/mL BSA, and 2.5 units of Immolase DNA polymerase (Bioline). The PCR reactions were initially denatured for 9 min at 95°C, followed by 35 cycles of 1 min at 95°C, 1 min at 64°C, and 1 min at 72°C and a final extension at 72°C for 10 min. The PCR products were purified by a QIA quick PCR purification kit (Qiagen). Sequencing reactions were carried out in a 10 μL reaction consisting of 1x BigDye terminator v1.1 ready reaction premix and 1.6 μM of the internal sequencing primer FXS-SEQF (5′-TCTGAGCGGGCGGCGGGCCGA-3′) for forward reactions or primer 571R for reverse reactions. Cycle sequencing conditions were performed in a GeneAmp PCR System 9700 thermal cycler with a temperature profile of 1 min at 96°C followed by 25 cycles of 10 sec at 96°C and 4 min at 60°C. The sequencing products were purified to remove unincorporated fluorescent dye terminator using a DyeEx 2.0 spin kit (Qiagen). All sequencing pellets were dissolved with 15 μL template suppressor reagent and separated by an ABI PRISM 310 genetic analyzer. The AGG interruption patterns were written in abbreviation, for example, 9A9A9, where 9 was (CGG)9 and A was AGG.

2.4. Data Analysis. The Haploview 4.2 program was used for SNP haplotypes analysis. We used Fisher's exact tests to examine the differences in haplotype frequencies among CGG repeat groups. The significant P value was assigned at 0.05.

3. Results

3.1. Haplotype Analysis. The high linkage disequilibrium found among the 8 SNPs studied is shown in Figure 1(b). Allele frequencies of all SNPs are shown in Table 2. When we analyzed the SNP haplotypes, three major haplotypes, GCGGATAA (Hap A), GCCGTTAA (Hap B), and TTCATCGC (Hap C), were found. The rare haplotypes (Hap D) included 11 different haplotypes with frequencies of less than 5% each. Hap A was similar to Hap B with 2 allele differences in the SNP loci (rs1805420 and rs25731) whereas Hap A was different from Hap C for all alleles in 8 SNPs.

3.2. Association of SNP Haplotypes and CGG Repeats. We divided the 176 samples into 6 groups based on the common and uncommon CGG repeats from small to large alleles (19–28, 29, 30, 31–35, 36, and 37–56) shown in Table 3. Strikingly, we found statistically significant associations between haplotypes and the common CGG repeat groups (Fisher's exact test < 0.001) but no statistical significance was found in other uncommon CGG repeat groups (Fisher's exact test = 0.0955). The 29-CGG-repeat group was associated with Hap A (41/55 or 74.5%), while the 30-CGG-repeat group was associated with Hap C (30/37 or 81.1%). In contrast, only one chromosome with Hap A and Hap C was observed in each of the 30- and 29-CGG-repeat groups. The 36-CGG-repeat group was associated with Hap B (27/32 or 84.4%). Hap B was not present in the 30-CGG-repeat group and only a few occurrences were noted in the 29-CGG-repeat group (5.5%).

TABLE 2: The allele frequencies of the 8 SNPs studied.

SNP	Major allele (%)	Minor allele (%)
WEX44 (rs1868140)	G (65.9)	T (34.1)
WEX82 (rs5904648)	C (65.3)	T (34.7)
WEX5 (rs1805420)	C (57.4)	G (42.6)
ATL1 (rs4949)	G (67.0)	A (33.0)
rs25731	T (58.0)	A (42.0)
IVS10 (rs25714)	T (64.2)	C (35.8)
rs25702	A (64.2)	G (35.8)
rs25723	A (64.8)	C (35.2)

The large CGG repeat (37–56) group was related to Hap A or Hap B (12/15 or 80%), while the 19–28- and 31–35-CGG-repeat groups had 44.4% (8/18) and 31.6% (6/19) of Hap A and Hap B, respectively.

3.3. Association of SNP Haplotypes and AGG Interruption Patterns. We randomly selected 95 X chromosomes from 176 samples (54%) for DNA sequencing, including uncommon and common alleles. The results revealed variety in both numbers of AGG and AGG interruption patterns in the CGG repeats of the *FMR1* gene (Figure 2). Most normal alleles had 2 AGG interruptions (48/95 or 50.5%). Alleles with a single or 3 AGG interruptions had the same frequencies of 20% (19/95). The no AGG and 4 AGG interruptions had frequencies of 4.2% (4/95) and 5.3% (5/95), respectively. The no AGG interruption was found in either low CGG repeats (21) or high CGG repeats (43 and 56) while the 4-AGG interruption was found in only high CGG repeats (43 and 45). The 3-AGG and 4-AGG interruptions were exclusively found in the Hap A and Hap B groups. However, no AGG and 2-AGG interruptions were found in all haplotypes. We also observed an allele possessing a 5′ tract with 20 CGG repeats (20A9). The 29 -CGG-repeat group with Hap A had an AGG configuration of 9A9A9 (10/17). The 30-CGG-repeat group with Hap C had an AGG configuration of 10A9A9 (16/18). The 36 CGG repeats with Hap B had an AGG configuration of 9A9A6A9 (13/18). This (CGG)6AGG pattern seemed specific to chromosomes with Hap B (i.e., 10A6A9 in 27 CGG repeats, 12A6A9 in 29 CGG repeats, 9A9A6A9 in 36 CGG repeats, 9A9A6A6A9 in 43 CGG repeats, and 9A9A6A8A9 in 45 CGG repeats). Only one chromosome with Hap B had the 9A23 pattern (33 CGG repeats) from 23 chromosomes with Hap B studied. Likewise, we observed that the 9A and 10A patterns at 5′ of the CGG repeats tract were related to Hap A and Hap C, respectively.

4. Discussion

The haplotype analysis using 8 SNPs in the present study provided more information than in previous studies [9, 17] which could not distinguish haplotypes with 29 CGG repeats from those with 36 CGG repeats (the third common allele exclusively found in Asians). Most chromosomes with 29

TABLE 3: SNP haplotypes frequencies in different CGG repeat groups.

Haplotype	Frequencies of the CGG groups (number)						
	19–28 CGG** (18)	29 CGG* (55)	30 CGG* (37)	31–35 CGG** (19)	36 CGG* (32)	37–56 CGG** (15)	Total number (176)
GCGGATAA (Hap A)	0.278 (5)	*0.745* *(41)*	0.027 (1)	0.263 (5)	0.156 (5)	0.467 (7)	**0.364** **(64)**
GCCGTTAA (Hap B)	0.167 (3)	0.055 (3)	0	0.053 (1)	*0.844* *(27)*	0.333 (5)	**0.222** **(39)**
TTCATCGC (Hap C)	0.333 (6)	0.018 (1)	*0.811* *(30)*	0.526 (10)	0	0.200 (3)	**0.284** **(50)**
Rare haplotypes (Hap D)	0.222 (4)	0.182 (10)	0.162 (6)	0.158 (3)	0	0	**0.130** **(23)**

Comparison based on CGG repeats groups.
*Common CGG repeat groups (29, 30, and 36; Fisher's exact test; P value < 0.001; statistical significance).
**Uncommon CGG repeat groups (19–28, 31–35, and 37–56; Fisher's exact test; P value = 0.0955; no statistical significance).

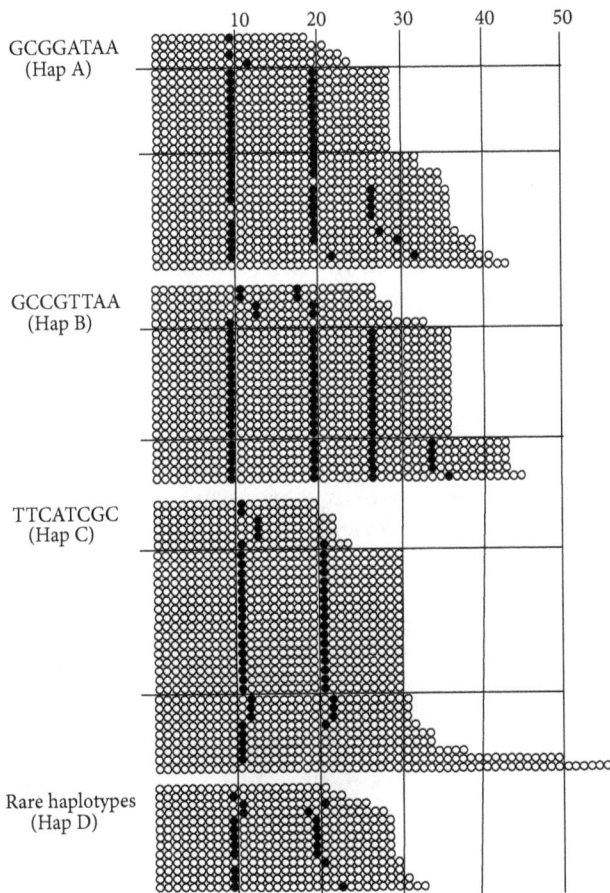

FIGURE 2: AGG interruption patterns of 95 X chromosomes. The CGG repeats are classified in haplotype groups. The AGG interruption patterns are shown from the 5′ to the 3′ ends of the CGG repeat tracts. A white circle represents a CGG and a black circle represents an AGG. The numbers of CGG repeats are indicated as numbers on top of the vertical lines (10, 20, 30, 40, and 50).

and 36 CGG repeats in Thai, Chinese, and Malay populations have G-T of the ATL1-IVS10 haplotype while the A-C haplotype was linked to chromosomes with 30 CGG repeats

in Thai, Malay, Chinese, and Indian populations [9, 17]. Table 2 shows that the 29 and 36 CGG repeat groups had different haplotypes from two SNPs (rs1805420, rs25731).

Analysis of haplotypes using 8 SNPs in our study showed significant associations between haplotypes and the common CGG repeats (29, 30, and 36). The 29-CGG-repeat group was associated with haplotype GCGGATAA (Hap A), the 30-CGG-repeat group was associated with haplotype TTCATCGC (Hap C), and the 36-CGG-repeat group was associated with haplotype GCCGTTAA (Hap B). The uncommon CGG repeats of the 19–28, 31–35, and 37–56 groups were not associated with any haplotype and had similar distributions of haplotypes. These findings suggest that uncommon CGG repeats randomly occur in all three common and rare haplotypes.

Most of chromosomes with 36 CGG repeats and Hap B had an AGG configuration of 9A9A6A9 that might be derived from chromosomes with 29 CGG repeats and Hap A (9A9A9) by 6A insertion [5]. This formation was also found in chromosomes with 43 CGG repeats and Hap B (9A9A6A6A9), which might be derived from chromosomes with 36 CGG repeats and Hap B by 6A insertion (Figure 3). However, a few Hap B-chromosomes with 27 and 29 CGG repeats had AGG configurations of 10A6A9 and 12A6A9 that might be derived from 20 (10A9) and 22 (12A9) CGG repeats of chromosomes with Hap C by insertion of 6A pattern (Figure 3).

Hap A and Hap C had different alleles in all SNPs. This suggests that Hap A and Hap C may have different evolutionary pathways. However, Hap A and Hap B are likely evolutionarily derived since they had similar SNP haplotypes (Table 3) and both haplotypes carried 9A pattern at 5′ of the CGG repeats tract (Figures 2 and 3). The evolution of CGG repeats is likely from primitive small to large CGG repeats. An evolutionary study of the CGG repeats of the FMR1 gene showed that most nonprimate mammals have a small number of uninterrupted CGG repeats with a mean of ~8 repeats, while the repeats of primates are larger with a mean of ~20 repeats and more highly specific interruptions [20]. Therefore, we hypothesize that there may be two distinct

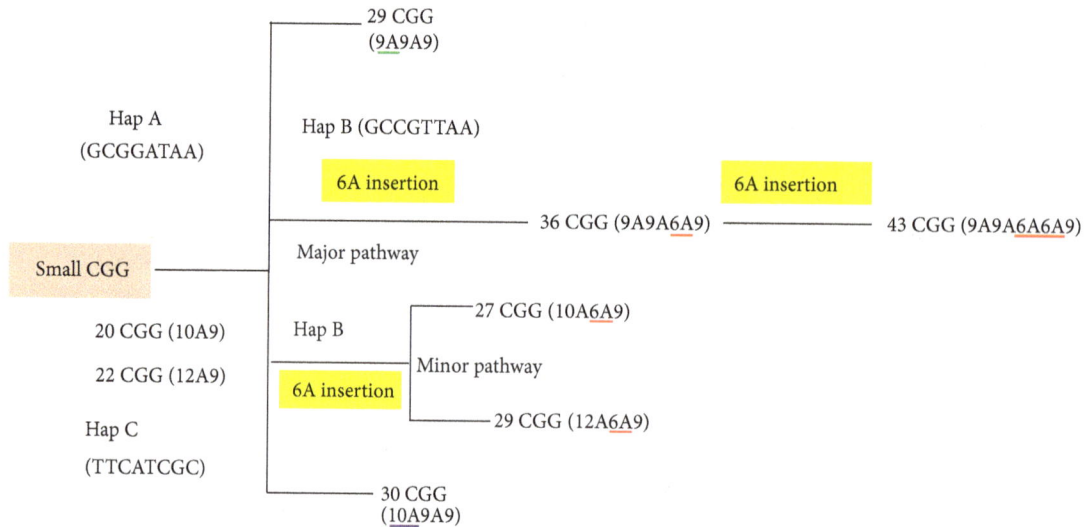

FIGURE 3: Simplified evolutionary pathways of the hypothesis. Chromosomes with 29 and 30 CGG repeats may have different evolutionary pathways since they had different haplotypes and AGG interruption patterns. The 6A pattern was linked to Hap B possibly derived from chromosomes with Hap A (major pathway) or Hap C (minor pathway).

pathways in our findings. First, chromosomes with 29 and 30 CGG repeats may independently arise from Hap A and Hap C by gradual replication slippage or recombination via the smaller alleles [20] and were stable by the 9A9A9 and 10A9A9 patterns, respectively [11, 21]. Second, the 6A pattern was linked to chromosomes with Hap B possibly derived from chromosomes with Hap A (major pathway) or Hap C (minor pathway). Simplified pathways of the hypothesis are shown in Figure 3. In addition, perhaps the 6A pattern enhances the stability of CGG repeat tracts [22, 23]. Thus, chromosomes with 36 CGG repeats linked to the 6A pattern have become the third most common allele in only Asian populations. It is also relevant to note that, to date, the 6A pattern has been exclusively found in Asians [5–11]. A study based on an Eskimo population indicated that the 6A pattern has been stably conserved for 15,000–30,000 years, since this group migrated from Asia to North America [7].

It has been proposed that AGG interruptions play a crucial role in maintaining the stability of the CGG repeats since premutation alleles often contain only one AGG or no AGG interruptions [3, 4, 24–26]. Haplotypes analysis using microsatellites near the *FMR1* gene (DXS548-FRAXAC1-FRAXAC2) found that specific haplotypes were associated with the loss of AGG interruptions of the CGG repeats in Caucasians [27] and Jewish Tunisians [28]. In contrast, the findings in African Americans using those three microsatellites and the SNP, ATL1 did not show a haplotype association with CGG repeats instability [29]. Also, our findings in this study support earlier studies where the SNP haplotype association between nearby SNPs and AGG interruption patterns in CGG repeats of the *FMR1* gene likely reflects linkage disequilibrium in each population [9, 17, 30]. Therefore, it is difficult to determine if an associated haplotype is a real factor for CGG repeats instability or a linkage disequilibrium in a specific population [31].

5. Conclusion

Our study showed new evidence that the specific haplotype (Hap B) was strongly linked to the 6A pattern in Thai subjects since almost all chromosomes with Hap B had at least one 6A configuration, regardless of CGG repeats (i.e., 10A6A9, 12A6A9, 9A9A6A6A9, and 9A9A9A6A8A9). The 6A pattern and Hap B may have originally occurred in the ancestors of Asian populations. However, we could not completely exclude that the findings may be by chance or sample selection bias. Further studies of SNP haplotypes and AGG interruption patterns in other Asian populations would be warranted, to confirm and expand on our findings.

Acknowledgments

The authors would like to thank Ms. Charunee Maharat, Ms. Supaporn Yangngam, and Ms. Oradawan Plong-On for technical assistance. This work was supported by the Graduate School, Prince of Songkla University (EC. 48/364-010), and was partly supported by the National Center for Genetic Engineering and Biotechnology (BIOTEC) Grant no. BT-B-01-MG-18-4814.

References

[1] Y.-H. Fu, D. P. A. Kuhl, A. Pizzuti et al., "Variation of the CGG repeat at the fragile X site results in genetic instability: resolution of the sherman paradox," *Cell*, vol. 67, no. 6, pp. 1047–1058, 1991.

[2] M. Pieretti, F. Zhang, Y.-H. Fu et al., "Absence of expression of the FMR-1 gene in fragile X syndrome," *Cell*, vol. 66, no. 4, pp. 817–822, 1991.

[3] E. E. Eichler, J. J. A. Holden, B. W. Popovich et al., "Length of uninterrupted CGG repeats determines instability in the FMR1 gene," *Nature Genetics*, vol. 8, no. 1, pp. 88–94, 1994.

[4] C. B. Kunst and S. T. Warren, "Cryptic and polar variation of the fragile X repeat could result in predisposing normal alleles," *Cell*, vol. 77, no. 6, pp. 853–861, 1994.

[5] S.-H. Chen, J. M. Schoof, N. E. Buroker, and C. R. Scott, "The identification of a (CGG)6AGG insertion within the CGG repeat of the FMR1 gene in Asians," *Human Genetics*, vol. 99, no. 6, pp. 793–795, 1997.

[6] M. C. Hirst, T. Arinami, and C. D. Laird, "Sequence analysis of long FMR1 arrays in the Japanese population: insights into the generation of long $(CGG)_n$ tracts," *Human Genetics*, vol. 101, no. 2, pp. 214–218, 1997.

[7] L. A. Larsen, J. S. M. Armstrong, K. Gronskov et al., "Analysis of $FMR1(CGG)_n$ alleles and FRAXA microsatellite haplotypes in the population of Greenland: implications for the population of the New World from Asia," *European Journal of Human Genetics*, vol. 7, no. 7, pp. 771–777, 1999.

[8] S. M. H. Faradz, J. Leggo, A. Murray, P. R. L. Lam-Po-Tang, M. F. Buckley, and J. J. A. Holden, "Distribution of FMR1 and FMR2 alleles in Javanese individuals with developmental disability and confirmation of a specific AGG-interruption pattern in Asian populations," *Annals of Human Genetics*, vol. 65, no. 2, pp. 127–135, 2001.

[9] Y. Zhou, K. Tang, H.-Y. Law, I. S. L. Ng, C. G. L. Lee, and S. S. Chong, "FMR1 CGG repeat patterns and flanking haplotypes in three Asian populations and their relationship with repeat instability," *Annals of Human Genetics*, vol. 70, no. 6, pp. 784–796, 2006.

[10] H.-H. Chiu, Y.-T. Tseng, H.-P. Hsiao, and H.-H. Hsiao, "The AGG interruption pattern within the CGG repeat of the FMR1 gene among Taiwanese population," *Journal of Genetics*, vol. 87, no. 3, pp. 275–277, 2008.

[11] C. M. Yrigollen, S. Sweha, B. Durbin-Johnson et al., "Distribution of AGG interruption patterns within nine world populations," *Intractable & Rare Diseases Research*, vol. 3, no. 4, pp. 153–161, 2014.

[12] S. Ennis, A. Murray, G. Brightwell, N. E. Morton, and P. A. Jacobs, "Closely linked cis-acting modifier of expansion of the CGG repeat in high risk FMR1 haplotypes," *Human Mutation*, vol. 28, no. 12, pp. 1216–1224, 2007.

[13] G. Brightwell, R. Wycherley, and A. Waghorn, "SNP genotyping using a simple and rapid single-tube modification of ARMS illustrated by analysis of 6 SNPs in a population of males with FRAXA repeat expansions," *Molecular and Cellular Probes*, vol. 16, no. 4, pp. 297–305, 2002.

[14] B. Xu, J. M. Schoof, N. E. Buroker, C. R. Scott, and S. H. Chen, "High frequency of the FMR1 IVS10+14C/T polymorphism in Asians, and its association with the fragile X syndrome in Caucasians," *The American Journal of Human Genetics*, vol. 65, supplement, abstract 2282, 1999.

[15] P. Limprasert, N. Ruangdaraganon, T. Sura, P. Vasiknanonte, and U. Jinorose, "Molecular screening for fragile X syndrome in Thailand," *The Southeast Asian Journal of Tropical Medicine and Public Health*, vol. 30, supplement 2, pp. 114–118, 1999.

[16] C. Charalsawadi, T. Sripo, and P. Limprasert, "Multiplex methylation specific PCR analysis of fragile X syndrome: experience in Songklanagarind Hospital," *Journal of the Medical Association of Thailand*, vol. 88, no. 8, pp. 1057–1061, 2005.

[17] P. Limprasert, V. Saechan, N. Ruangdaraganon et al., "Haplotype analysis at the FRAXA locus in Thai subjects," *American Journal of Medical Genetics*, vol. 98, no. 3, pp. 224–229, 2001.

[18] G. Brightwell, R. Wycherley, G. Potts, and A. Waghorn, "A high-density SNP map for the FRAX region of the X chromosome," *Journal of Human Genetics*, vol. 47, no. 11, pp. 567–575, 2002.

[19] S. S. Chong, E. E. Eichler, D. L. Nelson, and M. R. Hughes, "Robust amplification and ethidium-visible detection of the fragile X syndrome CGG repeat using Pfu polymerase," *American Journal of Medical Genetics*, vol. 51, no. 4, pp. 522–526, 1994.

[20] E. E. Eichler, C. B. Kunst, K. A. Lugenbeel et al., "Evolution of the cryptic FMR1 CGG repeat," *Nature Genetics*, vol. 11, no. 3, pp. 301–308, 1995.

[21] G. J. Latham, J. Coppinger, A. G. Hadd, and S. L. Nolin, "The role of AGG interruptions in fragile X repeat expansions: a twenty-year perspective," *Frontiers in Genetics*, vol. 5, no. 7, article 244, Article ID Article 244, 2014.

[22] P. Weisman-Shomer, E. Cohen, and M. Fry, "Interruption of the fragile X syndrome expanded sequence $d(CGG)_n$ by interspersed d(AGG) trinucleotides diminishes the formation and stability of $d(CGG)_n$ tetrahelical structures," *Nucleic Acids Research*, vol. 28, no. 7, pp. 1535–1541, 2000.

[23] C. B. Volle and S. Delaney, "AGG/CCT interruptions affect nucleosome formation and positioning of healthy-length CGG/CCG triplet repeats," *BMC Biochemistry*, vol. 14, no. 1, article 33, 2013.

[24] M. C. Hirst, P. K. Grewal, and K. E. Davies, "Precursor arrays for triplet repeat expansion at the fragile X locus," *Human Molecular Genetics*, vol. 3, no. 9, pp. 1553–1560, 1994.

[25] K. Snow, D. J. Tester, K. E. Kruckeberg, D. J. Schaid, and S. N. Thibodeau, "Sequence analysis of the fragile X trinucleotide repeat: implications for the origin of the fragile X mutation," *Human Molecular Genetics*, vol. 3, no. 9, pp. 1543–1551, 1994.

[26] N. Zhong, W. Yang, C. Dobkin, and W. T. Brown, "Fragile X gene instability: anchoring AGGs and linked microsatellites," *The American Journal of Human Genetics*, vol. 57, no. 2, pp. 351–361, 1995.

[27] E. E. Eichler, J. N. Macpherson, A. Murray, P. A. Jacobs, A. Chakravarti, and D. L. Nelson, "Haplotype and interspersion analysis of the FMR1 CGG repeat identifies two different mutational pathways for the origin of the fragile X syndrome," *Human Molecular Genetics*, vol. 5, no. 3, pp. 319–330, 1996.

[28] T. C. Falik-Zaccai, E. Shachak, M. Yalon et al., "Predisposition to the fragile X syndrome in Jews of Tunisian descent is due to the absence of AGG interruptions on a rare Mediterranean haplotype," *The American Journal of Human Genetics*, vol. 60, no. 1, pp. 103–112, 1997.

[29] D. C. Crawford, C. E. Schwartz, K. L. Meadows et al., "Survey of the fragile X syndrome CGG repeat and the short-tandem-repeat and single-nucleotide-polymorphism haplotypes in an African American population," *American Journal of Human Genetics*, vol. 66, no. 2, pp. 480–493, 2000.

[30] M. Barasoain, G. Barrenetxea, E. Ortiz-Lastra et al., "Single nucleotide polymorphism and FMR1 CGG repeat instability in two Basque valleys," *Annals of Human Genetics*, vol. 76, no. 2, pp. 110–120, 2012.

[31] S. Ennis, A. Murray, and N. E. Morton, "Haplotypic determinants of instability in the FRAX region: concatenated mutation or founder effect?" *Human Mutation*, vol. 18, no. 1, pp. 61–69, 2001.

13

Altered Body Weight Regulation in *CK1ε* Null and *tau* Mutant Mice on Regular Chow and High Fat Diets

Lili Zhou,[1,2] **Keith C. Summa,**[1,2] **Christopher Olker,**[1,2] **Martha H. Vitaterna,**[1,2] **and Fred W. Turek**[1,2]

[1]*Center for Sleep and Circadian Biology, Northwestern University, Evanston, IL 60208, USA*
[2]*Department of Neurobiology, Northwestern University, Evanston, IL 60208, USA*

Correspondence should be addressed to Fred W. Turek; fturek@northwestern.edu

Academic Editor: Francine Durocher

Disruption of circadian rhythms results in metabolic dysfunction. Casein kinase 1 epsilon (*CK1ε*) is a canonical circadian clock gene. Null and *tau* mutations in *CK1ε* show distinct effects on circadian period. To investigate the role of *CK1ε* in body weight regulation under both regular chow (RC) and high fat (HF) diet conditions, we examined body weight on both RC and HF diets in $CK1\varepsilon^{-/-}$ and $CK1\varepsilon^{tau/tau}$ mice on a standard 24 hr light-dark (LD) cycle. Given the abnormal entrainment of $CK1\varepsilon^{tau/tau}$ mice on a 24 hr LD cycle, a separate set of $CK1\varepsilon^{tau/tau}$ mice were tested under both diet conditions on a 20 hr LD cycle, which more closely matches their endogenous period length. On the RC diet, both $CK1\varepsilon^{-/-}$ and $CK1\varepsilon^{tau/tau}$ mutants on a 24 hr LD cycle and $CK1\varepsilon^{tau/tau}$ mice on a 20 hr LD cycle exhibited significantly lower body weights, despite similar overall food intake and activity levels. On the HF diet, $CK1\varepsilon^{tau/tau}$ mice on a 20 hr LD cycle were protected against the development of HF diet-induced excess weight gain. These results provide additional evidence supporting a link between circadian rhythms and energy regulation at the genetic level, particularly highlighting *CK1ε* involved in the integration of circadian biology and metabolic physiology.

1. Introduction

The coordination of daily rhythms in feeding behavior, body temperature, and energy storage and utilization across the 24 hr light-dark (LD) cycle is critical in maintaining homeostasis. It has been well established that circadian rhythms are controlled by endogenous circadian clock genes [1]. Core clock genes, such as *Clock*, *Bmal1*, *Period* (*Per*), and *Cryptochrome* (*Cry*), are key components of the transcriptional/translational feedback loop, which is considered to be the major mechanism underlying the generation of circadian rhythms [2]. Based on gene expression profiling studies, approximately 3%–20% of the transcriptome in any given tissue exhibits circadian oscillation, and a large proportion of the rhythmically regulated transcripts are involved in metabolic function [3]. In addition, the discovery that clock genes function in peripheral tissues involved in metabolism, such as liver, fat, heart, and muscle, provides further support

that circadian and metabolic processes are tightly linked [4, 5].

Accumulating evidence from both human and animal studies has strongly supported an important role for circadian rhythms in the regulation of metabolism. Shift workers have a higher incidence of diabetes, obesity, cancer, and cardiovascular diseases [6–8]. In clinical studies, forced misalignment of behavioral (i.e., sleep/wake) and circadian cycles in human subjects causes metabolic and endocrine abnormalities, including decreases in leptin and increases in glucose and insulin [9]. In rodents, genetic disruption of the molecular clock system leads to a variety of metabolic abnormalities, including obesity and metabolic syndrome in *Clock* mutant mice [10] and *Per2* deficient mice [11]. In contrast, mutant mice lacking the core clock gene *Bmal1*, either globally or exclusively in liver, have a lean phenotype [5, 12, 13]. Chronic circadian disruption, achieved by housing wild-type mice in a 20 hr LD cycle that is incongruous with their endogenous ~24 hr circadian period, results in accelerated weight gain and obesity, as well as dysregulation of metabolic hormones [14].

Casein kinase 1 epsilon (*CK1ε*), a member of the serine/threonine protein kinase family, is a canonical circadian gene which regulates circadian rhythms through posttranslational modification of the PER and CRY proteins [15]. In the kinase domain of *CK1ε*, a C to T single nucleotide transition results in a missense point mutation (*tau* mutation) of a conserved amino acid residue 178 (R178C), which causes profound changes in circadian organization in hamsters and mice, with a 4-hour decrease in the free-running period in homozygous mutants in constant darkness (DD) [16, 17]. Furthermore, the intrinsic 20 hr period length in *tau* mutant mice prevents stable entrainment to a conventional 24 hr LD cycle [15]. Another mouse model for *CK1ε* is the null mutant, which was developed using the loxP-Cre strategy to create a premature codon induced by frameshift in the *CK1ε* gene. In contrast to *tau* mutant mice, null mutant mice exhibited a significant but very mild lengthening of the circadian period [18].

Despite extensive evidence supporting a tight relationship between circadian rhythms and metabolism, studies examining the role of *CK1ε* in metabolic function remain limited. Previous studies in hamsters have demonstrated a significant impact of the *CK1ε tau* mutation on body weight regulation in males, with homozygous mutants weighing significantly less than wild-types [19–21]. Both *tau* mutant hamsters and *tau* mutant mice exhibit increased metabolic rates [18, 20]. Despite these findings, it is unknown whether *tau* mutant mice exhibit a reduction in body weight. In addition, no previous study has examined other mutations in *CK1ε*; thus, no comparison between different *CK1ε* mutants has been made. Furthermore, it is unclear whether the metabolic changes in *CK1ε tau* mutant hamsters are caused by an accelerated circadian pacemaker, as opposed to other pleiotropic effects of the mutant allele. It is also unknown whether *tau* mutants demonstrate altered body weight regulation on a high fat diet (HF) as they do on regular chow (RC).

Therefore, we used both $CK1ε^{-/-}$ and $CK1ε^{tau/tau}$ mutant mice as genetic tools to investigate the impact of these two mutations on body weight regulation in a 24 hr LD cycle under two different diet conditions: RC and HF. In addition, because of the abnormal entrainment in $CK1ε^{tau/tau}$ mice (not $CK1ε^{-/-}$ mice) on a 24 hr LD cycle, a separate set of $CK1ε^{tau/tau}$ mice were tested in a 20 hr LD cycle, which more closely matches their endogenous period length and permits patterns of entrainment comparable to those of wild-types on a standard 24 hr LD cycle. This was done to investigate whether the phenotype observed in $CK1ε^{tau/tau}$ mice on a 24 hr LD cycle was simply an artifact of the altered entrainment. Our results indicate that, on a RC diet, both *CK1ε* null and *tau* mutations on a 24 hr LD cycle, as well as the *CK1ε tau* mutation on a 20 hr LD cycle, exhibit significant effects on body weight, with mutant mice weighing less than wild-types. In contrast, on the HF diet, neither mutation on a 24 hr LD cycle led to a significant difference from wild-types.

Remarkably, a 20 hr LD cycle, which restores normal light entrainment in $CK1ε^{tau/tau}$ mice, provides resistance to excess body weight gain induced by a HF diet.

2. Materials and Methods

2.1. Animals and Experimental Protocol. All mutant animals used in this experiment were coisogenic C57BL/6J mice. The generation of these mutants has been described previously [18]. For all experiments, male wild-type ($CK1ε^{+/+}$), *CK1ε* null mutant ($CK1ε^{-/-}$), and *tau* mutant ($CK1ε^{tau/tau}$) mice were maintained in standard mouse cages with food and water available *ad libitum* on a conventional 12 hr light : 12 hr dark LD cycle (LD 12 : 12; lights on at 0600, lights off at 1800; 300 lux and 22–24°C ambient temperature). Until the beginning of the study, mice were group-housed and fed a RC diet (16% kcal from fat, 27% kcal from protein, and 57% kcal from carbohydrate; 7012 Teklad LM-485 Mouse/Rat Sterilizable Diet, Harlan Laboratories, Inc., Indianapolis, IN). At the age of 10 weeks, mice were transferred to individual cages under the same lighting and environmental conditions described above. The animals were randomized into experimental groups and fed either RC ($CK1ε^{+/+}$, $n = 18$; $CK1ε^{-/-}$, $n = 17$; $CK1ε^{tau/tau}$, $n = 13$) or HF diet (45% kcal from fat, 20% kcal from protein, and 35% kcal from carbohydrate; D12451, Research Diets, Inc. New Brunswick, NJ; $CK1ε^{+/+}$, $n = 16$; $CK1ε^{-/-}$, $n = 13$; $CK1ε^{tau/tau}$, $n = 16$).

Body weight was recorded weekly for 6 weeks. Food consumption was measured daily for 7 consecutive days in the second week of the experiment. A glucose tolerance test and an insulin tolerance test were performed on the seventh and eighth weeks, respectively, as described below. On the ninth week, serum samples, obtained by tail bleed, were collected 6 hours after light onset (by convention, referred to as Zeitgeber time (ZT) 6) from mice that had been fasted for 6 hours. At the end of the experiment, mice were euthanized (without fasting) at ZT6, and gonadal fat pads were harvested for analysis. A separate group of age-matched $CK1ε^{tau/tau}$ mutant mice were individually housed and fed either RC ($n = 10$) or HF ($n = 8$) diet. All procedures for this group were the same as above, except they were maintained on a 20 hr LD cycle (LD 10 : 10) for the duration of the experimental protocol. All procedures and protocols were approved in advance by the Institutional Animal Care and Use Committee of Northwestern University.

2.2. Locomotor Activity. Five or six mice of each genotype from a separate group of mice were singly housed in individual cages outfitted for locomotor activity analysis via detection of infrared beam breaks. These mice were fed either RC or HF diet under LD 12 : 12 or LD 10 : 10 cycle for 10 days and were used exclusively for locomotor activity analysis (i.e., they were not included in the body weight measurements or other metabolic analyses). Beam breaks were recorded in 6 min bins using the Chronobiology Kit (Stanford Software Systems, Stanford, CA, USA) and were analyzed using the ClockLab software (Actimetrics, Wilmette, IL, USA).

2.3. Body Temperature. Body temperatures were measured in 8–16-week-old male mice with a rectal thermometer (4600 thermometer, Measurement Specialties, Beavercreek, OH) inserted 1.7 cm in the middle of day time.

2.4. Blood Collection and Serum Insulin Analysis. Mice were transferred to clean cages at ZT0 to fast for 6 hours, during which access to water was unrestricted. At ZT6, tails were clipped to obtain blood samples for analysis and serum isolation. Briefly, a small (1 mm) cut was made at the end of the tail and about 50 μL of blood was obtained by gentle massage using tail blood collection tubes (BD Vacutainer Plus plastic serum tube, 2 mL, red top, #367820, BD Diagnostics, Franklin Lakes, NJ). Mice were then returned to their home cages. At room temperature, blood was incubated in an upright position for 45 min and then spun in a centrifuge for 6 min at 2000 ×g. The supernatant (serum) was removed, frozen, and stored at −80°C until future analysis. Serum insulin levels were determined by ELISA (Ultra Sensitive Mouse Insulin ELISA Kits, Crystal Chem Inc., Downers Grove, IL) according to the manufacturer's instructions. Blood glucose was determined as described below.

2.5. Glucose and Insulin Tolerance Tests. For the glucose tolerance test (GTT), mice were fasted for 6 hours (ZT0 to ZT6), after which baseline blood glucose levels were determined from a blood sample obtained from the tail of each mouse. A small (1 mm) cut was made at the end of the tail and a drop of blood was deposited onto a glucometer strip (Abbott Laboratories, Abbott Park, IL) by gentle massage for assessment of blood glucose level. Mice were then immediately injected intraperitoneally with 1.0 g/kg body weight of glucose (G8769, Sigma-Aldrich, St. Louis, MO). Additional blood samples and blood glucose readings were obtained via massage of the tail nick at 30, 60, and 120 min after the injection. Upon completion of the experiment, mice were returned to their home cages. Results are expressed as percentage of baseline glucose. Area under the curve (AUC) values were calculated by the trapezoidal rule.

For the insulin tolerance test (ITT), mice were fasted for 2 hours (ZT4 to ZT6), after which baseline blood glucose was measured as described above. Immediately after the baseline blood glucose sample was obtained, 0.75 units/kg body weight of regular human insulin Humulin® R U-100 (Eli Lilly, Indianapolis, IN; insulin was diluted to 1 : 1000 (0.1 units/mL) with sterile diluent) was injected intraperitoneally. After the insulin injection, blood glucose was sampled at 30, 60, 90, and 120 min, as described above. Upon completion of the experiment, animals were returned to their home cages. Results are expressed as percentage of baseline glucose. Area under the curve (AUC) values were calculated by the trapezoidal rule.

2.6. Statistics. All statistical analysis was performed using R software (http://www.r-project.org/) [22]. Time course data of body weight, GTT, and ITT on each diet were analyzed using repeated measures ANOVA with genotype as the between-subject factor and time as the within-subject

variable. Following a significant result on repeated measures ANOVA, single time point comparisons were made by Benjamini-Hochberg multiple comparison tests. All the other comparisons between genotypes and diets were conducted via two-way ANOVA, with Benjamini-Hochberg post hoc tests performed where appropriate for multiple comparisons. Group values are expressed as mean ± SEM. Significant differences were defined as $p < 0.05$.

3. Results

3.1. Altered Body Weight Regulation in CK1ε Mutants. Individually caged young adult (10-week) male mice were given either regular chow (RC) or high fat (HF) diet for the entire experimental protocol. After 6 weeks on RC, both $CK1\varepsilon^{-/-}$ and $CK1\varepsilon^{tau/tau}$ mice exhibited a significantly lower body weight than $CK1\varepsilon^{+/+}$ mice, approximately 15% less (Figure 1(a); $CK1\varepsilon^{+/+}$ = 29.38 ± 1.39 g, $CK1\varepsilon^{-/-}$ = 24.91 ± 0.98 g, and $CK1\varepsilon^{tau/tau}$ = 24.69 ± 1.67 g). Because the body weight of $CK1\varepsilon^{+/-}$ and $CK1\varepsilon^{+/tau}$ mice did not differ from wild-type mice (data not shown), the present study only focuses on results from homozygous mutants. Additional analyses indicated that the body weight differences between $CK1\varepsilon^{+/+}$ mice and both $CK1\varepsilon$ mutant mice were as early as the age of 3 weeks immediately after weaning (see Figure S1 in Supplementary Material available online at http://dx.doi.org/10.1155/2016/4973242). $CK1\varepsilon^{-/-}$ mice had significantly lower body weight than wild-type mice throughout the entire period of experiment. However, the stable body weight changes of $CK1\varepsilon^{tau/tau}$ mice compared to wild-type mice were only observed after week 8. To avoid this developmental fluctuation, we focused on the body weight only during adulthood.

Due to abnormal entrainment patterns and the mismatch between endogenous period length and the 24 hr LD cycle in homozygous *tau* mutants [15], we were interested in whether restoration of entrainment and resonance between the environmental cycle length and endogenous period in *tau* mutants would impact body weight regulation. $CK1\varepsilon^{-/-}$ mice, whose endogenous period is very close to 24 hr, have normal entrainment, so their phenotype was less likely to be affected by the entrainment. Therefore, we focused on only $CK1\varepsilon^{tau/tau}$ mice and maintained two separate groups of *tau* mutant mice on RC and HF diet, respectively, on a 20 hr LD cycle. We observed that the $CK1\varepsilon^{tau/tau}$ mice on a 20 hr LD cycle maintain significantly reduced body weight compared to wild-type mice on a 24 hr LD cycle on RC diet (Figure 1(a)).

On the HF diet, mice of all genotypes gained significantly more weight than the ones on the RC diet, as expected (Figure 1(c)). Significant differences were not evident between mutant and wild-type mice on a 24 hr LD cycle (Figure 1(b)). Intriguingly, when HF-fed $CK1\varepsilon^{tau/tau}$ mice were housed on a 20 hr LD cycle, the rate of body weight gain was reduced compared to other genotypes, resulting in a significantly lower weight gain after 6 weeks on the diet (Figure 1(c); $CK1\varepsilon^{+/+}$ = 15.5 ± 1.2 g, $CK1\varepsilon^{-/-}$ = 14.0 ± 0.8 g, $CK1\varepsilon^{tau/tau}$ = 16.4 ± 1.4 g, and $CK1\varepsilon^{tau/tau}$ on 20 hr LD = 9.6 ± 1.1 g).

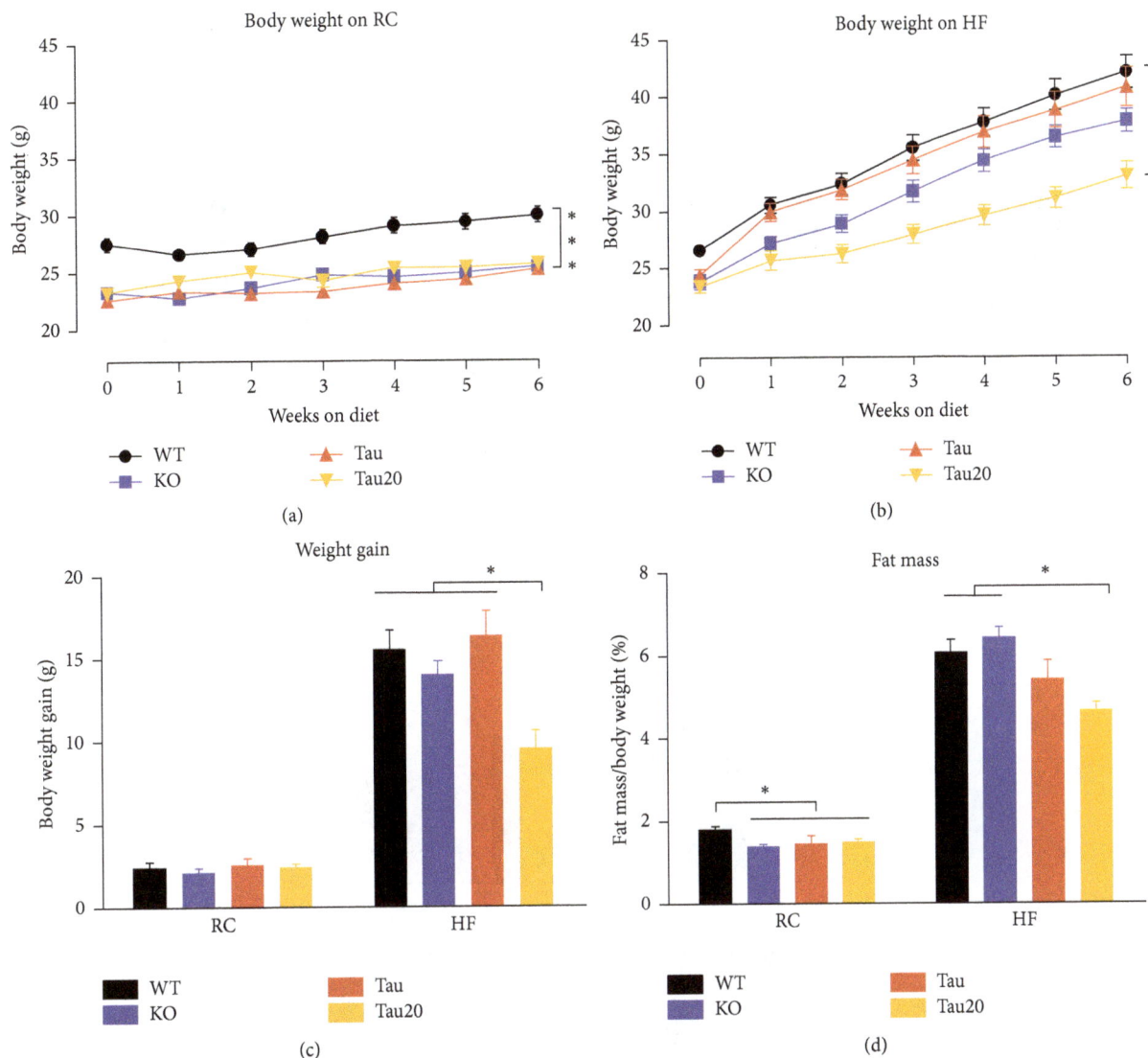

FIGURE 1: Altered body weight and fat mass in *CK1ε* mutant mice. (a) Growth curves of mice on RC diet. (b) Growth curves of mice on HF diet. (c) Total body weight gain after six weeks of either RC or HF diet. Body weight gain was calculated by subtracting the baseline weight at the beginning of the diet experiment from weight at the end of the diet experiment. (d) The gonadal fat pads from mice fed either RC or HF diet were weighed at the end of study. *CK1ε*$^{+/+}$ (black), *CK1ε*$^{-/-}$ (blue), and *CK1ε*$^{tau/tau}$ (red) mice were on a 24 hr LD cycle; *CK1ε*$^{tau/tau}$ mice kept on a 20 hr LD cycle are represented in yellow. Mean values are presented for each group, with error bars representing SEM. Asterisks indicate significant differences between groups ($p < 0.05$).

In agreement with the observed effects on body weight, we also observed reduced gonadal fat pad weight in *CK1ε* mutant mice. As shown in Figure 1(d), on RC, both *CK1ε*$^{-/-}$ and *CK1ε*$^{tau/tau}$ mutant mice on a 24 hr LD cycle, as well as *CK1ε*$^{tau/tau}$ on a 20 hr LD cycle, had a significantly reduced proportion of gonadal fat pad mass to total body weight (*CK1ε*$^{+/+}$ = 1.79 ± 0.09%, *CK1ε*$^{-/-}$ = 1.37 ± 0.05%, *CK1ε*$^{tau/tau}$ = 1.44 ± 0.19%, and *CK1ε*$^{tau/tau}$ on 20 hr LD = 1.48 ± 0.07%). On the HF diet (Figure 1(d)), mice of all genotypes exhibited a pronounced increase in percentage of gonadal fat pad weight, as expected; however, no significant differences were observed between *CK1ε*$^{-/-}$, *CK1ε*$^{tau/tau}$, and *CK1ε*$^{+/+}$ mice on a 24 hr LD cycle. Interestingly, as with body

weight, HF-fed *CK1ε*$^{tau/tau}$ mutant mice on a 20 hr LD cycle had a significant reduction in the percentage of gonadal fat pad weight, compared to HF-fed wild-type mice (*CK1ε*$^{+/+}$ = 6.05 ± 0.32% and *CK1ε*$^{tau/tau}$ on 20 hr LD = 4.64 ± 0.21%). With respect to the absolute mass of the gonadal fat pad, identical results were observed (data not shown).

3.2. Altered Diurnal Feeding Behavior and Locomotor Activity in *CK1ε* Mutants.
To determine whether the reduced body weight in *CK1ε* mutants was due to decreased food intake, we examined food consumption under RC and HF conditions. Total daily overall caloric intake did not differ between genotypes on 24 hr LD cycle (Figure 2(a)). However,

(a)

(b)

(c)

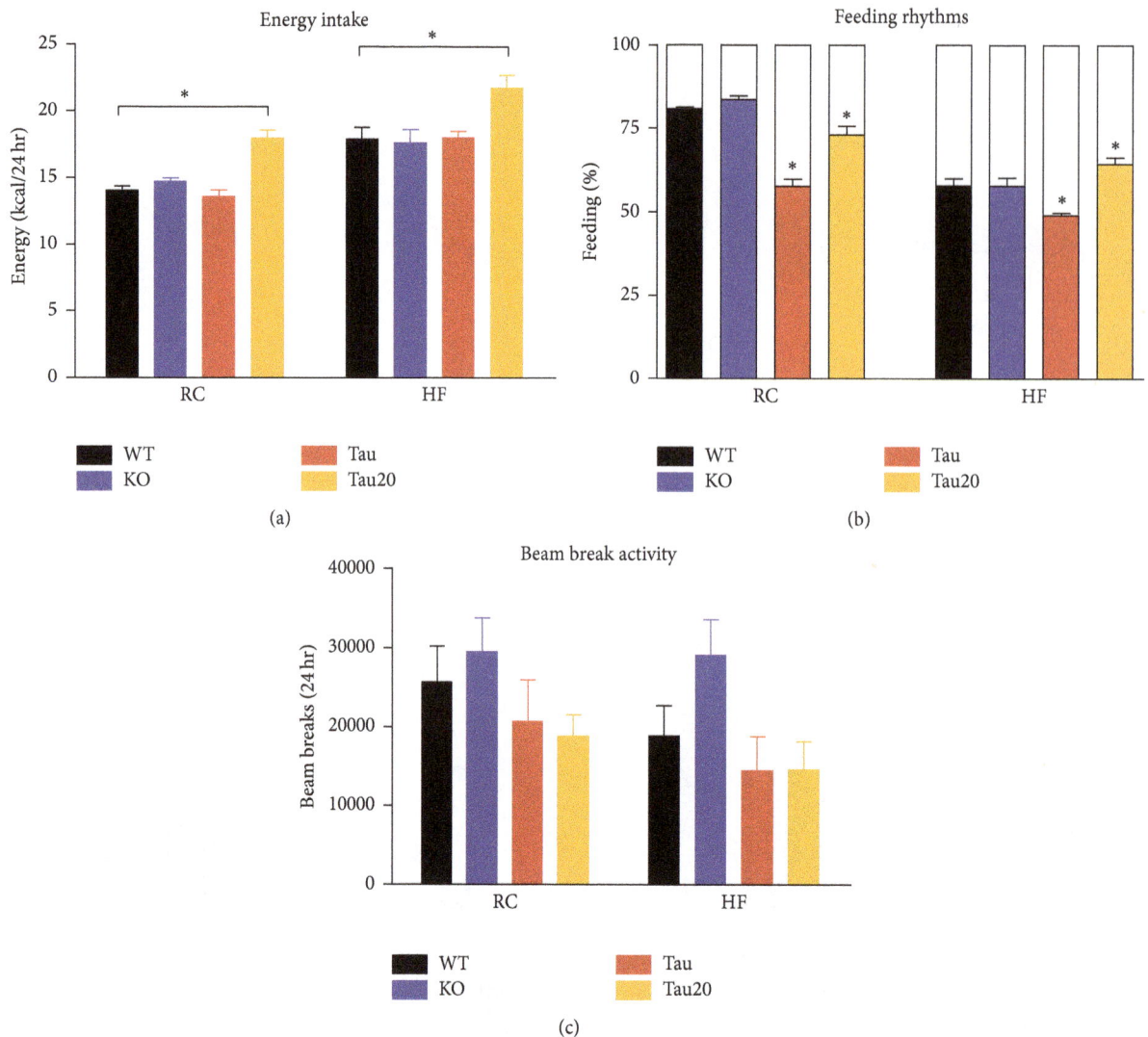

FIGURE 2: Altered diurnal feeding behaviors in $CK1\varepsilon$ mutant mice. (a) The diet, but not the genotype, has a significant effect on daily calorie intake. Energy intake was expressed as the average kilocalories consumed during each 24 hr period. (b) Percentage of calorie intake in light and dark periods. The top proportion, in white, represents the percentage of calorie intake in the light period. The bottom proportion, in different colors, represents the percentage of calorie intake in the dark period. Results were compared between mutants and wild-type controls on the same diet. (c) Locomotor activity of mice fed either RC or HF diet. The activity was expressed as the average beam breaks during each 24 hr period. $CK1\varepsilon^{+/+}$ (black), $CK1\varepsilon^{-/-}$ (blue), and $CK1\varepsilon^{tau/tau}$ (red) mice were on a 24 hr LD cycle; $CK1\varepsilon^{tau/tau}$ mice represented in yellow were on a 20 hr LD cycle. Mean values are presented for each group, with error bars representing SEM. Asterisks indicate significant differences between groups ($p < 0.05$).

$CK1\varepsilon^{tau/tau}$ mice on a 20 hr LD cycle consumed more energy than wild-type mice every 24 hr (RC: $CK1\varepsilon^{+/+}$ = 14.0 ± 0.3 kcal, $CK1\varepsilon^{tau/tau}$ on 20 hr LD = 17.9 ± 0.6 kcal; HF: $CK1\varepsilon^{+/+}$ = 17.9 ± 0.9 kcal, $CK1\varepsilon^{tau/tau}$ on 20 hr LD = 21.7 ± 1.0%). The distribution of food intake during the light versus dark periods was altered in the $CK1\varepsilon$ mutants compared to wild-types (Figure 2(b)). On RC, $CK1\varepsilon^{-/-}$ mice consumed a greater proportion of their total daily calories during the dark period. In contrast, $CK1\varepsilon^{tau/tau}$ mice consumed much less diet during the dark period (Figure 2(b); $CK1\varepsilon^{+/+}$ = 80.9 ± 1.0%, $CK1\varepsilon^{-/-}$ = 83.8 ± 1.0%, and $CK1\varepsilon^{tau/tau}$ = 57.8 ± 2.2%). Thus, the diurnal rhythm in energy intake in $CK1\varepsilon^{tau/tau}$ mice

was greatly attenuated on a 24 hr LD cycle. Interestingly, RC-fed $CK1\varepsilon^{tau/tau}$ mice housed on a 20 hr LD cycle consumed 73.1% of their total daily calories during dark period. This remained significantly lower than that of wild-type control mice but was improved compared to that of $CK1\varepsilon^{tau/tau}$ mice on a 24 hr LD cycle (Figure 2(b)). On the HF diet, diurnal rhythms of food intake in mice of all three genotypes on a 24 hr LD cycle were attenuated (Figure 2(b)). In particular, $CK1\varepsilon^{tau/tau}$ mice on a 24 hr LD cycle exhibited the greatest attenuation of diurnal feeding rhythms, consuming 49.1% calories during dark period. Surprisingly, HF-fed $CK1\varepsilon^{tau/tau}$ mice on a 20 hr LD cycle displayed improved diurnal rhythms

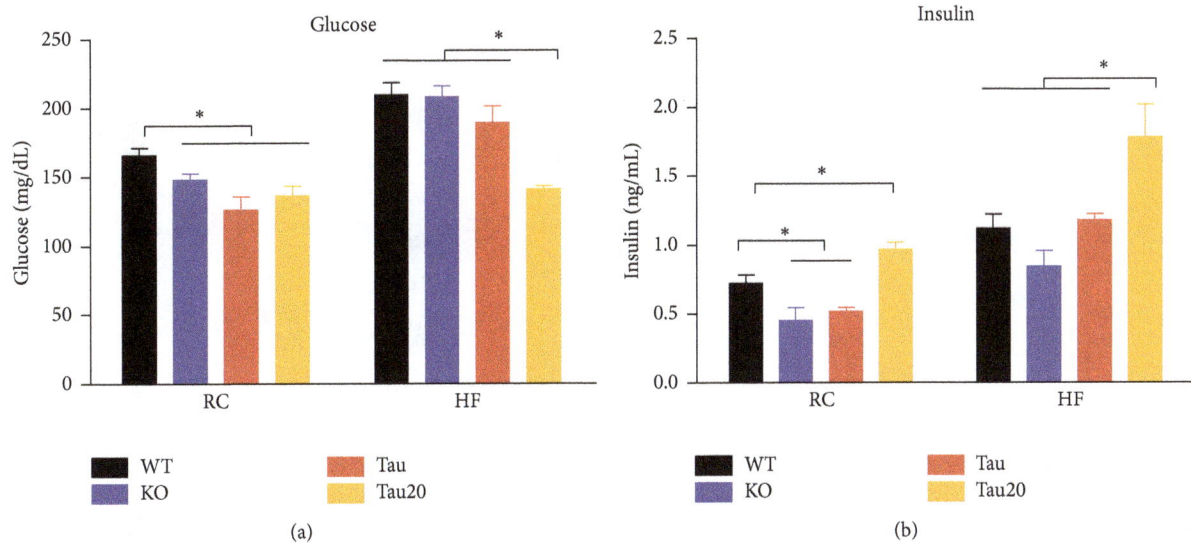

FIGURE 3: Altered glucose and insulin levels in *CK1ε* mutant mice. (a) Fasting serum glucose of mice fed either RC or HF diet. (b) Fasting serum insulin of mice fed either RC or HF diet. $CK1ε^{+/+}$ (white), $CK1ε^{-/-}$ (blue), and $CK1ε^{tau/tau}$ (red) mice were on a 24 hr LD cycle; $CK1ε^{tau/tau}$ mice represented in yellow were on a 20 hr LD cycle. Mean values are presented for each group, with error bars representing SEM. Asterisks indicate significant differences between groups ($p < 0.05$).

of energy intake compared to the other groups on HF diet, consuming 64.6% of their total daily calories during the dark phase. No significant differences in absolute beam break activity levels (Figure 2(c)) and the activity patterns (Figure S2) were evident between wild-type mice and mutants on both RC and HF diet. Additionally, no differences in body temperature between genotype groups were observed under each diet condition (Figure S3).

3.3. Altered Fasting Glucose and Insulin Levels in CK1ε Mutants.
We then examined fasting blood glucose and serum insulin levels in samples collected during the light phase. On the RC diet, both $CK1ε^{-/-}$ and $CK1ε^{tau/tau}$ mice on a 24 hr LD cycle, as well as $CK1ε^{tau/tau}$ mice on 20 hr LD cycle, had slight, but significant, reductions in fasting glucose compared to wild-type mice (Figure 3(a); $CK1ε^{+/+}$ = 165.7 ± 5.4 mg/dL, $CK1ε^{-/-}$ = 148.3 ± 4.3 mg/dL, $CK1ε^{tau/tau}$ = 126.4 ± 9.3 mg/dL, and $CK1ε^{tau/tau}$ on 20 hr LD = 136.6 ± 7.1 mg/dL). On the HF diet, $CK1ε^{+/+}$, $CK1ε^{-/-}$, and $CK1ε^{tau/tau}$ mice on a 24 hr LD cycle exhibited increased fasting glucose levels that did not significantly differ from one another. $CK1ε^{tau/tau}$ mutant mice on a 20 hr LD cycle exhibited reduced fasting glucose levels compared to wild-types (Figure 3(a); $CK1ε^{+/+}$ = 209.4 ± 8.8 mg/dL and $CK1ε^{tau/tau}$ on 20 hr LD = 141.1 ± 2.5 mg/dL).

Complex changes in fasting insulin levels were also observed. As shown in Figure 3(b), on RC diet and a 24 hr LD cycle, both $CK1ε^{-/-}$ and $CK1ε^{tau/tau}$ mice had lower levels of insulin than wild-type mice, whereas $CK1ε^{tau/tau}$ mice on a 20 hr LD cycle exhibited higher fasting insulin levels than wild-type mice ($CK1ε^{+/+}$ = 0.72 ± 0.06 ng/mL,

$CK1ε^{-/-}$ = 0.45 ± 0.09 ng/mL, $CK1ε^{tau/tau}$ = 0.52 ± 0.03 ng/mL, and $CK1ε^{tau/tau}$ on 20 hr LD = 0.97 ± 0.05 ng/mL). On HF diet, both $CK1ε^{-/-}$ and $CK1ε^{tau/tau}$ mice on a 24 hr LD cycle exhibited similar insulin levels. However, $CK1ε^{tau/tau}$ mice on 20 hr LD cycle exhibited a higher insulin level than wild-type mice on a 24 hr LD cycle (Figure 3(b); $CK1ε^{+/+}$ = 1.12 ± 0.10 ng/mL and $CK1ε^{tau/tau}$ on 20 hr LD = 1.78 ± 0.24 ng/mL).

3.4. Altered Glucose Tolerance in CK1ε Mutants.
To evaluate glucose utilization in the mutant mice, we performed both a glucose tolerance test and an insulin tolerance test. Because mice of different genotypes had different baseline levels, the data presented here were normalized by dividing the observed glucose value from the basal level of each genotype under each diet condition. On RC diet, $CK1ε^{tau/tau}$ mice on a 24 hr LD cycle had a slower rate of glucose uptake than the other groups (Figure 4(a)), and the area under the curve (AUC) was higher in $CK1ε^{tau/tau}$ mice than $CK1ε^{+/+}$ mice (Figure 4(c); $CK1ε^{+/+}$ = 100.0 ± 4.9% and $CK1ε^{tau/tau}$ = 141.8 ± 11.7%).

On HF diet, mice of all genotypes exhibited a reduced rate of glucose uptake compared to mice on RC diet (Figure 4(c)), and, among the groups, $CK1ε^{tau/tau}$ mice on a 24 hr LD cycle were slightly slower than wild-type mice, and $CK1ε^{tau/tau}$ mice on a 20 hr LD cycle were the most altered, having a significant and sustained reduction in glucose clearance (Figure 4(c); $CK1ε^{+/+}$ = 126.7 ± 6.9%, $CK1ε^{tau/tau}$ = 148.5 ± 7.5%, and $CK1ε^{tau/tau}$ on 20 hr LD = 182.6 ± 6.0%). No significant differences were observed under either diet condition from mice of any genotype during the ITT (Figures 4(d)–4(f)).

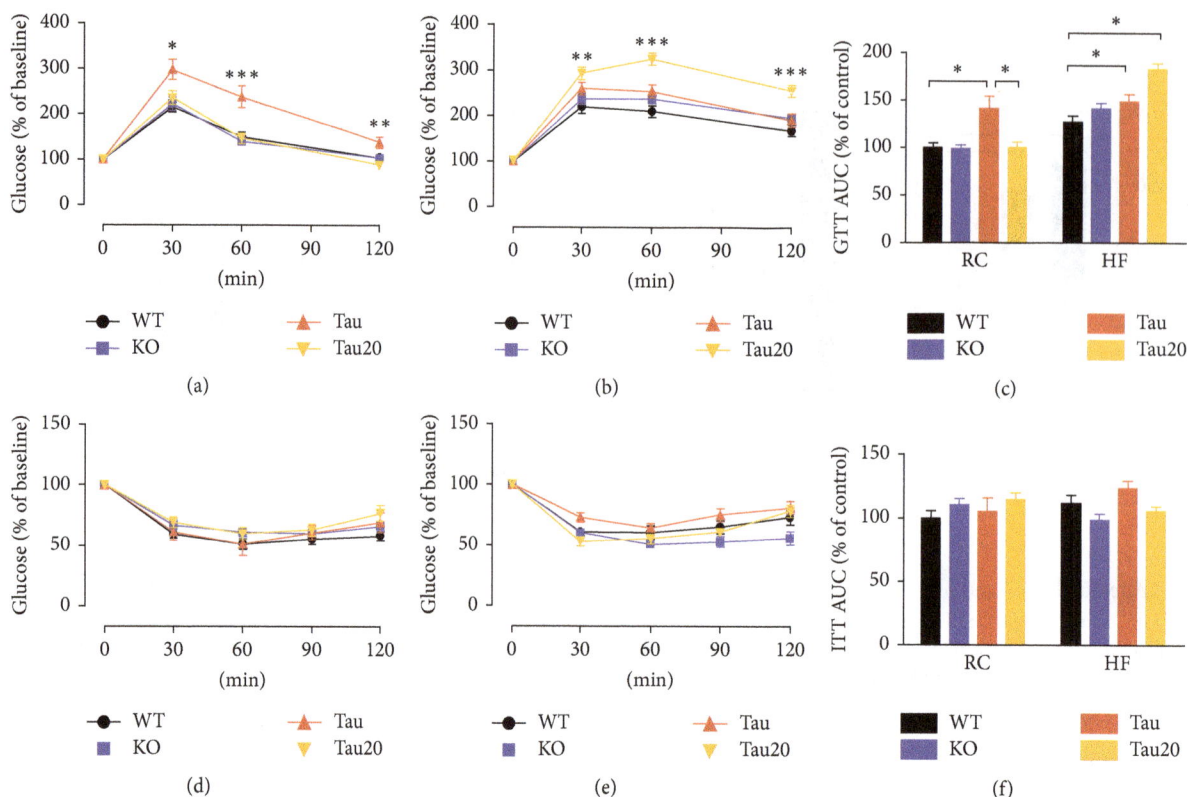

FIGURE 4: Alterations in GTT in $CK1\varepsilon$ mutant mice. Top panel: GTT. Intraperitoneal GTT was performed in mice fed either RC (a) or HF (b) diet. (c) Area under the curve (AUC) of the GTT performed in (a) and (b). Bottom panel: ITT. Intraperitoneal ITT was performed in mice fed either RC (d) or HF (e) diet. (f) Area under the curve (AUC) of the ITT performed in (d) and (e). $CK1\varepsilon^{+/+}$ (white), $CK1\varepsilon^{-/-}$ (blue), and $CK1\varepsilon^{tau/tau}$ (red) mice were on a 24 hr LD cycle; $CK1\varepsilon^{tau/tau}$ mice represented in yellow were kept on a 20 hr LD cycle. Mean values are presented for each group, with error bars representing SEM. Asterisks denote significant differences between mutant genotype and wild-type controls on the same diet ($^{*}p < 0.05$; $^{**}p < 0.01$; $^{***}p < 0.001$).

4. Discussion

Using two genetic mouse models of $CK1\varepsilon$ disruption (i.e., knock-out $CK1\varepsilon^{-/-}$ null mice and knock-in $CK1\varepsilon^{tau/tau}$ mutant mice) under different diet conditions, we have demonstrated distinct effects of the circadian clock gene $CK1\varepsilon$ on body weight regulation and susceptibility to excess weight gain induced by HF diet. In particular, by maintaining $CK1\varepsilon^{tau/tau}$ mice on a 20 hr LD cycle, which more closely corresponds to their endogenous circadian period length and enables normal entrainment, we have generated evidence suggesting that proper entrainment and synchrony between internal circadian rhythms and the external environment may limit, or even protect against, the development of excess body weight gain induced by HF diet.

We found that both homozygous null and tau mutant mice had a reduced body weight, approximately 15% lower than wild-type mice on RC diet. The magnitude was close to the percentage (18%) of reduced body mass reported in tau hamsters. But we believe that the body weight change in each group of $CK1\varepsilon$ mutants on RC diet is not caused by an accelerated circadian rate constant over physiological processes due to the discrepancy between impacts on body mass

and circadian rhythms phenotype in $CK1\varepsilon^{-/-}$ and $CK1\varepsilon^{tau/tau}$ mice. We also noted that altered energy expenditure or intake does not appear to be the primary reason for the reduced body weight in $CK1\varepsilon$ mutant mice, because we did not observe increased activity levels or reduced food intake in mutants compared to the wild-type mice. We also did not observe increased body temperature in mutants compared to the wild-type mice, which is consistent with previous results [20].

The cause of the reduced body weight in $CK1\varepsilon$ mutant mice is still unclear, but there are some possible mechanisms worth testing in the future. First of all, higher metabolic rates may be the main factor to determine the low body mass, which have been shown in both homozygous tau mutant hamsters and mice [19–21], although no study has been done in the $CK1\varepsilon$ null mutant. Second, it should not be excluded that alterations in development and cell growth might also contribute to the low body weight in $CK1\varepsilon$ mutants, if considering the slow-growth phenotype in yeast with a deletion of a $CK1\varepsilon$ homolog gene [23], as well as the known function of $CK1\varepsilon$ in promoting cell growth [24, 25]. Therefore, a further analysis of pathways involved in cell growth, such as Wnt and its intracellular effector β-catenin, in $CK1\varepsilon^{-/-}$ and $CK1\varepsilon^{tau/tau}$ mice will help in testing this hypothesis.

Studies have consistently demonstrated that misalignment of feeding behavior and circadian rhythms or a disrupted circadian clock can cause altered body weight regulation and result in the development of abnormalities consistent with the metabolic syndrome [10, 11, 26, 27]. In particular, a recent study demonstrated the harmful effects of chronic circadian disruption on metabolism in wild-type mice [14]. The mice were housed on a 20 hr LD cycle, incongruous with their endogenous 24 hr circadian period, and displayed significantly increased weight gain after 6 weeks on the altered LD cycle. Complementing this previous study, we took a different approach by studying $CK1\varepsilon^{tau/tau}$ mice, which have an endogenous 20 hr circadian period, in both a 24 hr LD cycle and a 20 hr LD cycle. Remarkably, we observed that the endogenous circadian period matched 20 hr LD cycle protected against the HF diet-induced weight gain in $CK1\varepsilon^{tau/tau}$ mice. We found that HF-fed $CK1\varepsilon^{tau/tau}$ mice in a 24 hr LD cycle were no longer leaner than wild-types as the *tau* mutant mice were on RC diet, and many of their metabolic parameters, such as absolute body weight, body weight gain, gonadal fat pad mass, and fasting blood glucose, were similar to those of wild-type mice on HF. However, when the LD cycle was adjusted to match their shortened endogenous circadian period, all these parameters were restored toward levels of wild-type mice on RC diet. Additionally, although the diurnal rhythms of food intake in HF-fed mice of all the genotypes were attenuated, which is similar to what was shown previously [28], HF-fed *tau* mice on a 20 hr LD cycle displayed the least attenuation in diurnal rhythms of energy intake. However, unlike the altered body weight in wild-type mice on the shortened LD cycle [14], we did not observe any difference in body weight in $CK1\varepsilon^{tau/tau}$ mice on RC diet between the 24 hr LD cycle and endogenous circadian period matched 20 hr LD cycle. The different responses to RC and HF diets in *tau* mice may be due to increased sensitivity to diet-induced weight gain on a metabolically "challenging" HF diet, compared to the RC diet, in the $CK1\varepsilon^{tau/tau}$ mice. Another possibility is that we only monitored body weight in $CK1\varepsilon^{tau/tau}$ mice in a 20 hr LD cycle for 6 weeks during adulthood; we do not know if a longer exposure to an endogenous circadian period matched LD cycle or if rearing in 20 hr LD cycle from birth would restore the body weight in RC-fed $CK1\varepsilon^{tau/tau}$ mice. Further experiments are needed to address these questions.

It has recently been reported that CK1ε *tau* mutant hamsters are protected against the development of cardiomyopathy and renal disease by adjusting the environmental LD cycle to match their shortened endogenous circadian period [29]. Both the present study and previous experiments [14, 29] have demonstrated that certain metabolic and pathological abnormalities may be restored or prevented by optimizing the LD cycle and suggest that strategies designed to synchronize and match internal circadian cycles with the external environment may be useful in limiting or preventing the development of metabolic abnormalities. Synchronizing internal circadian cycles with the external environment has at least two beneficial effects on circadian organization: proper entrainment and resonance between environmental and internal circadian period length, which need not be mutually exclusive. The present study could not distinguish between these two effects, and future work utilizing entrainment-specific or period-specific mutants would be necessary to do so.

In the present study, we had some discrepant observations. For example, HF-fed $CK1\varepsilon^{tau/tau}$ mice on a 20 hr LD cycle had significantly lower body weight, but higher energy intake. This might be an interacting effect of a higher metabolic rate and a restoration of the LD cycles matching the endogenous period on the HF diet-induced weight gain. An improved alignment of the endogenous rhythms with the environmental LD cycle may improve the temporal coordination between feeding and metabolism, expending the energy intake at the correct time more efficiently. High metabolic rate alone or correct LD cycle alone may not act as effectively as in HF-fed $CK1\varepsilon^{tau/tau}$ mice on a 20 hr LD cycle. Another discrepancy in the results is that the HF-fed $CK1\varepsilon^{tau/tau}$ mice on a 20 hr LD cycle have lower fasting glucose but higher AUG in GTT. Although in many cases a reduced glucose level is associated with improved GTT, or a high level of glucose is associated with impaired GTT, a coexistence of both low glucose level and impaired GTT sometimes happens. One example is gene pancreatic-derived factor (PANDER), which was recently found to be a novel hormone regulating glucose levels via interaction with both the liver and the endocrine pancreas. Although still glucose intolerant, PANDER-deficient mice fed a HF diet are protected from HF diet-induced hyperglycemia because of the decreased expression of the gluconeogenic genes PEPCK and G6Pase and the reduced glucose production in the liver [30]. Although it is still unclear what the exact mechanism is for HF-fed $CK1\varepsilon^{tau/tau}$ mice on a 20 hr LD cycle showing both low glucose level and GTT intolerance in the present study, it is possible that the *tau* mutation affects certain gene(s) which function similarly as PANDER or its receptor.

5. Conclusions

In conclusion, we have demonstrated that differences in body weight regulation and the response to a HF diet challenge exist in two different mouse mutants of *CK1ε* on a 24 hr LD cycle, as well as $CK1\varepsilon^{tau/tau}$ mice on a 20 hr LD cycle that matches their endogenous circadian period. Both $CK1\varepsilon^{-/-}$ and $CK1\varepsilon^{tau/tau}$ mice had reduced body weights on RC diet despite similar overall caloric consumption and daily activity levels. On a HF diet, however, $CK1\varepsilon^{tau/tau}$ mice on a 20 hr LD cycle were protected against the development of excess body weight gain induced by HF diet. These findings may provide unique insights for future strategies of obesity management, which involve the nutrient composition of the diet, the properties and principles of the circadian clock system, and the interactions between these two factors in determining the metabolic responses.

Competing Interests

The authors wish to disclose the absence of financial and pharmaceutical company support and off-label or investigational use of drugs or devices for this study. Martha H. Vitaterna has participated in research supported in part by Merck & Co., Inc., and in a study funded by Institut de Recherches Internationales Servier, but the experiments and results reported in this paper were not supported by either of these. Fred W. Turek has received consultant fees from Vanda Pharmaceuticals, Inc., and the Ingram Barge Company. The other authors have disclosed the absence of any financial competing interests.

Acknowledgments

Keith C. Summa was supported in part by the National Center for Research Resources (NCRR) and the National Center for Advancing Translational Sciences (NCATS), National Institutes of Health (NIH), through a Northwestern University Clinical and Translational Sciences Institute Predoctoral Training Grant (8UL1TR000150).

References

[1] J. S. Takahashi, H.-K. Hong, C. H. Ko, and E. L. McDearmon, "The genetics of mammalian circadian order and disorder: implications for physiology and disease," *Nature Reviews Genetics*, vol. 9, no. 10, pp. 764–775, 2008.

[2] C. H. Ko and J. S. Takahashi, "Molecular components of the mammalian circadian clock," *Human Molecular Genetics*, vol. 15, no. 2, pp. R271–R277, 2006.

[3] C. B. Green, J. S. Takahashi, and J. Bass, "The meter of metabolism," *Cell*, vol. 134, no. 5, pp. 728–742, 2008.

[4] S. Panda, M. P. Antoch, B. H. Miller et al., "Coordinated transcription of key pathways in the mouse by the circadian clock," *Cell*, vol. 109, no. 3, pp. 307–320, 2002.

[5] R. D. Rudic, P. McNamara, A.-M. Curtis et al., "BMAL1 and CLOCK, two essential components of the circadian clock, are involved in glucose homeostasis," *PLoS Biology*, vol. 2, no. 11, article e377, 2004.

[6] L. Di Lorenzo, G. De Pergola, C. Zocchetti et al., "Effect of shift work on body mass index: results of a study performed in 319 glucose-tolerant men working in a Southern Italian industry," *International Journal of Obesity*, vol. 27, no. 11, pp. 1353–1358, 2003.

[7] B. Karlsson, A. Knutsson, and B. Lindahl, "Is there an association between shift work and having a metabolic syndrome? Results from a population based study of 27,485 people," *Occupational and Environmental Medicine*, vol. 58, no. 11, pp. 747–752, 2001.

[8] Y. Suwazono, M. Dochi, K. Sakata et al., "A longitudinal study on the effect of shift work on weight gain in male Japanese workers," *Obesity*, vol. 16, no. 8, pp. 1887–1893, 2008.

[9] F. A. J. L. Scheer, M. F. Hilton, C. S. Mantzoros, and S. A. Shea, "Adverse metabolic and cardiovascular consequences of circadian misalignment," *Proceedings of the National Academy of Sciences of the United States of America*, vol. 106, no. 11, pp. 4453–4458, 2009.

[10] F. W. Turek, C. Joshu, A. Kohsaka et al., "Obesity and metabolic syndrome in circadian clock mutant mice," *Science*, vol. 308, no. 5724, pp. 1043–1045, 2005.

[11] S. Yang, A. Liu, A. Weidenhammer et al., "The role of mPer2 clock gene in glucocorticoid and feeding rhythms," *Endocrinology*, vol. 150, no. 5, pp. 2153–2160, 2009.

[12] K. A. Lamia, K.-F. Storch, and C. J. Weitz, "Physiological significance of a peripheral tissue circadian clock," *Proceedings of the National Academy of Sciences of the United States of America*, vol. 105, no. 39, pp. 15172–15177, 2008.

[13] S. Shimba, N. Ishii, Y. Ohta et al., "Brain and muscle Arnt-like protein-1 (BMAL1), a component of the molecular clock, regulates adipogenesis," *Proceedings of the National Academy of Sciences of the United States of America*, vol. 102, no. 34, pp. 12071–12076, 2005.

[14] I. N. Karatsoreos, S. Bhagat, E. B. Bloss, J. H. Morrison, and B. S. McEwen, "Disruption of circadian clocks has ramifications for metabolism, brain, and behavior," *Proceedings of the National Academy of Sciences of the United States of America*, vol. 108, no. 4, pp. 1657–1662, 2011.

[15] A. S. I. Loudon, Q. J. Meng, E. S. Maywood, D. A. Bechtold, R. P. Boot-Handford, and M. H. Hastings, "The biology of the circadian CK1ε tau mutation in mice and Syrian hamsters: a tale of two species," *Cold Spring Harbor Symposia on Quantitative Biology*, vol. 72, pp. 261–271, 2007.

[16] P. L. Lowrey, K. Shimomura, M. P. Antoch et al., "Positional syntenic cloning and functional characterization of the mammalian circadian mutation tau," *Science*, vol. 288, no. 5465, pp. 483–491, 2000.

[17] M. R. Ralph and M. Menaker, "A mutation of the circadian system in golden hamsters," *Science*, vol. 241, no. 4870, pp. 1225–1227, 1988.

[18] Q.-J. Meng, L. Logunova, E. S. Maywood et al., "Setting clock speed in mammals: the CK1ε tau mutation in mice accelerates circadian pacemakers by selectively destabilizing PERIOD proteins," *Neuron*, vol. 58, no. 1, pp. 78–88, 2008.

[19] R. J. Lucas, J. A. Stirland, Y. N. Mohammad, and A. S. I. Loudon, "Postnatal growth rate and gonadal development in circadian tau mutant hamsters reared in constant dim red light," *Journal of Reproduction and Fertility*, vol. 118, no. 2, pp. 327–330, 2000.

[20] M. Oklejewicz, R. A. Hut, S. Daan, A. S. I. Loudon, and A. J. Stirland, "Metabolic rate changes proportionally to circadian frequency in tau mutant Syrian hamsters," *Journal of Biological Rhythms*, vol. 12, no. 5, pp. 413–422, 1997.

[21] M. Oklejewicz, I. Pen, G. C. R. Durieux, and S. Daan, "Maternal and pup genotype contribution to growth in wild-type and tau mutant Syrian hamsters," *Behavior Genetics*, vol. 31, no. 4, pp. 383–391, 2001.

[22] R Development Core Team, *R: A Language and Environment for Statistical Computing*, R Foundation for Statistical Computing, Vienna, Austria, 2011.

[23] K. J. Fish, A. Cegielska, M. E. Getman, G. M. Landes, and D. M. Virshup, "Isolation and characterization of human casein kinase I epsilon (CKI), a novel member of the CKI gene family," *The Journal of Biological Chemistry*, vol. 270, no. 25, pp. 14875–14883, 1995.

[24] J. S. Boehm, J. J. Zhao, J. Yao et al., "Integrative genomic approaches identify IKBKE as a breast cancer oncogene," *Cell*, vol. 129, no. 6, pp. 1065–1079, 2007.

[25] W. S. Yang and B. R. Stockwell, "Inhibition of casein kinase 1-epsilon induces cancer-cell-selective, PERIOD2-dependent growth arrest," *Genome Biology*, vol. 9, no. 6, article R92, 2008.

[26] D. M. Arble, J. Bass, A. D. Laposky, M. H. Vitaterna, and F. W. Turek, "Circadian timing of food intake contributes to weight gain," *Obesity*, vol. 17, no. 11, pp. 2100–2102, 2009.

[27] M. Garaulet, P. Gómez-Abellán, J. J. Alburquerque-Béjar, Y. Lee, J. M. Ordovás, and F. A. Scheer, "Timing of food intake predicts

weight loss effectiveness," *International Journal of Obesity*, vol. 37, no. 4, pp. 604–611, 2013.

[28] A. Kohsaka, A. D. Laposky, K. M. Ramsey et al., "High-fat diet disrupts behavioral and molecular circadian rhythms in mice," *Cell Metabolism*, vol. 6, no. 5, pp. 414–421, 2007.

[29] T. A. Martino, G. Y. Oudit, A. M. Herzenberg et al., "Circadian rhythm disorganization produces profound cardiovascular and renal disease in hamsters," *American Journal of Physiology—Regulatory Integrative and Comparative Physiology*, vol. 294, no. 5, pp. R1675–R1683, 2008.

[30] C. E. Robert-Cooperman, C. G. Wilson, and B. R. Burkhardt, "PANDER KO mice on high-fat diet are glucose intolerant yet resistant to fasting hyperglycemia and hyperinsulinemia," *FEBS Letters*, vol. 585, no. 9, pp. 1345–1349, 2011.

14

The Noncell Autonomous Requirement of Proboscipedia for Growth and Differentiation of the Distal Maxillary Palp during Metamorphosis of *Drosophila melanogaster*

Anthony Percival-Smith, Gabriel Ponce, and Jacob J. H. Pelling

Department of Biology, University of Western Ontario, London, ON, Canada N6A 5B7

Correspondence should be addressed to Anthony Percival-Smith; aperciva@uwo.ca

Academic Editor: Norman A. Doggett

The *Drosophila* maxillary palpus that develops during metamorphosis is composed of two elements: the proximal maxillary socket and distal maxillary palp. The HOX protein, Proboscipedia (PB), was required for development of the proximal maxillary socket and distal maxillary palp. For growth and differentiation of the distal maxillary palp, PB was required in the cells of, or close to, the maxillary socket, as well as the cells of the distal maxillary palp. Therefore, PB is required in cells outside the distal maxillary palp for the expression, by some mechanism, of a growth factor or factors that promote the growth of the distal maxillary palp. Both *wingless (wg)* and *hedgehog (hh)* genes were expressed in cells outside the distal maxillary palp in the lancinia and maxillary socket, respectively. Both *wg* and *hh* were required for distal maxillary palp growth, and *hh* was required noncell autonomously for distal maxillary palp growth. However, expression of *wg-GAL4* and *hh-GAL4* during maxillary palp differentiation did not require PB, ruling out a direct role for PB in the regulation of transcription of these growth factors.

1. Introduction

The life cycle of *Drosophila* has two distinct free-living forms: the larva and adult. During embryogenesis a larva is formed, and during the larval stages and metamorphosis the imaginal cells proliferate and differentiate to form an adult. The head of the larva and adult fly are highly derived relative to the archetypical insect head [1]. The important function of the mouthparts in adapting to distinct ecological niches [2] explains the large diversity of morphology of mouthparts in insects. The morphogenesis of the adult *Drosophila* mouthparts, the maxillary palpus and proboscis, requires four *Hox* genes: *labial (lab)*, *Deformed (Dfd)*, *pb*, and *Sex combs reduced (Scr)* [3–7]. The diversity of the structure and function of insect mouthparts observed during evolution of the lineages leading to *Drosophila*, *Tribolium*, and *Oncopeltus* is reflected in distinct requirements of HOX proteins for mouthpart development. The requirements of LAB, PB, DFD, and SCR in maxillary palpus development and the maxillary palpus phenotype due to the loss of these HOX proteins are distinct in *Drosophila*, *Tribolium*, and *Oncopeltus* [5, 6, 8, 9].

Even within the *Drosophila* life cycle, the requirements of HOX proteins for mouthpart development are distinct [10]. During embryogenesis PB is expressed in, but not required for, mouthpart development; SCR patterns the labial segment and DFD patterns the maxillary segment [11]. In adults, PB is required for patterning the maxillary palpus and PB with SCR is required for patterning the proboscis [12, 13].

The *Drosophila* maxillary palpus is a highly derived sensory appendage. The establishment of the adult maxillary palpus developmental field requires temporal regulation of wingless (WG) expression during the larval stages [14]. Although DFD expression during second and third stadium larvae defines a maxillary field, it is the delayed expression of WG that specifies maxillary palpus versus antennal identity. Precocious expression of WG in the maxillary primordia results in a maxillary palpus to antenna homeotic transformation. The maxillary palpus has a proximal-distal axis. Proximal-distal axis formation of the legs is well described in *Drosophila* [15, 16]. In the first step, the anterior and posterior compartments are established by the expression of Engrailed (EN) and Hedgehog (HH) in the posterior compartment.

TABLE 1: Stocks.

Name	Genotype	Origin
APS303	$y\ w;\ P\{hspFLP\};\ pb^{20}/TM6B,\ P\{walLy\}$	[12]
APS304	$y\ w,\ P\{w^{+},\ pb^{a}>y^{+}>Tub\alpha1\}B;\ pb^{27}/TM6B,\ P\{walLy\}$	[12]
APS202	$y\ w;\ P\{ry^{+},\ neo^{r},\ FRT\}82B\ pb^{27}/TM6B,\ P\{walLy\}$	[12]
APS205	$y\ w;\ P\{ry^{+},\ neo^{r},\ FRT\}82B\ Sb^{63b}\ M(3)95A^{2}\ P\{y^{+},\ ry^{+}\}96E/TM6B,\ P\{walLy\}$	This work
APS201	$y\ w;\ P\{ry^{+},\ neo^{r},\ FRT\}82B\ pb^{27}\ Scr^{2}\ P\{w^{+},\ ry^{+}\}90E/TM6B,\ P\{walLy\}$	[12]
APS121	$y\ w;\ P\{ry^{+},\ neo^{r},\ FRT\}82B\ pb^{20}\ Sb^{63b}\ M(3)95A^{2}\ P\{y^{+},\ ry^{+}\}96E/TM6B,\ P\{walLy\}$	[12]
DJ103	$y\ w;\ P\{ry^{+},\ neo^{r},\ FRT\}82B\ M(3)95A^{2}\ P\{y^{+},\ ry^{+}\}96E\ P\{exd^{+},\ w^{+}\}/TM6B,\ P\{walLy\}$	[19]
GS902	$y\ w,\ P\{hspFLP\}^{122};\ P\{UAStrc^{S292A\ T453A},w^{+}\},\ hh\text{-}GAL4/TM2$	G. Struhl
APS402	$y\ w;\ P\{UAStrc^{S292A\ T453A},w^{+}\},\ P\{ry^{+},\ neo^{r},\ FRT\}82B\ pb^{27}\ hh\text{-}GAL4/TM6B,\ P\{walLy\}$	This work
APS403	$y\ w;\ P\{UASmyr\text{-}mRFP,\ w^{+}\}/CyO;\ P\{UAStrc^{S292A\ T453A},w^{+}\},\ P\{ry^{+},\ neo^{r},\ FRT\}82B\ pb^{27}$ $hh\text{-}GAL4/TM6B,\ P\{walLy\}$	This work
APS404	$y\ w;\ P\{UASEGFP,\ w^{+}\},\ pb^{20}/TM6B,\ P\{walLy\}$	This work
APS405	$y\ w;\ P\{ry^{+},\ neo^{r},\ FRT\}82B\ pb^{27}\ Scr^{2}\ e\ hh^{9}/TM6B,\ P\{walLy\}$	This work
CB10	$w1118;\ P\{pb\text{-}GAL4,\ w^{+}\},\ P\{UASlacZ,\ w^{+}\}/CyO$	[20]
GS30	$w^{1118};\ P\{ry^{+},\ neo^{r},\ FRT\}82B\ e\ hh^{9}/TM2$	G. Struhl
GFP	$w^{1118};\ P\{UASEGFP,\ w+\}$	Bloomington stock center
VDRC60010	$w^{1118},\ P\{UASdicer2,\ w^{+}\};\ Pin/CyO$	[21]
GP1	$y\ w;\ P\{UASYFP,\ w^{+}\};\ P\{Ubi\ GFP\}$	This work
APS454	$y\ w;\ P\{UAS\ YFP,\ w^{+}\};\ P\{ry^{+},\ neo^{r},\ FRT\}82B\ pb^{27}/TM6B,\ P\{walLy\}$	This work
APS455	$y\ w;\ pb^{20},\ P\{Ubi\ GFP,\ w^{+}\}$	This work
KB1	$y\ w,\ P\{hspFLP\};\ wgGAL4^{270}$	[22]
S491	$w^{1118};P\{dpp\text{-}GAL4,\ w+\}$	[23]

HH activates the expression of Decapentaplegic (DPP) in a sector of dorsal cells and the expression of wingless (WG) in a sector of ventral cells. The expression of the DPP and WG morphogens patterns the proximal-distal axis by regulating the expression of genes such as *Distalless (Dll)* and *homothorax (hth)* [16].

Determination and differentiation is easy to observe during embryogenesis and larval imaginal disc development but not during metamorphosis, because the pupae are opaque, the larval tissue is undergoing histolysis and the developing imaginal tissue is fragile. Although easy to identify body parts that have undergone overt differentiation in fixed pupal material, undifferentiated cells are hard to assign an origin and future. Finally dynamic temporal changes in gene expression are hard to identify by comparing one static, fixed and dissected pupal stage against another. The development of live imaging of metamorphosis allows access to the events of metamorphosis [17, 18]. In this paper, we show that PB is required noncell autonomously for growth of the distal maxillary palp but not by regulation of the transcription of two growth factor genes *wingless (wg)* and *hedgehog (hh)*.

2. Materials and Methods

2.1. Drosophila Stocks and Crosses. The fly strains were maintained on standard medium. All genotypes were generated by standard *Drosophila* crosses. The stocks used in this study are listed in Table 1.

2.2. Immunolocalization of PB/SCR and Detection of GFP/ RFP. Staged prepupae and pupae were dissected from the pupal case in *Drosophila* Ringer's solution, and the pupal membrane was torn along the dorsal side of the thorax to allow penetration of the fixative. The prepupae and pupae were fixed for 20 min in PBS and 4% formaldehyde. Pupae were dissected further to remove more of the pupal membrane and histolysed larval tissue, and refixed for 20 min. For immunolocalization, rabbit anti-PB E9 polyclonal antibody and mouse anti-SCR monoclonal antibody were used to detect PB and SCR expression [24, 25]. The primary antibodies were visualized with donkey FITC conjugated anti-rabbit and Texas-red conjugated anti-mouse antibodies (Jackson laboratories). In the case of detection of GFP and RFP, fixed material was counterstained with DAPI. Images were collected on a Zeiss confocal microscope in the Biotron integrated microscopy facility. Detection of β-galactosidase activity was performed with the X-gal substrate using conditions as described in [26].

2.3. Live Imaging of Metamorphosis. Similar methods were used as described in [17] with one important modification: white prepupae were suspended in a small drop of halocarbon

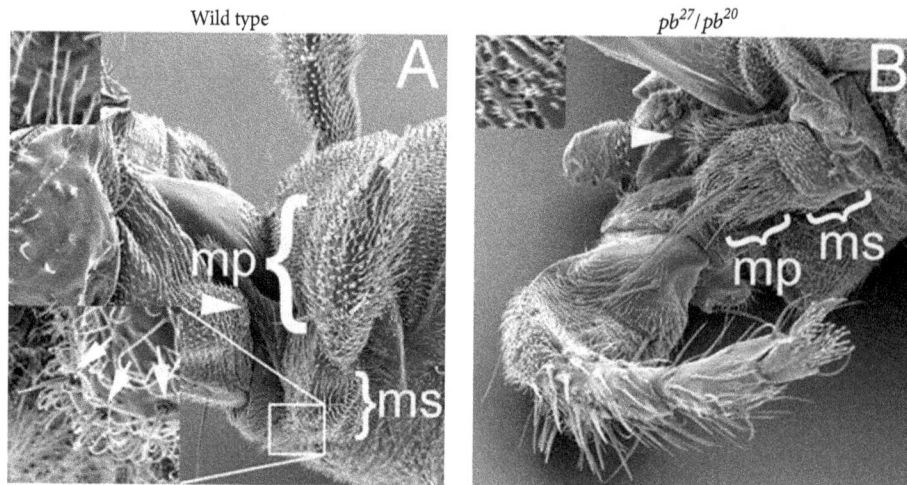

FIGURE 1: The proboscipedia phenotype. In both panels the ventral side is on the left. Panel (A) is wild type mouthparts. The mp bracket indicates the distal maxillary palp and the ms bracket indicates the proximal maxillary socket. The insert at the bottom left is a close-up of the portion of the maxillary socket with three proximal palpus bristles indicated by the arrows. Panel (B) is a pb^{27}/pb^{20} transformed mouthpart with the reduced distal maxillary palp (mp) and reduced proximal maxillary socket (ms) indicated with brackets. The arrowheads indicate the lancinia in both panels. The inserts on the top left of each panel show the long tricombs found on the maxillary socket of wild type (panel A) and the short tricombs found on the maxillary socket of pb mutants (panel B).

oil on a coverslip of a humidity chamber to improve greatly the live imaging of metamorphosis. Images of the time lapse were collected at approximately 2.53 minute intervals with a Hamamatsu digital camera mounted on a Leica DMRBE microscope and 2.5 minute intervals with a Zeiss confocal microscope. All images were exported as.tiff files, and to preserve relative intensities, all images of the time lapse were simultaneously adjusted for brightness, contrast and size with Adobe Photoshop. For the experiments with hh-GAL4 and wg-GAL4 in wild type and pb^{27}/pb^{20} pupae, the pupae were mounted side by side during image capture, and after adjustment for intensity, the image of each pupa of the set was separated and used to make a movie in Adobe After Effects. These movies were synchronized in Adobe After Effects such that head eversion of the pupa occurred in the same frame. A movie of the appropriate time stamp was imported into Adobe After Effects and also synchronized to head eversion. The final movie was rendered and compressed in Adobe After Effects and exported as a QuickTime (MPEG4) file.

2.4. Phenotypic Analysis of Drosophila Heads.
Heads were dissected from eclosed or pharate adults. For bright field microscopy the heads were incubated with 80% acetic acid 20% glycerol overnight at 60°C. The heads were mounted on slides in 1 : 1 Hoyer's mountant : lactic acid [27]. For scanning electron microscopy, the heads were critical point dried and sputter gold coated. Images were collected on a Hitachi 3400-N variable pressure scanning electron microscope in the Biotron integrated microscopy facility.

2.5. Mosaic Analyses.
Flip-mediated mitotic recombination was used to generate all clones of mutant tissue [28]. Larvae were heat shocked for 1 h at 36.5°C. For marking the clones on the adult cuticle, either the $FRT82B\ Sb\ M\ y^+$,

$FRT82B\ pb^{20}\ Sb\ M\ y^+$ or $FRT82B\ M\ y^+exd^+$ chromosome were used screening for $Sb^+\ M^+\ y^+$ cells (Table 4) [12, 19]. The pb^{27} clones in pupae expressing RFP driven by the hh-GAL4 driver were generated in the genotype $y\ w;\ P\{UASRFP\}/P\{hspFLP\};\ P\{UAStrc^{S292A\ T453A}\},\ FRT82B\ pb^{27}hh$-GAL4/$FRT82B\ P\{UbiGFP\}$ and homozygous pb^{27} cells identified by lack of expression of GFP.

2.6. Marking Adult Cuticle for Expression of hh-GAL4.
The reporter $P\{UAStrc^{S292A\ T453A}\}$ was used [29]. Cells that express $TRC^{S292A\ T453A}$ had multiple stunted tricombs that were detected with a scanning electron microscope.

2.7. RNAi Reduction of Expression.
RNAi lines were obtained from Vienna Drosophila RNAi Center and virgin females crossed with w^{1118}, $P\{UASdicer2,\ w^+\};\ P\{pb$-GAL4, $w^+\},\ P\{UASlacZ,\ w^+\}/CyO$ males (VDRC60010 X CB10) [20, 21]. The crosses were reared at 29°C. The heads were mounted in Hoyer's mountant [27], and the length of the maxillary palps measured in Openlab 3.1. Five independent biological replicates were set up, and the mean lengths of male and female distal maxillary palps from the replicates were analyzed with an ANOVA for statistical significance in SSPS v. 16.0.

3. Results

3.1. The Requirement of PB for Maxillary Palpus Development.
The maxillary palpus and the proboscis constitute the mouthparts of $Drosophila$. The maxillary palpus is composed of two pieces: the distal maxillary palp, the mobile sensory appendage, and the proximal maxillary socket into which the distal maxillary palp is inserted (Figure 1(A)). The formation of the adult mouthparts required the Hox gene pb (Figure 1), and PB protein was specifically expressed in the developing

maxillary palpus and proboscis (Figures 2(C) and 2(D)). The HOX protein SCR, required for proboscis development but not required for maxillary palpus development, was expressed in the developing proboscis (Figure 2(D)). PB was expressed in both the cells of the distal maxillary palp and the cells surrounding the distal maxillary palp. This latter expression of PB outside the distal maxillary palp initiated a close examination of the *pb* null phenotype.

The cells surrounding and outside the distal maxillary palp primordium give rise to the proximal maxillary socket and lancinia. In null pb^{27}/pb^{20} mutant adults both the distal maxillary palp and the proximal maxillary socket were reduced, but the lancinia was unaffected (Figure 1(B)). The reduction of the proximal maxillary socket was also associated with the loss of the proximal palpus bristles and the 7.5 fold reduction of tricomb length from 9.64 ± 0.77 μm to 1.28 ± 0.09 μm (±SEM, $n = 12$) on maxillary socket cells [30] (Figures 1(A) and 1(B)). Therefore, cells surrounding the distal maxillary palp were also affected by loss of PB expression.

3.2. Live Imaging of Maxillary Palpus Development. Live imaging of metamorphosis was employed to observe mouthpart development (Figure 2; Supplemental Data Movies 1 and 2 in Supplementary Material available online at https://doi.org/10.1155/2017/2624170) [17]. The prepupal and planerocephalic stages of metamorphosis were recorded, and head eversion is the boundary between the two stages. For all live imaging the first image after head eversion (AHE) is time zero of the pupal planerocephalic stage, and the time during the prepupal stage leading up to head eversion is before head eversion (BHE) (Figures 2(H) and 2(I)). To mark mouthpart development, a *pb-GAL4* driver was used to drive expression of GFP or YFP [20]. *pb-GAL4* has many sites of ectopic expression in addition to expression in the mouthparts [20]. In wandering third stadium larvae, *pb-GAL4* is expressed in a small ring of cells in the aristal primordia, which was observed in the early prepupae (Figure 2(F)). In addition, *pb-GAL4* is expressed in the wings, legs, larval salivary glands, brain, and peripheral nervous system (Figures 2(F)–2(M)); Supplemental Data Movie 1).

Live imaging revealed that the levels and pattern of *pb-GAL4* expression were dynamic during metamorphosis. During the prepupal stage, *pb-GAL4* was expressed strongly in the salivary glands, brain, PNS (Keilin's organs and an anterior sensory complex potentially the labial sensory organ), and labial imaginal discs (Figure 2(F)). The major event observed with *pb-GAL4* during the prepupal stage important for mouthpart development was the fusion of the labial imaginal discs, which was associated with strong expression of *pb-GAL4* (Figures 2(A), 2(G), and 2(H)). Before head eversion, the imaginal tissue of the eye antennal, clypeolabral, and labial discs had fused, and *pb-GAL4* was expressed strongly in the fused labial discs, in the antenna, and in the primordia of the maxillary palpus (Figure 2(A)). However, we were unable to detect *pb-GAL4* expression in the maxillary palpus during the prepupal stage in live imaging, and therefore, do not know exactly when *pb-GAL4* is first expressed in the maxillary palpus.

After head eversion, differentiation of the mouthparts continues forming readily identifiable mouthparts (Figure 2(M)). Expression of *pb-GAL4* in the mouthpart primordia was obscured for a few hours after head eversion by the histolysis of the salivary glands in the movie using fluorescence optics (Figure 2(J)). Using confocal microscopy, *pb-GAL4* expression was detected in the maxillary palp and proboscis primordia after head eversion (Figures 2(N)–2(Q) Supplemental Movie 2). As the GFP/YFP signal expressed from the larval salivary glands degraded, two bright spots of *pb-GAL4* expression appeared in the mouthpart primordia (Figure 2(K), Supplemental Movies 1 and 2). The expression of *pb-GAL4* intensified in the two spots and the cells of the developing proboscis became more visible (Figure 2(l)). A bright spot of *pb-GAL4* expression appeared to be pushed dorsally during the differentiation of distal maxillary palps. Distal maxillary palp growth occurred between 7:37 and 27:40 h AHE. The bright spot of GFP expression was in cells of the maxillary socket (Figure 2(B)). In the mouthparts, the *pb-GAL4* driver reproduces the expression pattern of PB well (Figures 2(B), 2(C), and 2(M)). Importantly both *pb-GAL4* and PB are expressed in cells surrounding the distal maxillary palp as well as the cells of the developing distal maxillary palp.

Antennal differentiation was illuminated by ectopic expression of *pb-GAL4*. Just before head eversion the expression of *pb-GAL4* in the antenna went from a small circle in the arista primordia to throughout the antenna becoming more intense (Figure 2(H), Supplemental Movie 1). After head eversion the expression of *pb-GAL4* was very strong (Figure 2(I)). Between 3:45 and 17:55 h AHE, the antenna continued differentiation and migrated toward the centerline with the ongoing differentiation of the head.

3.3. Noncell Autonomous Requirement of PB for Growth of the Distal Maxillary Palp. Using FLP-mediated mitotic recombination to generate genetically mosaic flies with clones of pb^{27} mutant cells showed that PB is required noncell autonomously in cells of, or close to, the proximal maxillary socket for distal maxillary palp growth, as well as being required in the cells of the distal maxillary palp for growth [12, 28]. All pb^{27} clones (Sb+ M+) in the distal maxillary palp were reduced (Figure 3(A)) (Table 2), suggesting that PB is required in the distal maxillary palp for growth. Interestingly though, one-quarter of the Sb M (*pb+*) distal maxillary palps were also reduced, suggesting that PB is also required noncell autonomously in cells outside the distal maxillary palp for growth (Figure 3(B)) (Table 2). To determine which cells outside the distal maxillary palp PB was required in, a second mosaic analysis using *FRT pb²⁷ Scr²* and *FRT M y⁺ exd⁺* chromosomes was performed scoring the y⁺ phenotype of the tricombs and maxillary socket bristles [19]. In vestigial maxillary palps the tricombs of the socket cells were reduced 7.5 fold in length and were too small to assess the y⁺/⁻ phenotype (Figures 1(A) and 1(B)). Of the 189 maxillary palps examined in genetically mosaic flies, 22 had y⁺ wild type distal maxillary palps (Figure 3(C)). In all 22 of these examples, the maxillary socket cells were (*pb+*), as both the tricombs and maxillary socket bristles had the y⁺ phenotype, suggesting that PB expressed in the ectoderm cells of, or very

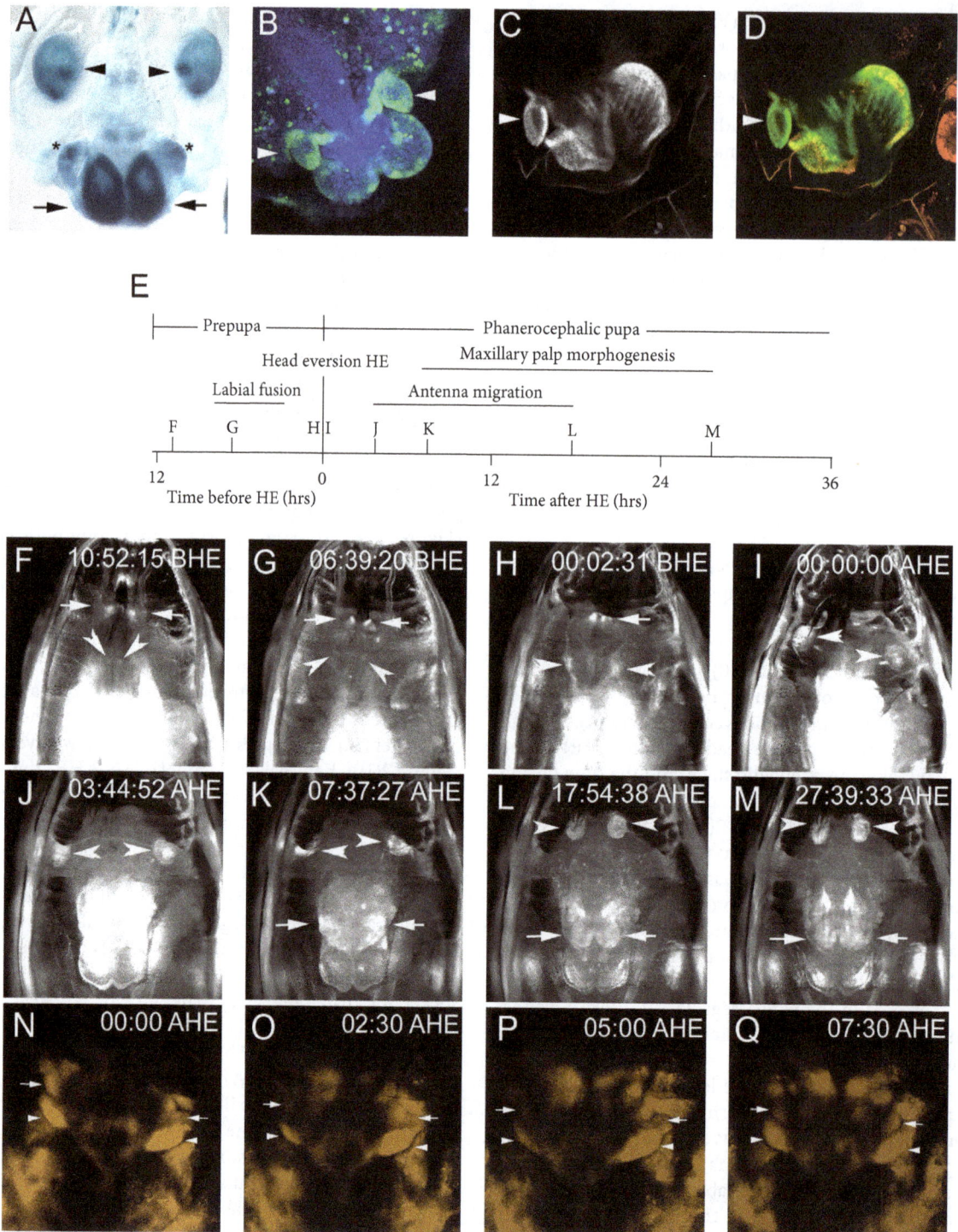

FIGURE 2: The expression of PB and *pb-GAL4* during metamorphosis. Panel (A) is the expression of *pb-GAL4* recorded using the *UASlacZ* reporter gene fixed just before head eversion. The arrows indicate the fused labial discs, asterisks indicate the maxillary palp primordia, and the arrowheads show the antennal primordia. Panel (B) is the expression of *pb-GAL4* in developing mouthparts (approximately 18 h AHE) using *UASEGFP* as the reporter gene (green). The arrowheads indicate the distal maxillary palps. The tissue is stained with DAPI (blue). Panels (C) and (D) are the expression of PB (C) and the expression of PB (green) and SCR (red) (D) at approximately 36 h AHE. The arrowheads indicate the distal maxillary palp. Panel (E) is the time line of metamorphosis indicating the stages and major events observed. The start and stop point for labial fusion, antenna migration, and maxillary palp morphogenesis are estimates based on first evidence of movement. The letters indicate the relative time of the images shown in panels (F)–(M). Panels (F)–(M) are individual frames from live imaging shown in Supplemental Data Movie 1. The time the image was recorded is indicated (h: min: sec BHE or AHE) and the arrows indicate the developing labial segment and the arrowheads indicate the aristal primordia and antennal primordia expression. Panels (N)–(Q) are the first 7.5 minutes AHE of YFP expression driven by *pb-GAL4* shown in Supplemental Data Movie 2. The arrowheads indicate the proboscis primordia and the arrows the maxillary palpus primordia. In panels (A)–(D) and (I)–(Q) the dorsal side of the head is at the top and the ventral is at the bottom.

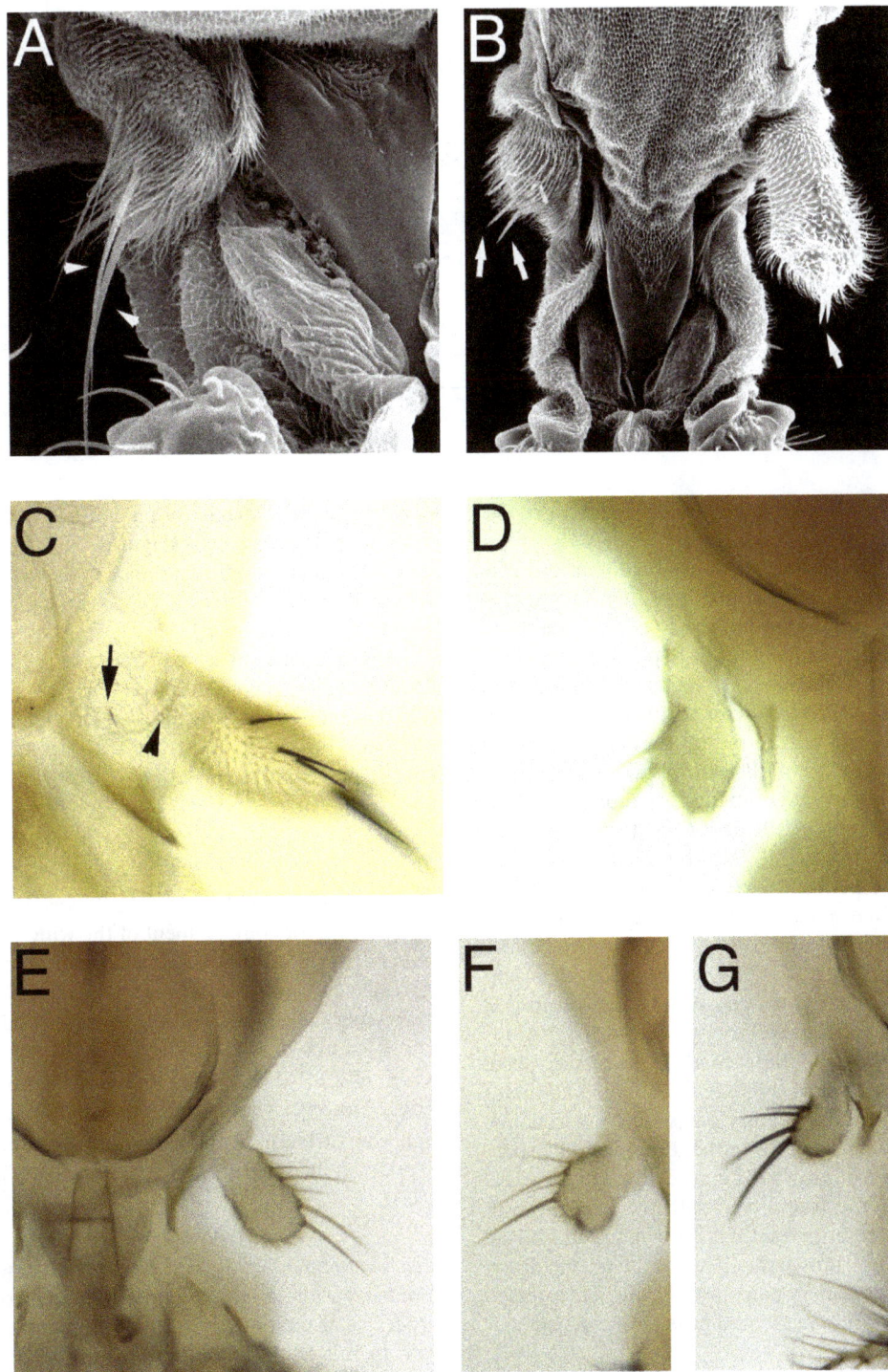

FIGURE 3: The maxillary palpus phenotypes of three independent mosaic analyses. Panels (A) and (B) are scanning electron micrographs of the effects of clonal loss of PB function generated in flies with the genotype $y\ w;\ P\{hspFLP\}/+;\ P\{ry^+,\ neo^r,\ FRT\}82B\ pb^{27}/P\{ry^+,\ neo^r,\ FRT\}82B$ $Sb^{63b}\ M(3)95A^2\ P\{y^+,\ ry^+\}96E$. In panel (A), a $Sb^+\ pb^{27}$ clone in the distal maxillary palp is shown and the maxillary palp is reduced. In panel (B), $Sb\ pb^+$ distal maxillary palps are shown; the right is wild type, and the left is reduced indicated by the arrow. In panels (A) and (B), the arrows indicate Sb bristles and the arrowheads Sb^+ bristles. Panels (C) and (D) are bright field micrographs of clonal loss of PB function generated in the genotype $y\ w;\ P\{hspFLP\}/+;\ P\{ry^+,\ neo^r,\ FRT\}82B\ pb^{27}\ Scr^2\ P\{w^+,\ ry^+\}90E/P\{ry^+,\ neo^r,\ FRT\}82B\ M(3)95A^2\ P\{y^+,\ ry^+\}96E$ $P\{exd^+,\ w^+\}$. Panel (C) is one of the 22 $pb^{27}/+$ wild type distal maxillary palps with the y^+ proximal palpus bristle indicated by the arrow and y^+ maxillary socket tricombs indicated with the arrowhead. Panel (D) shows a $y^-\ pb^{27}Scr^2$ maxillary palpus. Panels (E)–(G) are bright field micrographs of distal maxillary palps from the clonal ectopic expression of PB in a pb^{27}/pb^{20} mutant background generated in the genotype $y\ w,\ P\{w^+,\ pb^a{>}y^+{>}Tub\alpha1\}B;\ P\{hspFLP\}/+;\ pb^{27}/pb^{20}$. Panel (E) is a rescued y^- and PB expressing ($pb^a{>}Tub\alpha1$) maxillary palpus, panel (F) is a reduced y^- and PB expressing ($pb^a{>}Tub\alpha1$) maxillary palpus, and panel (G) is a reduced $y^+\ pb^-$ ($pb^a{>}y^+{>}Tub\alpha1$) maxillary palpus.

TABLE 2: Distribution of phenotypes in the *pb* loss-of-function mosaic analysis.

Sb M y^+ (pb^+)		Sb$^+$ M y^- (pb^-)	
Wild type maxillary palps	Reduced maxillary palps	Wild type maxillary palps	Reduced maxillary palps
40[a]	13	0	96

[a]The numbers only include distal maxillary palps that were completely Sb$^+$ or Sb; distal maxillary palps that were a mix of genotypes were not included.

TABLE 3: Distribution of phenotypes in PB ectopic expression rescue mosaic analysis.

y^- (pb^+)		y^+ (pb^-)	
Rescued maxillary palps	Reduced maxillary palps	Rescued maxillary palps	Reduced maxillary palps
63[a]	10	0	51

[a]The numbers only include distal maxillary palps that were completely y^+ or y^-; distal maxillary palps that were a mix of genotypes were not included.

close to, the maxillary socket is required for growth of the distal maxillary palp.

A potential hypothesis for the noncell autonomous role of PB is that PB expressed in the proximal maxillary socket cells is required for transcription of the *pb* gene in the distal maxillary palp. A simple model for this hypothesis is that PB expression in the proximal maxillary socket cells is required for the expression of a secreted factor that binds and acts on the distal maxillary palp cells to induce transcription of the *pb* gene, and this expression of PB in the distal maxillary palp cells directs growth and differentiation. A mosaic analysis with PB expressed from a *Tubulin αl pb* fusion gene was used to test this hypothesis [12]. Flip recombinase was used to excise the y^+ gene from a *Tubαl>y^+>pb^a* construct (>*FRT* site) to create a *Tubαl>pb* fusion gene expressing PB in a pb^{27}/pb^{20} mutant background. In all y^+ distal maxillary palps, which do not express PB from the *Tubαl pb* fusion gene, the distal maxillary palp was reduced indicating that PB expression in the distal maxillary palp cells is required for rescue (Figure 3(G)). 87% of the y^- and PB expressing distal maxillary palps were rescued (Figure 3(E)). But 13% of the y^- and PB expressing distal maxillary palps were not rescued, suggesting that expression of PB in the distal maxillary palp is not sufficient to rescue growth. (Figure 3(F)) (Table 3). This was the same phenomena observed with the generation of pb^{27} clones, and more importantly if PB is required noncell autonomously for *pb* transcription in the distal maxillary palp, then all distal maxillary palps with PB being expressed from the *Tubulin αl* promoter would have been rescued.

3.4. Expression of Wingless, Decapentaplegic, and Hedgehog during Maxillary Palpus Differentiation. The three secreted proteins WG, DPP, and HH are required for establishing the proximal-distal axis of the leg, wing, and antenna. These three proteins are potential candidates for a PB-regulated factor secreted from cells within or close to the maxillary socket that promotes growth of the distal maxillary palp along the proximal-distal axis. The expression patterns of these secreted factors were assessed using GAL4 driver lines during metamorphosis (Figure 4) [22, 23, 31]. Both *hh-GAL4* and *wg-GAL4* were expressed strongly in cells outside the

developing distal maxillary palp and less so in some of the distal maxillary palp cells (Figures 4(B) and 4(C)). However, *dpp-GAL4* was strongly expressed in the distal maxillary palp cells ruling out DPP as a candidate for the PB-regulated growth factor (Figure 4(D)).

Although the site of expression of *wg-GAL4* outside the distal maxillary palp is the developing lancinia, the site of *hh-GAL4* expression outside the distal maxillary palp is unclear. To mark the adult cells that had strongly expressed *hh-GAL4*, the *UAStricornered*$^{S292A\ T453A}$ fusion gene was used [29]. The TRC$^{S292A\ T453A}$ protein inhibits TRC$^+$ protein activity resulting in multiple short tricombs on each cell that express TRC$^{S292A\ T453A}$. This was most easily observed when expression of TRC$^{S292A\ T453A}$ was driven by *hh-GAL4* in the posterior compartment of the wing (Figure 4(E)). The cells of the anterior compartment had long single tricombs on each cell, but the cells of the posterior compartment expressing GAL4 had multiple short tricombs (Figure 4(E)). The tricombs of the distal maxillary palp were unaffected when TRC$^{S292A\ T453A}$ was expressed using the *hh-GAL4* driver. However, ventral cells of the maxillary socket had multiple short tricombs indicating high levels of TRC$^{S292A\ T453A}$ expression had occurred in these cells (Figures 4(F) and 4(G)).

3.5. Requirement of wg and hh for the Growth of the Distal Maxillary Palp. Both WG and HH are required for many processes at many stages of development. Particularly relevant to this study is the importance of WG expression for the establishment of the maxillary palp field during larval development [14]. Therefore, to target reduction of expression of WG and HH to the developing maxillary palpus, the *pb-GAL4* driver and *UASRNAi* lines were used. The use of the *pb-GAL4* driver restricted expression of RNAi molecules to the maxillary palpus during pupal development, and in *Drosophila* RNAi mediated reduction of expression is cell autonomous [21]. To increase the activity of GAL4 expressed from *pbGAL4*, the flies were grown at 29°C [32]. Three *HH RNAi* lines were obtained: two of which (ID# 1402 and 1403) were predicted to have one off target (CG4637); and one line (ID# 43255) was predicted to have five off targets

TABLE 4: The effect of mouthpart-specific, RNAi mediated inhibition of components of the WG and HH pathways on distal maxillary palp length.

RNAi line	Construct	Targeted mRNA	Length of the distal maxillary palp (μm) $\pm$ SEM[*]	
			Female (*UASdicer2*[+])	Male (*UASdicer2*[−])
y w	—	—	160 ± 3^a (5)[#]	139 ± 1^a (5)
1402	193	*HH*	144 ± 3^b (5)	141 ± 2^a (5)
1403	193	*HH*	126 ± 1^c (5)	127 ± 3^b (5)
43255	6242	*HH*	138 ± 2^b (5)	136 ± 3^a (5)
y w	—	—	156 ± 3^a (5)	134 ± 1^a (5)
13351	5007	*WG*	134 ± 3^b (5)	135 ± 4^a (5)
39676	5007	*WG*	130 ± 6^b (2)	127 ± 2^a (3)
7767	1372	*ARM*	138 ± 1^b (5)	141 ± 1^a (4)
107344	102545	*ARM*	— (0)	108 ± 8^b (4)
25940	10429	*PAN*	143 ± 3^b (5)	138 ± 2^a (5)
9542	577	*SMO*	136 ± 2^b (4)	134 ± 2^a (5)

[*]Data in the same column with the same letter are not significantly different ($P > 0.05$).
[#]Number of biological replicates.

(CG17450, CG32819, CG32820, CG8665, CG9934) [21, 33, 34]. The predicted off targets were not shared between the two constructs. As a result of the crossing scheme, females expressed Dicer from *UASdicer2*, but males did not. In Dicer expressing females, all *HH RNAi* lines exhibited a significant reduction in the length of the distal maxillary palp (Table 4). The only significant reduction observed in males, which do not express Dicer, was with RNAi line 1403, which showed the strongest effect in females. Two *WG RNAi* line were obtained (ID# 13351 and 39676). Both carried the same construct with no predicted off targets and showed a significant reduction in the length of the distal maxillary palp. Using RNAi lines to target the reduction of expression of components of the WG and HH signal transduction pathways, Armidillo (ARM) and Pangolin (PAN) of the WG pathway and Smoothened (SMO) of the HH pathway were shown to be required for distal maxillary palp growth (Table 4).

3.6. The Requirement of HH in the Growth of the Distal Maxillary Palp. The *hh* gene encodes a secreted ligand, and therefore, *hh* mutant alleles behave noncell autonomously in a mosaic analysis [35]. To determine whether HH is required noncell autonomously for maxillary palpus development as expected from *hh-GAL4* expression (Figure 4), we induced *hh*[9] mutant clones using FLP-mediated mitotic recombination. The right palp in Figure 5(A) had *hh*[9] mutant clone of cells in the distal maxillary palp marked by the Sb[+] bristles and exhibited a wild type phenotype indicating that *hh* was not required in the cells of the distal maxillary palp for growth. All other distal maxillary palps shown in Figures 5(A) and 5(B) were shortened or absent confirming the RNAi results that HH was required for distal maxillary palp growth. These two observations also show that HH was required noncell autonomously for distal maxillary palp growth.

Although both *pb* and *hh* were required for the growth of the distal maxillary palp, the pb and hh phenotypes were

distinct: loss of PB expression resulted in a vestigial maxillary palpus; whereas, loss of HH expression resulted sometimes in a complete deletion of the maxillary palpus (Figures 5(A) and 5(B)). Using *FRT82 pb*[27] *Scr*[2] *hh*[9] and *FRT82 pb*[20] *Sb M y*[+] chromosomes, *hh*[9] clones were induced in a *pb* mutant background (Figure 5(C)). As observed with *hh*[9] clones in a wild type background, *hh*[9] clones in a *pb* mutant background also resulted in loss of the vestigial palp indicating that growth of the vestigial palp is HH-dependent (Figure 5(C)).

3.7. Expressions of wg-GAL4 and hh-GAL4 Were Not PB-Dependent. The expression of *wg-GAL4* and *hh-GAL4* were assessed in parallel live imaging experiments where both wild type and *pb*[27]/*pb*[20] prepupae were mounted side by side and allowed to undergo metamorphosis. The expression of GAL4 was detected with a *UASYFP* reporter gene and all cells of the pupae were marked with GFP expressed from *UbiGFP*. In parallel live imaging of *wg-GAL4* expression in wild type and *pb*[27]/*pb*[20] prepupae and pupae, *wg-GAL4* was strongly expressed in lancinia of both the wild type and *pb* mutant (Figure 6; Supplemental Data Movie 3) indicating that *wg* is not regulated by PB. In parallel live imaging of *hh-GAL4* expression in wild type and *pb*[27]/*pb*[20] prepupae, *hh-GAL4* was strongly expressed in the salivary glands of wild type but not *pb*[27]/*pb*[20] prepupae indicating that *hh* expression in the salivary gland is PB-dependent (Supplemental Data Movie 4). However, in wild type pupae *hh-GAL4* was expressed strongly in the cells of the maxillary socket and *hh-GAL4* expression was only expressed a little less in the maxillary palp socket cells of the vestigial maxillary palpus of *pb* mutants (Figure 6 Supplemental Data Movie 4). Although in other repeat experiments, a greater difference between expression of *hh-GAL4* in wild type and *pb* mutants was observed, *hh-GAL4* is still expressed in *pb* mutants. The variation observed between experiments could be due to the *pb* mutant mouthpart cells not being very healthy resulting

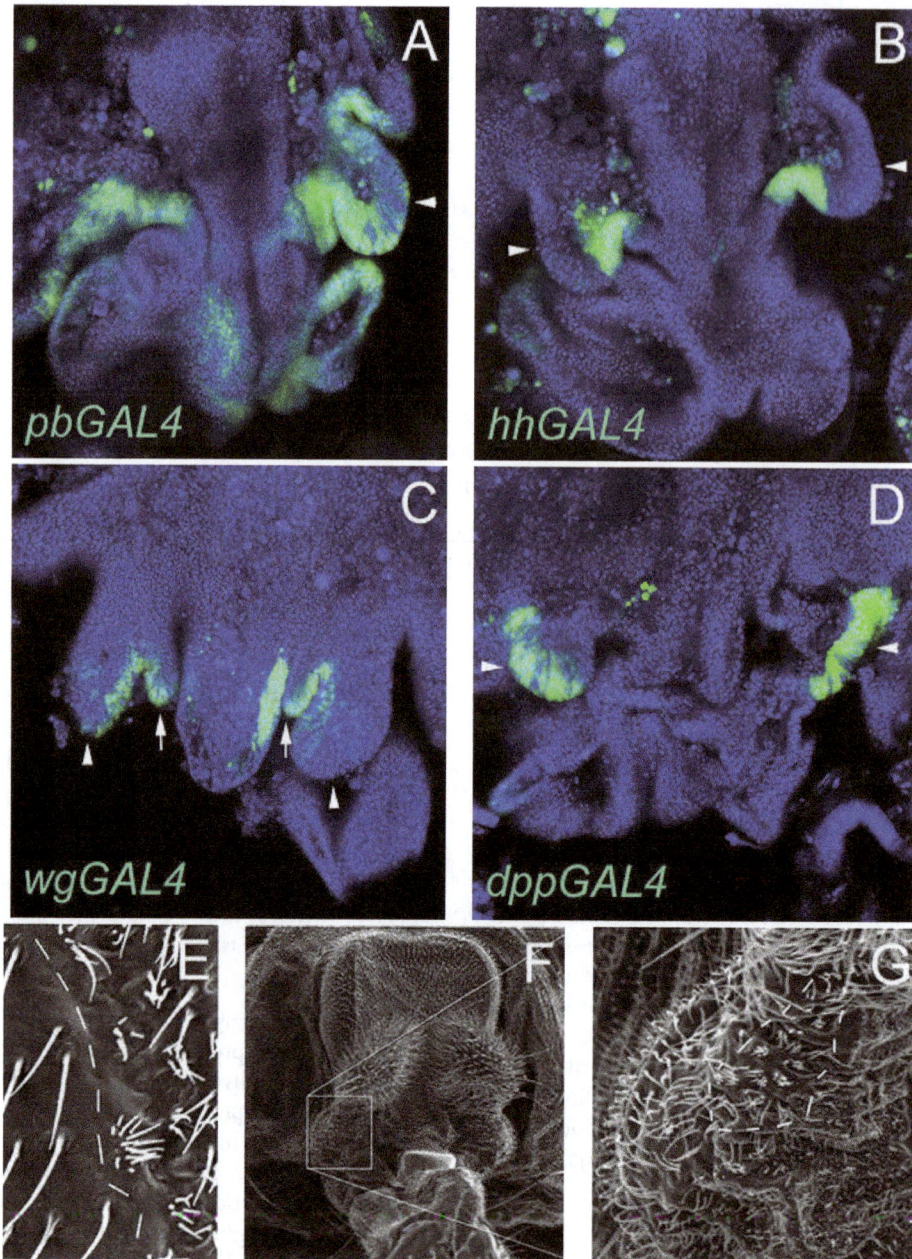

FIGURE 4: Expression of *pb*, *hh*, *wg*, and *dpp-GAL4* drivers during maxillary palpus development. In panels (A)–(D), the driver is indicated on the bottom lefthand corner. The arrowheads indicate the developing maxillary palpus and the arrows in panel (C) indicate the developing lancinia. The tissue is stained with DAPI (blue), and the expression of the drivers was detected with a *UASEGFP* reporter (green). Panel (E) is the expression of *hh-GAL4* in the wing marked by expression of TRC$^{S292A\ T453A}$ from the *UAStrc*$^{S292A\ T453A}$ reporter. The dotted line indicates the anterior-posterior compartment boundary, and multiple short bristles are observed in the posterior compartment. Panels (F) and (G) are the expression of *hh-GAL4* in the maxillary palpus marked by expression of TRC$^{S292A\ T453A}$ from the *UAStrc*$^{S292A\ T453A}$ reporter. The box in (F) indicates the close-up shown in (G). The dotted line in (G) indicates the field of cells that have multiple short tricombs indicating expression of TRC$^{S292A\ T453A}$.

in a nonreproducible level of *hh-GAL4* expression. Or as clearly observed in the movies, once the maxillary palps of wild type and *pb* mutants start to differentiate they are very different from one another early in differentiation and the lower level of *hh-GAL4* expression may reflect divergence of the structure of the wild type and mutant palps. To investigate

further whether PB was required for *hh-GAL4* expression, a FLP-mediated mosaic analysis was performed. The *hh-GAL4* allele is an insertion of a GAL4 enhancer detector into the *hh* locus, and both the *pb* and *hh* loci are on the right arm of chromosome 3. Therefore, the *FRT82 pb*^{27} *hh-GAL4* chromosome created to perform the mosaic analysis resulted

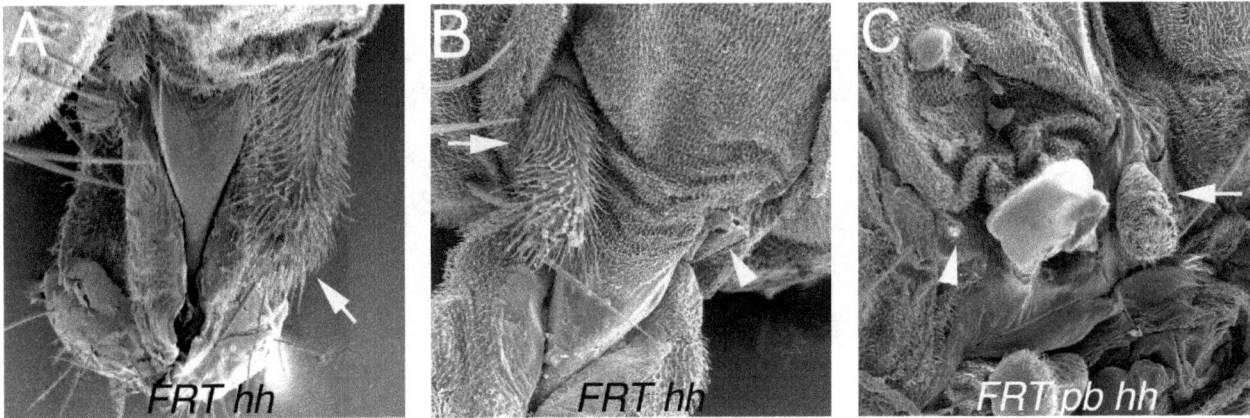

FIGURE 5: Genetic analysis of the requirement of *hh*. Panels (A) and (B) are scanning electron micrographs of the effects of clonal loss of HH function generated in flies with the genotype *y w; P{hspFLP}/+; P{ry⁺, neoʳ, FRT}82B e hh⁹/P{ry⁺, neoʳ, FRT}82B Sb⁶³ᵇ M(3)95A² P{y⁺, ry⁺}96E*. The arrow in panel (A) points to a *hh* clone in the distal maxillary palp that was marked with Sb⁺ M⁺ bristles and that did not affect growth; the growth of the other three distal maxillary palps in panels (A) and (B) were affected to varying degrees and lacked bristles. Panel (C) is a *hh⁹* genetic mosaic in a *pb* mutant background generated in the genotype *y w; P{hspFLP}/+; P{ry⁺, neoʳ, FRT}82B pb²⁷ Scr² e hh⁹/P{ry⁺, neoʳ, FRT}82B pb²⁰ Sb⁶³ᵇ M(3)95A² P{y⁺, ry⁺}96E*. The arrows indicate reduced maxillary palpus and the arrowheads indicate the loss of the maxillary palpus in panels (B) and (C). In panel (C) the left vestigial maxillary palpus was missing and remaining vestigial palpus on the right was Sb M (*hh⁹/+*).

in 3 distinct cellular genotypes that were most easily observed in the posterior compartment of the wing (Figures 7(A)–7(D)): parental RFP and GFP expressing cells, loss of GFP expression (*pb²⁷hh-GAL4*) but not RFP expression, and gain of GFP expression but loss of RFP expression due to the loss of *hh-GAL4 (FRT UbiGFP)*. In *pb²⁷* clones in the maxillary socket cells, RFP, and therefore *hh-GAL4*, was still expressed at a high level (Figures 7(E)–7(H)). PB is not required cell autonomously for *hh-GAL4* expression.

4. Discussion

4.1. Noncell Autonomous Requirement of PB in Maxillary Palpus Development. PB is required for the development of both the adult proboscis and the maxillary palpus [3]. At the wandering third instar larval stage, PB is expressed in the labial imaginal disc but not in the eye antennal imaginal disc that harbors the primordia for the maxillary palpus [20]. During the prepupal stage *pb-GAL4* expression intensifies in the differentiating labial discs, and is expressed in the maxillary palpus primordia. During the first thirty hours of pupal development the mouthparts undergo major events of morphogenesis forming a structure that is easily recognized as adult mouthparts. During this stage of pupal development *pb-GAL4* expression is dynamic and intense. The expression of PB and *pb-GAL4* are not restricted to the proboscis and distal maxillary palp, but are also expressed in the cells of the maxillary socket and surrounding tissue. Expression of PB in the cells of the maxillary socket, or cells close to it, is required for the growth of the distal maxillary palp. This noncell autonomous requirement of PB in cells outside the distal maxillary palp is not due to the transcriptional regulation of genes that encode the growth factors HH and WG, even though WG and HH are required for growth of the distal maxillary palp.

Although the HOX protein PB is a transcription factor, and is expected to have a cell autonomous role in regulation of PB-regulated genes, these regulated genes can function on pathways involved in cell-cell communication. This phenomenon is well described in a number of HOX systems in *Drosophila*. In morphogenesis of the embryonic gut, Ultrabithorax (UBX) is required for the expression of the growth factor DPP [36]. Also SCR and other HOX proteins are required noncell autonomously for induction of ectopic tarsi, and Antennapedia is required noncell autonomously for leg determination [12, 37, 38]. PB is required in a complex combination of cell autonomy and noncell autonomy in the regulation of WG and HH pathways during proboscis determination [39, 40]. For regulation of the growth of the haltere UBX is required noncell autonomously and UBX regulated genes involved in mediating this noncell autonomy are identified [41, 42].

PB is also required in the distal maxillary palps cells for growth. In our mosaic analysis we were unable to assess the growth phenotype of *pb⁺* and *pb⁻* cells in palps of mixed *pb⁺/pb⁻* genotypes, and therefore, we were unable to assess whether PB is required cell autonomously in the developing distal maxillary palp cells. If PB is required cell autonomously in the distal maxillary palp cells, then it is possible that PB also regulates the expression of the components (receptor, signal transduction, etc.) that receive the noncell autonomous PB-regulated signal coming from cells outside the distal maxillary palp.

4.2. PB and Proximal-Distal Axis Formation. The major maxillary palpus phenotype caused by loss of PB expression is the loss of growth of the distal maxillary palp along the proximal-distal axis resulting in a vestigial stump. This phenotype suggested the possibility that PB regulates the expression of

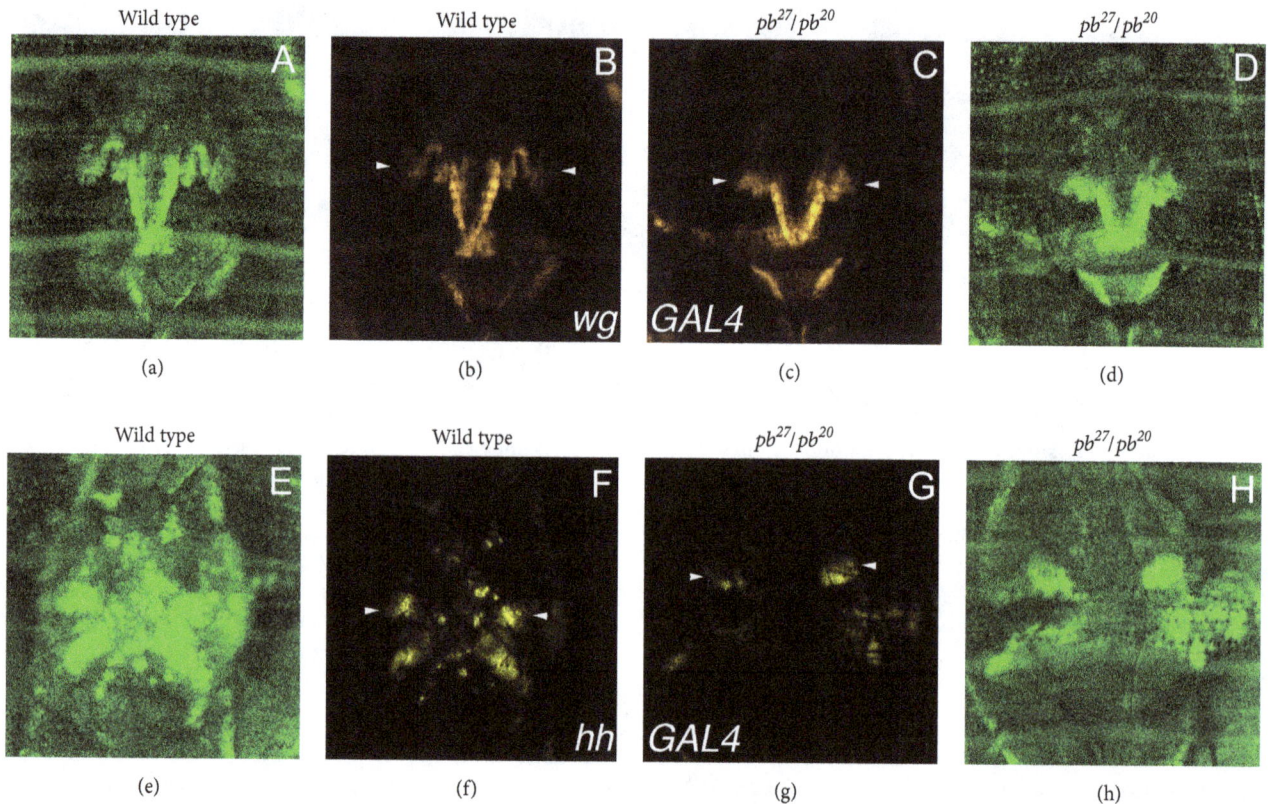

FIGURE 6: Expression of *wg-GAL4* and *hh-GAL4* in wild type and *pb²⁷/pb²⁰* mutants. Panels (A)–(D) are *wg-GAL4*; panels (E)–(H) are *hh-GAL4*. Panels (A), (D), (E), and (H) show expression of *UbiGFP* (green). Panels (B), (C), (F), and (G) show expression of YFP (yellow) from a *UASYFP* reporter gene. Panels (A), (B), (E), and (F) are wild type and panels (C), (D), (G), and (H) are *pb²⁷/pb²⁰* mutants. The arrowheads indicate expression of YFP in the maxillary palps of wild type and *pb* mutants at 16 h AHE.

genes required for the formation of the proximal-distal axis of the leg, wing and antennal appendages. *wg* and *hh* are transcribed in cells outside the distal maxillary palp, and are required for the growth of the distal maxillary palp, but the transcription of these genes does not require PB. These results may suggest that the system PB regulates for proximal-distal axis formation is independent of the system that WG and HH function in for proximal-distal axis formation of the distal maxillary palp. Although this may be the case, our results really only suggest that transcription of the genes that encode the secreted ligands WG and HH are not PB-regulated. UBX is required to suppress the growth of haltere cells, and UBX does not do this by suppressing DPP expression directly but through components involved in the interpretation of the DPP gradient [41, 42]. It is possible that PB functions in a similar manner during distal maxillary palp growth. For example, PB may regulate the expression of a gene in cells outside the distal maxillary palp that is required for a specific posttranslational modification of the secreted factor WG or HH, and this modified form of WG or HH is important for proximal-distal axis formation [43]. In a second explanation, PB is required for repression of expression of a secreted inhibitor of a growth factor. Therefore, PB may have a role in regulating the activity of HH, WG, or DPP in proximal-distal axis formation.

4.3. The Derived Drosophila Maxillary Palpus. The maxillary palpus of *Drosophila* is a highly derived structure relative to that proposed for the archetypical insect head [1, 44, 45]. This high level of derivation may be reflected in two other observations. First, analysis of mitotic clones did not detect anterior-posterior compartment formation in the maxillary palpus even though HH is expressed in a spatially restricted domain during maxillary palpus differentiation [30, 46]. Second, *Dll-GAL4* expression in the distal maxillary palp and maxillary socket [17] suggests that the derived maxillary palpus may be of telopodite origin. These two observations may suggest that the compartmental boundaries are established during metamorphosis and the maxillary palpus is homologous to the distal arista and tarsus of the antenna and leg, respectively. Therefore, PB may be regulating the Epidermal Growth Factor Receptor (EGFR) pathway, which is important for determining the proximal-distal axis of the distal segments of the legs and antennae [47, 48]. This is supported by the identification of components of the EGFR signal transduction pathway as being important for the antenna to maxillary palp transformation caused by ectopic expression of PB [49]. In addition, analysis of the pathways involved in proximal-distal axis formation of *Tribolium castaneum* mouthparts has shown an involvement of the EGFR pathway [45]. Although an interesting possibility, when considering the ligands of

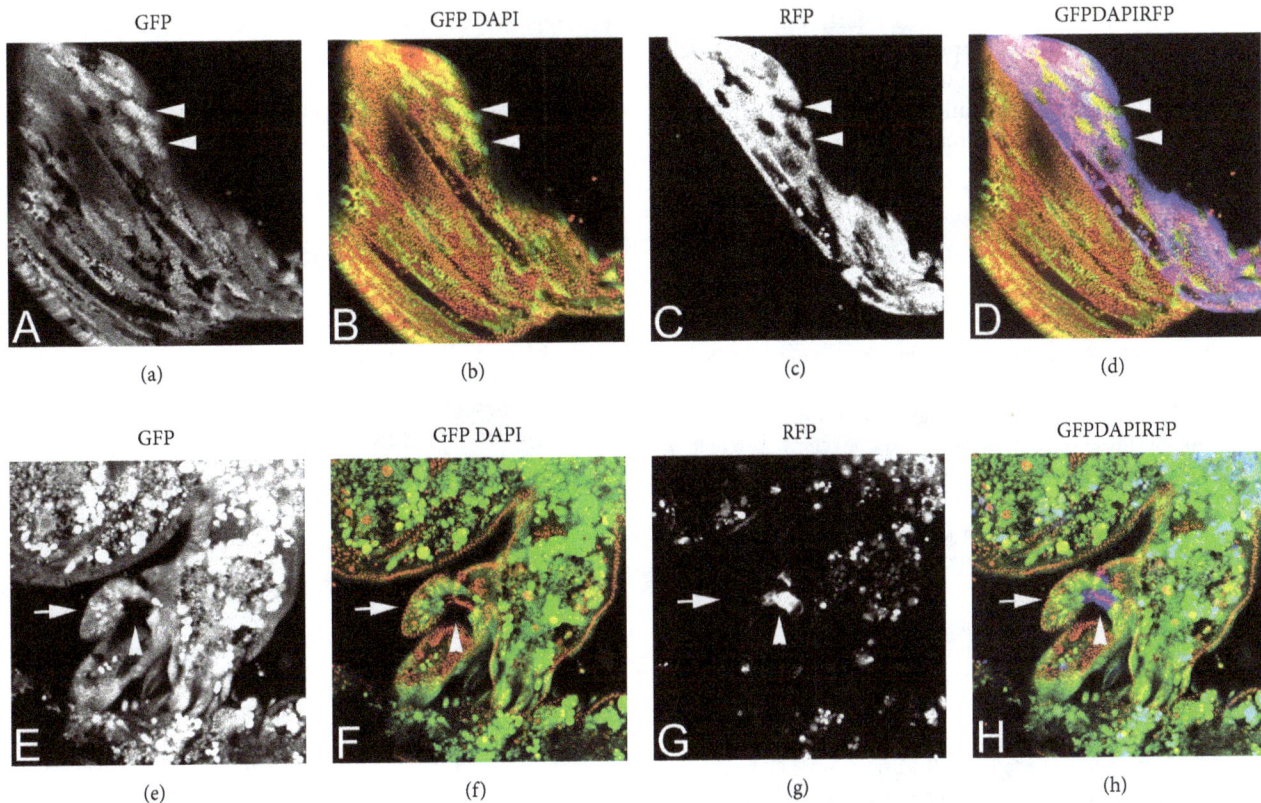

FIGURE 7: Clonal analysis of the requirement of PB in *hh-GAL4* expression. All clones were generated in the genotype *y w; P{UASRFP}/P{hspFLP}; P{UAStrc^{S292A T453A}}, FRT82B pb^{27}hh-GAL4/FRT82B P{UbiGFP}*. Panels (A)–(D) are a pupal wing and panels (E)–(H) are a pupal maxillary palp. Panels (A) and (E) are GFP expression; panels (B) and (F) are GFP expression (green) and nuclei visualized with DAPI (red); panels (C) and (G) are RFP expression; and panels (D) and (H) are GFP (green) and RFP (blue) expression with the nuclei visualized with DAPI (red). The two arrowheads in panels (A)–(D) indicate clones of cells that are homozygous for *UbiGFP* and have lost RFP expression due to loss of *hh-GAL4*. In panels (E)–(H), the arrow indicates the developing distal maxillary palp, and the arrowhead indicates a *pb^{27}* mutant clone that shows strong expression of RFP indicating strong expression of *hh-GAL4*.

the conserved genetic toolkit, there is unlikely to be a single cell that is unaffected by the HH, WG, EGFR, Notch etc. pathways during their development, so it may be naïve to look for direct PB-dependent regulation of the genes that encode the secreted ligand of these pathways as HOX proteins may regulate growth by more subtle mechanisms [41, 42].

5. Conclusions

The HOX transcription factor PB is required both in the cells of the distal maxillary palp and in cells of, or close to, the adjacent maxillary socket for growth of the distal maxillary palp. Therefore, an important role of PB in the growth of the distal maxillary palp is the regulation a cell-cell communication pathway(s). The genes *wg* and *hh* are expressed in cells outside the distal maxillary palp and are required for growth of the maxillary palp. Although WG and HH are good candidates for mediating the noncell autonomous requirement of PB, transcription of the *wg* and *hh* genes is not directly regulated by PB. But the option remains that PB may be required for activation of either WG or HH protein activity, or that PB may regulate the expression of another signaling pathway altogether.

Competing Interests

The authors have no competing interests.

Authors' Contributions

Anthony Percival-Smith designed and performed the experiments presented in Figures 1, 2, 3, 4, 5, and 7 and wrote the first drafts of the manuscript. Gabriel Ponce developed the flies for dual live imaging, assisted with collection of the live imaging data, and revised the manuscript. Jacob J. H. Pelling processed the live imaging data to make the movies and revised the manuscript.

Acknowledgments

The authors thank Gary Struhl, the Bloomington Stock Center, and the Vienna Drosophila RNAi Center for fly stocks, and Anthony Percival-Smith thanks Markus Affolter for lab space during a sabbatical leave where some of this work was performed. The authors thank Gary Struhl for suggesting using *UAStrc^{S292A T453A}* to mark GAL4 expressing cells, Sheila Macfie for assistance with statistical analysis, and Dan Bath for critically reading an early version of the manuscript. This work was supported by a National Science and Engineering Research Council NSERC Discovery grant to Anthony Percival-Smith and a NSERC Undergraduate Summer Research award to Jacob J. H. Pelling.

References

[1] R. E. Snodgrass, *Principles of Insect Morphogenesis*, McGraw-Hill Book Co., New York, NY, USA, 1935.

[2] P. R. Grant and B. R. Grant, "Unpredictable evolution in a 30-year study of Darwin's finches," *Science*, vol. 296, no. 5568, pp. 707–711, 2002.

[3] T. C. Kaufman, "Cytogenetic analysis of chromosome 3 in Drosophila melanogaster: isolation and characterization of four new alleles of the proboscipedia (pb) locus," *Genetics*, vol. 90, no. 3, pp. 579–596, 1978.

[4] G. Struhl, "Genes controlling segmental specification in the Drosophila thorax," *Proceedings of the National Academy of Sciences of the United States of America*, vol. 79, no. 23 I, pp. 7380–7384, 1982.

[5] V. K. L. Merrill, F. R. Turner, and T. C. Kaufman, "A genetic and developmental analysis of mutations in the Deformed locus in Drosophila melanogaster," *Developmental Biology*, vol. 122, no. 2, pp. 379–395, 1987.

[6] V. K. L. Merrill, R. J. Diederich, F. R. Turner, and T. C. Kaufman, "A genetic and developmental analysis of mutations in labial, a gene necessary for proper head formation in Drosophila melanogaster," *Developmental Biology*, vol. 135, no. 2, pp. 376–391, 1989.

[7] D. L. Cribbs, C. Benassayag, F. M. Randazzo, and T. C. Kaufman, "Levels of homeotic protein function can determine developmental identity: evidence from low-level expression of the Drosophila homeotic gene proboscipedia under Hsp7O control," *EMBO Journal*, vol. 14, no. 4, pp. 767–778, 1995.

[8] R. W. Beeman, J. J. Stuart, M. S. Haas, and R. E. Denell, "Genetic analysis of the homeotic gene complex (HOM-C) in the beetle *Tribolium castaneum*," *Developmental Biology*, vol. 133, no. 1, pp. 196–209, 1989.

[9] C. L. Hughes and T. C. Kaufman, "RNAi analysis of deformed, proboscipedia and sex combs reduced in the milkweed bug Oncopeltus fasciatus: novel roles for Hox genes in the Hemipteran head," *Development*, vol. 127, no. 17, pp. 3683–3694, 2000.

[10] T. C. Kaufman, M. A. Seeger, and G. Olsen, "Molecular and genetic organization of the antennapedia gene complex of Drosophila melanogaster," *Advances in Genetics*, vol. 27, pp. 309–362, 1990.

[11] M. A. Pultz, R. J. Diederich, D. L. Cribbs, and T. C. Kaufman, "The proboscipedia locus of the Antennapedia complex: a molecular and genetic analysis," *Genes & Development*, vol. 2, no. 7, pp. 901–920, 1988.

[12] A. Percival-Smith, J. Weber, E. Gilfoyle, and P. Wilson, "Genetic characterization of the role of the two HOX proteins, Proboscipedia and Sex Combs Reduced, in determination of adult antennal, tarsal, maxillary palp and proboscis identities in Drosophila melanogaster," *Development*, vol. 124, no. 24, pp. 5049–5062, 1997.

[13] L. Sivanantharajah and A. Percival-Smith, "Analysis of the sequence and phenotype of Drosophila Sex combs reduced alleles reveals potential functions of conserved protein motifs

of the Sex combs reduced protein," *Genetics*, vol. 182, no. 1, pp. 191–203, 2009.

[14] G. Lebreton, C. Faucher, D. L. Cribbs, and C. Benassayag, "Timing of Wingless signalling distinguishes maxillary and antennal identities in Drosophila melanogaster," *Development*, vol. 135, no. 13, pp. 2301–2309, 2008.

[15] W. J. Brook, F. J. Diaz-Benjumea, and S. M. Cohen, "Organizing spatial pattern in limb development," *Annual Review of Cell and Developmental Biology*, vol. 12, pp. 161–180, 1996.

[16] T. Lecuit and S. M. Cohen, "Proximal-distal axis formation in the drosophila leg," *Nature*, vol. 388, no. 6638, pp. 139–145, 1997.

[17] R. E. Ward, P. Reid, A. Bashirullah, P. P. D'Avino, and C. S. Thummel, "GFP in living animals reveals dynamic developmental responses to ecdysone during Drosophila metamorphosis," *Developmental Biology*, vol. 256, no. 2, pp. 389–402, 2003.

[18] C. Vinegoni, C. Pitsouli, D. Razansky, N. Perrimon, and V. Ntziachristos, "In vivo imaging of Drosophila melanogaster pupae with mesoscopic fluorescence tomography," *Nature Methods*, vol. 5, no. 1, pp. 45–47, 2008.

[19] A. Percival-Smith and D. J. Hayden, "Analysis in Drosophila melanogaster of the interaction between sex combs reduced and extradenticle activity in the determination of tarsus and arista identity," *Genetics*, vol. 150, pp. 189–198, 1998.

[20] C. Benassayag, S. Plaza, P. Callaerts et al., "Evidence for a direct functional antagonism of the selector genes proboscipedia and eyeless in Drosophila head development," *Development*, vol. 130, no. 3, pp. 575–586, 2003.

[21] G. Dietzl, D. Chen, F. Schnorrer et al., "A genome-wide transgenic RNAi library for conditional gene inactivation in Drosophila," *Nature*, vol. 448, no. 7150, pp. 151–156, 2007.

[22] O. Gerlitz, D. Nellen, M. Ottiger, and K. Basler, "A screen for genes expressed in Drosophila imaginal discs," *International Journal of Developmental Biology*, vol. 46, no. 1, pp. 173–176, 2002.

[23] K. Staehling-Hampton, P. D. Jackson, M. J. Clark, A. H. Brand, and F. M. Hoffmann, "Specificity of bone morphogenetic protein-related factors: cell fate and gene expression changes in Drosophila embryos induced by decapentaplegic but not 60A," *Cell Growth and Differentiation*, vol. 5, no. 6, pp. 585–593, 1994.

[24] D. L. Cribbs, M. A. Pultz, D. Johnson, M. Mazzulla, and T. C. Kaufman, "Structural complexity and evolutionary conservation of the Drosophila homeotic gene proboscipedia," *EMBO Journal*, vol. 11, no. 4, pp. 1437–1449, 1992.

[25] M. A. Glicksman and D. L. Brower, "Expression of the *Sex combs reduced* protein in *Drosophila* larvae," *Developmental Biology*, vol. 127, no. 1, pp. 113–118, 1988.

[26] H. J. Bellen, C. J. O'Kane, C. Wilson, U. Grossniklaus, R. K. Pearson, and W. J. Gehring, "P-element-mediated enhancer detection: a versatile method to study development in Drosophila," *Genes & Development*, vol. 3, no. 9, pp. 1288–1300, 1989.

[27] E. Wieschaus and C. Nusslein-Volhard, "Looking at embryos," in *Drosophila: A Practical Approach*, D. B. Roberts, Ed., pp. 199–228, IRL Press, Oxford, UK, 1986.

[28] T. Xu and G. M. Rubin, "Analysis of genetic mosaics in developing and adult Drosophila tissues," *Development*, vol. 117, no. 4, pp. 1223–1237, 1993.

[29] Y. He, X. Fang, K. Emoto, Y.-N. Jan, and P. N. Adler, "The tricornered Ser/Thr protein kinase is regulated by phosphorylation and interacts with furry during Drosophila wing hair development," *Molecular Biology of the Cell*, vol. 16, no. 2, pp. 689–700, 2005.

[30] J. L. Haynie and P. J. Bryant, "Development of the eye-antenna imaginal disc and morphogenesis of the adult head in *Drosophila melanogaster*," *Journal of Experimental Zoology*, vol. 237, no. 3, pp. 293–308, 1986.

[31] H. Tanimoto, S. Itoh, P. Ten Dijke, and T. Tabata, "Hedgehog creates a gradient of DPP activity in Drosophila wing imaginal discs," *Molecular Cell*, vol. 5, no. 1, pp. 59–71, 2000.

[32] T. E. Haerry, O. Khalsa, M. B. O'Connor, and K. A. Wharton, "Synergistic signaling by two BMP ligands through the SAX and TKV receptors controls wing growth and patterning in Drosophila," *Development*, vol. 125, no. 20, pp. 3977–3987, 1998.

[33] M. M. Kulkarni, M. Booker, S. J. Silver et al., "Evidence of off-target effects associated with long dsRNAs in *Drosophila* melanogaster cell-based assays," *Nature Methods*, vol. 3, no. 10, pp. 833–838, 2006.

[34] Y. Ma, A. Creanga, L. Lum, and P. A. Beachy, "Prevalence of off-target effects in Drosophila RNA interference screens," *Nature*, vol. 443, no. 7109, pp. 359–363, 2006.

[35] J. Mohler and K. Vani, "Molecular organization and embryonic expression of the hedgehog gene involved in cell-cell communication in segmental patterning of Drosophila," *Development*, vol. 115, no. 4, pp. 957–971, 1992.

[36] K. Immerglück, P. A. Lawrence, and M. Bienz, "Induction across germ layers in Drosophila mediated by a genetic cascade," *Cell*, vol. 62, no. 2, pp. 261–268, 1990.

[37] A. Percival-Smith, W. A. Teft, and J. L. Barta, "Tarsus determination in Drosophila melanogaster," *Genome*, vol. 48, no. 4, pp. 712–721, 2005.

[38] G. Struhl, "A homoeotic mutation transforming leg to antenna in *Drosophila*," *Nature*, vol. 292, no. 5824, pp. 635–638, 1981.

[39] L. Joulia, H.-M. Bourbon, and D. L. Cribbs, "Homeotic proboscipedia function modulates hedgehog-mediated organizer activity to pattern adult Drosophila mouthparts," *Developmental Biology*, vol. 278, no. 2, pp. 495–510, 2005.

[40] L. Joulia, J. Deutsch, H.-M. Bourbon, and D. L. Cribbs, "The specification of a highly derived arthropod appendage, the Drosophila labial palps, requires the joint action of selectors and signaling pathways," *Development Genes and Evolution*, vol. 216, no. 7-8, pp. 431–442, 2006.

[41] M. A. Crickmore and R. S. Mann, "Hox control of organ size by regulation of morphogen production and mobility," *Science*, vol. 313, no. 5783, pp. 63–68, 2006.

[42] M. A. Crickmore and R. S. Mann, "Hox control of morphogen mobility and organ development through regulation of glypican expression," *Development*, vol. 134, no. 2, pp. 327–334, 2007.

[43] R. K. Mann and P. A. Beachy, "Novel lipid modifications of secreted protein signals," *Annual Review of Biochemistry*, vol. 73, pp. 891–923, 2004.

[44] D. R. Angelini and T. C. Kaufman, "Insect appendages and comparative ontogenetics," *Developmental Biology*, vol. 286, no. 1, pp. 57–77, 2005.

[45] D. R. Angelini, F. W. Smith, A. C. Aspiras, M. Kikuchi, and E. L. Jockusch, "Patterning of the adult mandibulate mouthparts in the red flour beetle, Tribolium castaneum," *Genetics*, vol. 190, no. 2, pp. 639–654, 2012.

[46] G. Morata and P. A. Lawrence, "Development of the eye-antenna imaginal disc of Drosophila," *Developmental Biology*, vol. 70, no. 2, pp. 355–371, 1979.

[47] G. Campbell, "Distalization of the *Drosophila* leg by graded EGF-receptor activity," *Nature*, vol. 418, no. 6899, pp. 781–785, 2002.

[48] M. I. Galindo, S. A. Bishop, and J. P. Couso, "Dynamic EGFR-Ras signalling in Drosophila leg development," *Developmental Dynamics*, vol. 233, no. 4, pp. 1496–1508, 2005.

[49] M. Boube, C. Benassayag, L. Seroude, and D. L. Cribbs, "Rasl-mediated modulation of Drosophila homeotic function in cell and segment identity," *Genetics*, vol. 146, no. 2, pp. 619–628, 1997.

15

Serotonin-Related Gene Polymorphisms and Asymptomatic Neurocognitive Impairment in HIV-Infected Alcohol Abusers

Karina Villalba,[1] Jessy G. Dévieux,[1] Rhonda Rosenberg,[1] and Jean Lud Cadet[2]

[1]*Department of Health Promotion and Disease Prevention, Robert Stempel College of Public Health and Social Work, Florida International University, Miami, FL 33181, USA*
[2]*Molecular Neuropsychiatry Research Branch, DHHS/NIH/NIDA Intramural Research Program, Baltimore, MD, USA*

Correspondence should be addressed to Karina Villalba; kvill012@fiu.edu

Academic Editor: Paul J. Lockhart

HIV-infected individuals continue to experience neurocognitive deterioration despite virologically successful treatments. While the cause remains unclear, evidence suggests that HIV-associated neurocognitive disorders (HAND) may be associated with neurobehavioral dysfunction. Genetic variants have been explored to identify risk markers to determine neuropathogenesis of neurocognitive deterioration. Memory deficits and executive dysfunction are highly prevalent among HIV-infected adults. These conditions can affect their quality of life and HIV risk-taking behaviors. Single nucleotide polymorphisms in the *SLC6A4*, *TPH2*, and *GALM* genes may affect the activity of serotonin and increase the risk of HAND. The present study explored the relationship between *SLC6A4*, *TPH2*, and *GALM* genes and neurocognitive impairment in HIV-infected alcohol abusers. A total of 267 individuals were genotyped for polymorphisms in *SLC6A4* 5-HTTLPR, *TPH2* rs4570625, and *GALM* rs6741892. To assess neurocognitive functions, the Short Category and the Auditory Verbal Learning Tests were used. *TPH2* SNP rs4570625 showed a significant association with executive function in African American males (odds ratio 4.8, 95% CI, 1.5–14.8; $P = 0.005$). Similarly, *GALM* SNP rs6741892 showed an increased risk with African American males (odds ratio 2.4, 95% CI, 1.2–4.9; $P = 0.02$). This study suggests that *TPH2* rs4570625 and *GALM* rs6741892 polymorphisms may be risk factors for HAND.

1. Introduction

The development of combination antiretroviral therapy (ART) has mitigated the severity of the human immunodeficiency virus (HIV) epidemic. Therapeutic advances have transformed HIV/AIDS from a life-threatening illness to a chronic condition [1]. Despite substantial improvements in life expectancy and a lower incidence of HIV-associated neurocognitive disorders (HAND), neuropsychological and neurocognitive deficits continue to be highly prevalent [2]. Clinical neurocognitive manifestations of HAND in the ART era differ from the typical AIDS dementia complex [1]. In the pre-ART era, a progressive subcortical dementia with motor and cognitive slowing was common.

However, today more cortical than subcortical involvement is often reported [3, 4]. HAND encompasses a range of cognitive impairments, including slowed processing and deficient memory and attention, decreased executive function, and behavioral changes, such as apathy or lethargy [5]. Although this type of impairment is much more subtle than the classic HIV dementia, it still affects daily function, quality of life, and antiretroviral adherence and can increase HIV risk behaviors [1, 6]. Asymptomatic or minimally symptomatic neurocognitive disorders are more prevalent in individuals in the current ART era than in the pre-ART era [7]. The causes of continuing high rates of HIV-associated neurocognitive disorder in the ART era are uncertain [7]. Comorbid disorders such as aging, coinfection with hepatitis C, and drug abuse may act as moderators of neurocognitive decline [7, 8]. However, there is a need to identify additional risk markers for the development of HAND [9].

Cognitive control processes regulating thought and action are multifaceted functions influenced by heritable genetic factors and environmental influences [10]. The consequences of cognitive impairment are seen by the large range of both neurologic and neuropsychiatric disorders that affect the quality of life [8, 11, 12]. Cognitive impairment is highly heritable, and individual differences in executive function and memory are influenced by genetic variations [13]. Several studies have demonstrated associations between serotonin polymorphisms with sustained attention, memory, and executive function phenotypes in both clinical and nonclinical populations [14–17]. Furthermore, cognitive neuroscience and pharmacology associate dopamine and serotonin as neuromodulators of executive function [10]. Executive function is part of a system that acts in a supervisory capacity in the brain through planning, decision making, and response control for purposeful goal-directed behavior [18]. Memory is a subclass of cognitive function, divided into long-term and short-term (working) memory. Long-term memory is further divided into explicit and implicit memory [19].

Serotonin pathways arise from the dorsal and ventral raphe nuclei to innervate cortical and subcortical brain regions, including the limbic forebrain, basal ganglia, frontal cortices, thalamus, and the hypothalamus [20]. The neurotransmitter serotonin, 5-hydroxytryptamine (5-HT), is implicated in the pathophysiology of several psychological, behavioral, and psychiatric disorders [21]. Molecular genetics studies on 5-HT have shown that single nucleotide polymorphisms in the TPH2 and SLC6A4 are associated with impaired memory and executive function [21–23]. SLC6A4 (solute carrier family 6, member 4) encodes the serotonin transporter that affects serotonergic neurotransmission by reuptake of synaptic serotonin, ending neurotransmission [24]. Serotonin reuptake variation is linked to a functional polymorphism in the promoter region of the SLC6A4 on chromosome 17q11.1–q12 [24]. Polymorphisms within SLC6A4 have influenced memory regulation, decision making, and response inhibition capabilities [21–23].

The TPH2 (tryptophan hydroxylase 2) gene encodes a member of the protein-dependent aromatic acid hydroxylase family [15]. It is the rate-limiting enzyme of 5-HT synthesis in the brain, which transforms tryptophan into 5-hydroxytryptophan, the direct precursor of 5-HT [16, 25]. The functional SNP rs4570625 is found within the transcriptional region of TPH2 on chromosome 12p21.1 [15]. Evidence of TPH2 variations playing a role in cognition comes from studies implicating TPH2 in the pathophysiology of ADHD and obsessive compulsive disorder [26–28]. Studies have shown that the homozygous TT genotype in SNP rs4570625 is associated with poorer executive function compared to GG and GT genotypes [15, 22]. These studies showed a compensatory adjustment of deficits in executive control functions [16, 29].

The GALM (galactose mutarotase) gene catalyzes the conversion of beta-D-galactose to alpha-D-galactose, which may affect regional neurophysiology, leading to local increases in serotonin release in the brain [30]. A genome-wide association study (GWAS) found a strong association between thalamic region and a coding SNP rs6741892 in GALM using the tracer [11C]DASB-BPND, used to measure brain serotonin transporter levels [31].

The study demonstrated that SNP rs6741892 accounted for about 50% of the variance in [11C]DASB-BPND binding potential in the thalamus, especially for the TT genotype [31]. There is accumulated evidence from genetic studies

suggesting that genetically determined polymorphisms in serotonin-related genes may amplify differences in cognitive performance measures in individuals with already impaired cognition [32, 33].

The present study explored potential associations with SLC6A4 5-HTTLPR, TPH2 rs4570625, and GALM rs6741892 polymorphism and cognitive functions in HIV-infected adults. As described in this section, the existing literature on serotonin genetics and cognition is in HIV-uninfected populations. The present study explored potential associations with SLC6A4 5-HTTLPR, TPH2 rs4570625, and GALM rs6741892 polymorphisms and cognitive functions in HIV-infected adults, using a number of cognitive measures reported to be valid and reliable [34, 35].

2. Methods

The participants in this study were previously outlined in the study by Villalba et al. [36].

2.1. Genotyping. DNA was extracted from whole blood by manual extraction using the QIAamp DNA Mini Kit (Valencia, CA). Genotyping for TPH2 and GALM SNPs was conducted using TaqMan® SNP Genotyping Assays (Foster City, CA) on Bio-Rad CFX96™ real-time PCR instrument (Hercules CA). Polymerase Chain Reaction (PCR) amplifications were performed by using the Probes Supermix. For the promoter variant called 5-HTTLPR, Bio-Rad CFX Manager software (version 3.0) was used for data acquisition and genotype assignment. The primer sequences used for the 5-HTTLPR amplification were obtained from a previous study [30]. The sequences were as follows: $5'$-GCTCTGAAT-GCCAGCACCTAAC-$3'$ (forward primer) and AGAGGG-ACTGAGCTGGACAACCAC-$3'$ (reverse primer) amplifying 522 bp for the 16-repeat allele and 478 bp for the 14-repeat allele [30]. All genotyping was performed blindly with unknown clinical status or background data on the samples.

2.2. Neurocognitive Assessment. All participants were assessed on the same battery of neurocognitive tests and in the same order. Verbal memory was measured with the Auditory Verbal Learning Test (AVLT), using the version World Health Organization/University of California Los Angeles (WHO/UCLA) [37], and executive function was measured by the Short Category Test (SCT) [38]. Alcohol use and other drugs of abuse were also measured using the Timeline Followback (TLFB) and the Alcohol Use Disorders Identification score [39]. There are no agreed criteria to measure HIV-associated neurocognitive disorders. However, the Frascati criteria have been validated and are widely used to classify HIV impairments [3]. The Frascati criteria were used to determine cognitive impairment for asymptomatic neurocognitive impairment (ANI), mild neurocognitive impairment (MND), and HIV-associated dementia (HAD) [40] (Table 1).

2.2.1. Auditory Verbal Learning Test. The AVLT assessment was based on a five-trial presentation of a 15-noun word list (list A) with a presentation rate of one word per second. On completion of trial 5, a single word presentation of a

TABLE 1: Categories of HIV-associated neurocognitive disorder according to Frascati criteria.

	Neurocognitive status[*]	Functional status
Asymptomatic neurocognitive impairment	1 SD below the mean in 2 cognitive domains	No impairment in activities of daily living
Mild neurocognitive impairment or disorder	1 SD below the mean in 2 cognitive domains	Impairment in activities of daily living
HIV-associated dementia	2 SD below the mean in 2 cognitive domains	Notable impairment in activities of daily living

SD: standard deviation.
[*]Neurocognitive testing should include an assessment of at least five domains, including attention-information processing, language, abstraction-executive, complex perceptual motor skills, memory (including learning and recall), simple motor skills, or sensory, perceptual skills.

TABLE 2: Genotype frequencies and cognitive scores for AVLT and SCT tests.

Chr.	Position	Gene	Variant	Minor allele	A/A	A/B	B/B	MAF	Cognitive T-scores Mean (SD)
12	12:71938143	*TPH2*	rs4570625	T	93	120	47	0.18	40.6 (15.1)
2	2:38689828	*GALM*	rs6741892	T	90	114	60	0.22	47.4 (9.8)

15-noun word interference list (list B) was presented. The test measured retention, learning, and recognition rates with higher scores representing better episodic memory [41, 42]. This instrument demonstrates high test-retest reliability, with alpha scores ranging from 0.51 to 0.72 [43].

2.2.2. Short Category Test. The SCT assessment consisted of five booklets, one for each subtest, with 20 cards per subtest. All of the cards within each subtest were organized according to a single principle. The test required the individual to formulate an organizing concept for each subtest. The number of errors on each booklet was added to determine impairment with lower scores representing better executive function [38]. Test-retest coefficients range from 0.60 to 0.96, depending upon the severity of impairment in the sample [38].

3. Analysis

Statistical analyses were performed using Stata v.11 (StataCorp, College Station, TX). Pearson's X^2, Student's t-test (for means), or median test (for medians) was used to compare characteristics between the participants and nonparticipants. To standardize cognitive measures for this study, standardized T-scores were developed by using multiple linear regression methods analyzing the influence of age, sex, education, and ethnicity on each cognitive test score. Each of the five cognitive domains was included as dependent variables: memory (recognition measures form AVLT) executive function (number of errors on the SCT). The continuous predictor was age, and categorical predictors were sex, education, and race/ethnicity. All the predictors in the model were included in each regression, retaining only the variables that significantly contributed to the prediction of cognitive test score. These predictive scores were subtracted from each individual's actual composite score to calculate residual scores. Finally, residual scores were converted to T-scores (mean = 50 and SD = 10) which were used to determine cognitive impairment for asymptomatic neurocognitive

impairment (ANI). Mild neurocognitive impairment (MND) and HIV-associated dementia (HAD) were not part of the analyses due to small number of participants with MND and HAD. Logistic methods were used to calculate crude and multifactorial (self-reported ethnicity/race, alcohol use severity, viral load, CD4 count, cannabis, and cocaine use) adjusted odds ratios (OR), including 95% confidence intervals (CIs). The ORs, with 95% CIs, were used as a measure of effect size. Test for interaction was performed. Bonferroni method was used for correction for multiple comparisons. All statistical tests were two-tailed, and the threshold for statistical significance was set at $P < 0.05$. Ethnic and gender-specific associations were calculated through stratified analyses. Whenever possible the STREGA reporting guidelines for genetic association studies were used. Genotyping counts were tested for Hardy-Weinberg equilibrium using an exact test. By default, the additive genetic model was used, but due to previous associations in the recessive model for *SCL6A4*, the recessive model was also used.

4. Results

A total of 267 HIV-infected alcohol abusers completed the study. Please refer to the study by Villalba et al. [36] for participant characteristics. The Frascati criteria were used to measure neurocognitive impairment. Asymptomatic neurocognitive impairment, greater than one standard deviation below the mean, was observed in 125 (47%) and mild neurocognitive impairment, greater than two standard deviations below the mean, was seen in 11 (4%). HIV-associated dementia was not observed (executive function: mean = 45.2, SD = 10.9; memory: mean = 40.0, SD = 9.1).

Genotyping results, including genotype frequencies, are presented in Tables 2 and 3. All SNPs were in Hardy-Weinberg equilibrium. Analyses yielded significant associations with executive function and *TPH2* and memory and *GALM* genetic polymorphisms.

Whereas 5-HTTLPR polymorphism did not show an association with cognitive flexibility as previously suggested

TABLE 3: *TPH2* and *GALM* associations with cognitive impairment stratified by sex and race/ethnicity (ORs and 95% CIs).

Gene	Variant	Domain	OR allele (95% CI)	P value	OR sex (95% CI)	P value	OR race (95% CI)	P value	OR race × sex (95% CI)	P value
TPH2	rs4570625	Executive function	2.5 (1.1 to 4.9)	$P = 0.02$	4.0 (1.6 to 10.5)	$P = 0.007$	3.3 (1.4 to 7.6)	$P = 0.007$	4.8 (1.5 to 14.8)	$P = 0.005$
GALM	rs6741892	Memory	1.9 (1.2 to 3.1)	$P = 0.006$	2.3 (1.2 to 4.2)	$P = 0.009$	1.9 (1.1 to 3.6)	$P = 0.02$	2.4 (1.2 to 4.9)	$P = 0.02$

OR stratified by sex, OR stratified by race, OR stratified by sex and race.

[27], the SNP rs4570625 of *TPH2* gene showed an overall association in the dominant model with impaired executive function (odds ratio = 2.5, 95% CI, 1.1–4.9; $P = 0.02$). Furthermore, the association showed an increased risk in males (odds ratio = 4.0, 95% CI, 1.6–10.5; $P = 0.007$), not in females ($P_{\text{interaction}} = 0.08$ for sex). Greater risk was observed in African American males (odds ratio 4.8, 95% CI, 1.514.8; $P = 0.005$). For the SNP rs6741892 of the *GALM* gene, a significant association with impaired memory (odds ratio = 1.9, 95% CI, 1.2–3.1; $P = 0.006$) was observed. The risk again was increased in African American males (odds ratio 2.4, 95% CI, 1.2–4.9; $P = 0.02$). Results from this study showed that the associations between serotonin genes and asymptomatic neurocognitive impairment are male-specific. When stratified by race/ethnicity, results were only significant in African Americans; Caucasians and Hispanics were nonsignificant. The interaction between *GALM* and *THP2* polymorphisms with alcohol use was nonsignificant ($P = 0.65$).

5. Discussion

This study provides further evidence for the role of 5-HT in cognition, where functional polymorphisms of two candidate genes in the serotonergic signaling pathway influence executive function and memory. Significant associations were found between *TPH2* SNP rs4570625 and executive dysfunction and *GALM* SNP rs6741892 and impaired memory. Previous studies have shown sex differences in cognition due to dopamine genes interacting with sex and impacting cognition. Similarly, we sought to analyze if serotonin genes were modulated by sex. Thus, stratification by potential effect modifiers (sex and race) showed an even greater effect in African American males but not in females. For the polymorphism 5-HTTLPR, no statistically significant associations were found with neurocognitive measures. Our data suggest that the 5-HTTLPR polymorphism is probably not a risk factor for executive dysfunction and supports previous studies that reported no association between this polymorphism and cognitive decline [44, 45]. However, results are mixed; other studies provided evidence of the influence of 5HTTLPR polymorphism on executive function [14, 46].

The findings of this study indicate that homozygous TT genotype in SNP rs4570625 showed higher error rates measured by the Short Category Test than TG and GG genotypes increasing the risk for executive dysfunction. These findings parallel and extend those of functional imaging and molecular genetic studies suggesting that polymorphism

rs4570625 is a risk marker for executive dysfunction [27, 47, 48]. SNP rs4570625 affects the transcription rate of *TPH2*, which may increase the activity of prefrontal cortex [15]. The prefrontal cortex plays a central role in top-down control of many higher-order executive tasks [49–51]. Additional evidence from a functional image study showed a significant association between SNP rs4570625 and increased activity in several prefrontal and parietal sites during updating of working memory [48]. The authors suggested that the effect of SNP rs4570625 was not specific for attention, impulse control, or working memory, rather it seemed to reflect one common basal cognitive process [48]. These studies showed a compensatory adjustment of deficits in executive control functions [15, 16]. Similarly, results in this study are in line with studies suggesting increased prefrontal activity due to serotonin dysregulation affecting executive function [14, 52–54].

Behavioral studies have demonstrated that executive dysfunction (i.e., poor learning) is central to HIV-neurocognitive impairment and most likely affects behaviors, including adherence to antiretroviral medication and unemployment (or underemployment) [7].

Heaton and colleagues found that medically asymptomatic HIV-infected adults with executive dysfunction were twice as likely to be unemployed and perceived greater vocational difficulties than their unimpaired counterparts [55]. Similarly, another study showed that, for recently diagnosed individuals, the key predictors for finding employment were learning and memory [56].

This study is the first to analyze the functionality of the *GALM* SNP rs6741892 in relation to 5-HT transporter. Thus, a significant association between rs6741892 and impaired memory measured by Auditory Verbal Learning Test was found in HIV-infected adults.

The Auditory Verbal Learning Test was used because of the repeated presentations of words and their successive testing at various time intervals which allowed for the analysis of different memory learning processes such as acquisition, retention, retrieval, and interference [41]. The thalamus plays a significant role in regulating higher level brain activity [57]. Several functional imaging studies showed a relationship between memory processes and the thalamus [58]. Results in this study showed an association with explicit memory and are consistent with neuroimaging reports of compromised thalamus and associated memory structures [57, 58]. Explicit memory is correlated with limited use of higher level encoding strategies, such as semantic clustering and strategic retrieval. This can lead to issues involving medication nonadherence and problematic work-related issues in HIV-infected adults [56, 59]. This study has several limitations. First, there was relatively low frequency of homozygous TT genotype of the *TPH2* SNP. However, it should be noted that there is relatively low occurrence of TT genotype within the general population. In fact, compared to previous studies, the current study included a rather high proportion of homozygous TT genotype carriers compared to others [15, 60]. Second, because of the low power of the study to detect smaller effect sizes, some important associations may not have emerged as statistically significant. Multiple comparisons were necessary due to the exploratory nature of the study, including the analysis of the SNP functionality in the *GALM* gene, as well as the use of all three genetic models. These results should be viewed with caution and should be replicated before a definitive conclusion can be drawn. In general the additive model is used to assess statistical associations of SNPs. While the additive model has sufficient power to detect associations in most situations, there may be occasions where statistical significance is not found, when, in fact, there is an association. Consequently, a strength in this study was the use of multiple genetic models to determine associations that may remain undetectable by the exclusive use of the additive model. Third, due to the vast number of HIV antiretroviral drugs used by study participants, we did not adjust for HIV medication type. Since certain HIV antiretroviral drugs may also affect cognition, this may potentially confound the results. Fourth, two main approaches are used to approximate individual ancestry in association with studies, self-reported race, and ancestry informative markers. We did not use ancestry informative markers due to DNA requirements. Instead, we used self-reported ancestry that may capture common environmental influences as well as ancestral background. However, self-identified racial categories may not always consistently predict ancestral population clusters. Finally, since this was a cross-sectional study stemming from a behavioral intervention trial of HIV-infected subjects, we did not have a healthy control group. Although we adjusted for alcohol and drug use, the results may not adequately explain whether impairments in memory and executive function were correlated with the presence of SNPs *TPH2* rs4570625 and *GALM* rs6741892 or mediated by HIV and alcohol/drug use. Alternatively, these results can serve as an initial point for future research in cognitive phenotypes for HAND in adults. Molecular genetics, as applied in the present study, offers further analytic insight beyond behavioral assessment and neuroimaging and may present a reasonable instrument for the dissociation of different executive control processes.

This study may pave the way for future research integrating the examination of genetic factors in behavioral prevention interventions with HIV-infected populations. Future studies are needed to further identify specific neurocognitive aspects that are especially relevant to HIV-associated neurocognitive disorders. Studies that incorporate genetic factors in combination with neurocognitive testing would benefit from also including the effects of genetic factors on cognitive functioning in healthy individuals since gene by disorder interactions might be expected. Furthermore, it would be beneficial to investigate haplotypes rather than genotypes in studies on cognitive performance in HAND. Since most of the polymorphisms have a small relative effect on cognition, a larger sample is optimal to detect an effect. In addition to the genes analyzed in this study, other genes related to cognitive function should be included. The available findings provide preliminary information for identifying targets for cognitive rehabilitation and other behavioral interventions.

6. Conclusion

The current study was the first to explore the relationship between serotonin-related polymorphisms and asymptomatic neurocognitive impairment in a sample of HIV-infected adults. The results showed a significant association between *TPH2* rs4570625 and individual differences in executive function and *GALM* rs6741892 with memory. The two associations were male-specific. The present study validates previous results pointing to genetic influences on executive function and memory. Moreover, a significant association between SNP rs6741892 and memory was demonstrated, which may imply SNP rs6741892 as a functional polymorphism in the *GALM* gene.

Competing Interests

There are no potential competing interests by any of the authors.

Acknowledgments

This work was supported by National Institute on Alcohol Abuse and Alcoholism (Grant R01AA017405). Karina Villalba was supported by National Institute of General Medical Sciences of the National Institutes of Health (Grant R25 GM061347).

References

[1] D. B. Clifford and B. M. Ances, "HIV-associated neurocognitive disorder," *The Lancet Infectious Diseases*, vol. 13, no. 11, pp. 976–986, 2013.

[2] S. P. Woods, D. J. Moore, E. Weber, and I. Grant, "Cognitive neuropsychology of HIV-associated neurocognitive disorders," *Neuropsychology Review*, vol. 19, no. 2, pp. 152–168, 2009.

[3] A. Antinori, G. Arendt, J. T. Becker et al., "Updated research nosology for HIV-associated neurocognitive disorders," *Neurology*, vol. 69, no. 18, pp. 1789–1799, 2007.

[4] B. A. Navia, B. D. Jordan, and R. W. Price, "The AIDS dementia complex: I. Clinical features," *Annals of Neurology*, vol. 19, no. 6, pp. 517–524, 1986.

[5] R. K. Heaton, D. R. Franklin, R. J. Ellis et al., "HIV-associated neurocognitive disorders before and during the era of combination antiretroviral therapy: differences in rates, nature, and predictors," *Journal of NeuroVirology*, vol. 17, no. 1, pp. 3–16, 2011.

[6] F. Gray, H. Adle-Biassette, F. Chretien, G. Lorin de la Grandmaison, G. Force, and C. Keohane, "Neuropathology and neurodegeneration in human immunodeficiency virus infection. Pathogenesis of HIV-induced lesions of the brain, correlations with HIV-associated disorders and modifications according to treatments," *Clinical Neuropathology*, vol. 20, no. 4, pp. 146–155, 2001.

[7] J. Foley, M. Ettenhofer, M. Wright, and C. H. Hinkin, "Emerging issues in the neuropsychology of HIV infection," *Current HIV/AIDS Reports*, vol. 5, no. 4, pp. 204–211, 2008.

[8] P. Anand, S. A. Springer, M. M. Copenhaver, and F. L. Altice, "Neurocognitive impairment and HIV risk factors: a reciprocal relationship," *AIDS and Behavior*, vol. 14, no. 6, pp. 1213–1226, 2010.

[9] A. J. Levine, E. J. Singer, and P. Shapshak, "The role of host genetics in the susceptibility for HIV-associated neurocognitive disorders," *AIDS and Behavior*, vol. 13, no. 1, pp. 118–132, 2009.

[10] J. J. M. Barnes, A. J. Dean, L. S. Nandam, R. G. O'Connell, and M. A. Bellgrove, "The molecular genetics of executive function: role of monoamine system genes," *Biological Psychiatry*, vol. 69, no. 12, pp. e127–e143, 2011.

[11] L. Boissé, M. J. Gill, and C. Power, "HIV infection of the central nervous system: clinical features and neuropathogenesis," *Neurologic Clinics*, vol. 26, no. 3, pp. 799–819, 2008.

[12] A. F. T. Arnsten and B.-M. Li, "Neurobiology of executive functions: catecholamine influences on prefrontal cortical functions," *Biological Psychiatry*, vol. 57, no. 11, pp. 1377–1384, 2005.

[13] M. J. Frank and J. A. Fossella, "Neurogenetics and pharmacology of learning, motivation, and cognition," *Neuropsychopharmacology*, vol. 36, no. 1, pp. 133–152, 2011.

[14] M. Bosia, S. Anselmetti, A. Pirovano et al., "HTTLPR functional polymorphism in schizophrenia: executive functions vs. sustained attention dissociation," *Progress in Neuro-Psychopharmacology and Biological Psychiatry*, vol. 34, no. 1, pp. 81–85, 2010.

[15] M. Reuter, Y. Kuepper, and J. Hennig, "Association between a polymorphism in the promoter region of the TPH2 gene and the personality trait of harm avoidance," *International Journal of Neuropsychopharmacology*, vol. 10, no. 3, pp. 401–404, 2007.

[16] M. Reuter, C. Esslinger, C. Montag, S. Lis, B. Gallhofer, and P. Kirsch, "A functional variant of the tryptophan hydroxylase 2 gene impacts working memory: a genetic imaging study," *Biological Psychology*, vol. 79, no. 1, pp. 111–117, 2008.

[17] A. Sarosi, X. Gonda, G. Balogh et al., "Association of the STin2 polymorphism of the serotonin transporter gene with a neurocognitive endophenotype in major depressive disorder," *Progress in Neuro-Psychopharmacology and Biological Psychiatry*, vol. 32, no. 7, pp. 1667–1672, 2008.

[18] T. E. Goldberg and D. R. Weinberger, "Genes and the parsing of cognitive processes," *Trends in Cognitive Sciences*, vol. 8, no. 7, pp. 325–335, 2004.

[19] S. Gupta, S. P. Woods, E. Weber, M. S. Dawson, and I. Grant, "Is prospective memory a dissociable cognitive function in HIV infection?" *Journal of Clinical and Experimental Neuropsychology*, vol. 32, no. 8, pp. 898–908, 2010.

[20] M. Luciana, P. F. Collins, and R. A. Depue, "Opposing roles for dopamine and serotonin in the modulation of human spatial working memory functions," *Cerebral Cortex*, vol. 8, no. 3, pp. 218–223, 1998.

[21] J. H. Borg, T. Saijo, M. Inoue et al., "Serotonin transporter genotype is associated with cognitive performance but not regional 5-HT1A receptor binding in humans," *International Journal of Neuropsychopharmacology*, vol. 12, no. 6, pp. 783–792, 2009.

[22] S. Enge, M. Fleischhauer, K.-P. Lesch, A. Reif, and A. Strobel, "Serotonergic modulation in executive functioning: linking genetic variations to working memory performance," *Neuropsychologia*, vol. 49, no. 13, pp. 3776–3785, 2011.

[23] A. A. Kehagia, G. K. Murray, and T. W. Robbins, "Learning and cognitive flexibility: frontostriatal function and monoaminergic modulation," *Current Opinion in Neurobiology*, vol. 20, no. 2, pp. 199–204, 2010.

[24] K.-P. Lesch, D. Bengel, A. Heils et al., "Association of anxiety-related traits with a polymorphism in the serotonin transporter gene regulatory region," *Science*, vol. 274, no. 5292, pp. 1527–1531, 1996.

[25] D. J. Walther, S. Bashammakh, H. Hörtnagl, M. Voits, H. Fink, and M. Bader, "Synthesis of serotonin by a second tryptophan hydroxylase isoform," *Science*, vol. 3, pp. 5603–5676, 2003.

[26] R. Mössner, S. Walitza, F. Geller et al., "Transmission disequilibrium of polymorphic variants in the tryptophan hydroxylase-2 gene in children and adolescents with obsessive-compulsive disorder," *International Journal of Neuropsychopharmacology*, vol. 9, no. 4, pp. 437–442, 2006.

[27] A. Strobel, G. Dreisbach, J. Müller, T. Goschke, B. Brocke, and K.-P. Lesch, "Genetic variation of serotonin function and cognitive control," *Journal of Cognitive Neuroscience*, vol. 19, no. 12, pp. 1923–1931, 2007.

[28] S. Walitza, T. J. Renner, A. Dempfle et al., "Transmission disequilibrium of polymorphic variants in the tryptophan hydroxylase-2 gene in attention-deficit/hyperactivity disorder," *Molecular Psychiatry*, vol. 10, no. 12, pp. 1126–1132, 2005.

[29] C. G. Baehne, A.-C. Ehlis, M. M. Plichta et al., "Tph2 gene variants modulate response control processes in adult ADHD patients and healthy individuals," *Molecular Psychiatry*, vol. 14, no. 11, pp. 1032–1039, 2009.

[30] H. Correa, A. C. Campi-Azevedo, L. De Marco et al., "Familial suicide behaviour: association with probands suicide attempt characteristics and 5-HTTLPR polymorphism," *Acta Psychiatrica Scandinavica*, vol. 110, no. 6, pp. 459–464, 2004.

[31] X. Liu, D. M. Cannon, N. Akula et al., "A non-synonymous polymorphism in galactose mutarotase (GALM) is associated with serotonin transporter binding potential in the human thalamus: results of a genome-wide association study," *Molecular Psychiatry*, vol. 16, no. 6, pp. 584–585, 2011.

[32] J. Borg, S. Henningsson, T. Saijo et al., "Serotonin transporter genotype is associated with cognitive performance but not regional 5-HT1A receptor binding in humans," *International Journal of Neuropsychopharmacology*, vol. 12, no. 6, pp. 783–792, 2009.

[33] C. M. L. Leone, J. C. de Aguiar, and F. G. Graeff, "Role of 5-hydroxytryptamine in amphetamine effects on punished and unpunished behaviour," *Psychopharmacology*, vol. 80, no. 1, pp. 78–82, 1983.

[34] J. R. Crossen and A. N. Wiens, "Comparison of the auditory-verbal learning test (AVLT) and California Verbal Learning Test (CVLT) in a sample of normal subjects," *Journal of Clinical and Experimental Neuropsychology*, vol. 16, no. 2, pp. 190–194, 1994.

[35] L. Wetzel and T. Boll, *Short Category Test*, Western Psychological Services, Los Angeles, Calif, USA, 1987.

[36] K. Villalba, J. G. Devieux, R. Rosenberg, and J. L. Cadet, "*DRD2* and *DRD4* genes releated to cognitive deficits in HIV-infected adults who abuse alcohol," *Behavioral and Brain Functions*, vol. 11, article 25, 2015.

[37] M. Maj, P. Satz, R. Janssen et al., "WHO Neuropsychiatric AIDS Study, cross-sectional phase II: neuropsychological and neurological findings," *Archives of General Psychiatry*, vol. 51, no. 1, pp. 51–61, 1994.

[38] L. Wetzel and T. J. Boll, *Short Category Test, Booklet Format*, W. P. Services, Ed., Western Psychological Services, Los Angeles, Calif, USA, 1987.

[39] S. A. Maisto, M. McNeil, K. Kraemer, M. E. Kelley, and J. Conigliaro, "An empirical investigation of the factor structure of the AUDIT," *Psychological Assessment*, vol. 12, no. 3, pp. 346–353, 2000.

[40] N. S. Gandhi, R. T. Moxley, J. Creighton et al., "Comparison of scales to evaluate the progression of HIV-associated neurocognitive disorder," *HIV Therapy*, vol. 4, no. 3, pp. 371–379, 2010.

[41] E. Vakil, Y. Greenstein, and H. Blachstein, "Normative data for composite scores for children and adults derived from the rey auditory verbal learning test," *Clinical Neuropsychologist*, vol. 24, no. 4, pp. 662–677, 2010.

[42] W. Van der Elst, M. P. J. Van Boxtel, G. J. P. Van Breukelen, and J. Jolles, "Rey's verbal learning test: normative data for 1855 healthy participants aged 24–81 years and the influence of age, sex, education, and mode of presentation," *Journal of the International Neuropsychological Society*, vol. 11, no. 3, pp. 290–302, 2005.

[43] M. D. Lezak, D. B. Howieson, and D. W. Loring, *Neuropsychological Assessment*, Oxford University Press, New York, NY, USA, 2004.

[44] S. B. Marini, V. Bessi, A. Tedde et al., "Implication of serotonin-transporter (5-HTT) gene polymorphism in subjective memory complaints and mild cognitive impairment (MCI)," *Archives of Gerontology and Geriatrics*, vol. 52, no. 2, pp. e71–e74, 2011.

[45] A. Payton, L. Gibbons, Y. Davidson et al., "Influence of serotonin transporter gene polymorphisms on cognitive decline and cognitive abilities in a nondemented elderly population," *Molecular Psychiatry*, vol. 10, no. 12, pp. 1133–1139, 2005.

[46] J. P. Roiser, R. D. Rogers, L. J. Cook, and B. J. Sahakian, "The effect of polymorphism at the serotonin transporter gene on decision-making, memory and executive function in ecstasy users and controls," *Psychopharmacology*, vol. 188, no. 2, pp. 213–227, 2006.

[47] M. Reuter, U. Ott, D. Vaitl, and J. Hennig, "Impaired executive control is associated with a variation in the promoter region of the tryptophan hydroxylase 2 gene," *Journal of Cognitive Neuroscience*, vol. 19, no. 3, pp. 401–408, 2007.

[48] M. E. Reuter, C. Esslinger, C. Montag, S. Lis, B. Gallhofer, and P. Kirsch, "A functional variant of the tryptophan hydroxylase 2 gene impacts working memory: a genetic imaging study," *Biological Psychology*, vol. 79, no. 1, pp. 111–117, 2008.

[49] A. Pasupathy and E. K. Miller, "Different time courses of learning-related activity in the prefrontal cortex and striatum," *Nature*, vol. 433, no. 7028, pp. 873–876, 2005.

[50] H. F. Clarke, J. W. Dalley, H. S. Crofts, T. W. Robbins, and A. C. Roberts, "Cognitive inflexibility after prefrontal serotonin depletion," *Science*, vol. 304, no. 5672, pp. 878–880, 2004.

[51] M. V. Puig and A. T. Gulledge, "Serotonin and prefrontal cortex function: neurons, networks, and circuits," *Molecular neurobiology*, vol. 44, no. 3, pp. 449–464, 2011.

[52] S. F. Enge, K.-P. Lesch, A. Strobel, M. Fleischhauer, and A. Reif, "Serotonergic modulation in executive functioning: linking genetic variations to working memory performance," *Neuropsychologia*, vol. 49, no. 13, pp. 3776–3785, 2011.

[53] J. E. Cattie, K. Doyle, E. Weber, I. Grant, S. P. Woods, and HIV Neurobehavioral Research Program (HNRP) Group, "Planning deficits in HIV-associated neurocognitive disorders: component processes, cognitive correlates, and implications for everyday functioning," *Journal of Clinical and Experimental Neuropsychology*, vol. 34, no. 9, pp. 906–918, 2012.

[54] E. Dahlin, L. Nyberg, L. Bäckman, and A. S. Neely, "Plasticity of executive functioning in young and older adults: immediate training gains, transfer, and long-term maintenance," *Psychology and Aging*, vol. 23, no. 4, pp. 720–730, 2008.

[55] R. K. Heaton, R. A. Velin, J. A. McCutchan et al., "Neuropsychological impairment in human immunodeficiency virus-infection: implications for employment," *Psychosomatic Medicine*, vol. 56, no. 1, pp. 8–17, 1994.

[56] W. G. Van Gorp, J. G. Rabkin, S. J. Ferrando et al., "Neuropsychiatric predictors of return to work in HIV/AIDS," *Journal of the International Neuropsychological Society*, vol. 13, no. 1, pp. 80–89, 2007.

[57] M. D. Johnson and G. A. Ojemann, "The role of the human thalamus in language and memory: evidence from electrophysiological studies," *Brain and Cognition*, vol. 42, no. 2, pp. 218–230, 2000.

[58] R. Fama, M. J. Rosenbloom, S. A. Sassoon, T. Rohlfing, A. Pfefferbaum, and E. V. Sullivan, "Thalamic volume deficit contributes to procedural and explicit memory impairment in HIV infection with primary alcoholism comorbidity," *Brain Imaging and Behavior*, vol. 8, no. 4, pp. 611–620, 2014.

[59] P. M. Maki, D. M. Little, D. Fornelli et al., "Impairments in memory and hippocampal function in HIV-positive vs HIV-negative women: a preliminary study," *Neurology*, vol. 72, no. 19, pp. 1661–1668, 2009.

[60] R. Osinsky, A. Schmitz, N. Alexander, Y. Kuepper, E. Kozyra, and J. Hennig, "TPH2 gene variation and conflict processing in a cognitive and an emotional Stroop task," *Behavioural Brain Research*, vol. 198, no. 2, pp. 404–410, 2009.

16

Induced Pluripotent Stem Cell as a New Source for Cancer Immunotherapy

Farzaneh Rami,[1] **Halimeh Mollainezhad,**[2] **and Mansoor Salehi**[1]

[1]*Department of Genetics and Molecular Biology, School of Medicine, Isfahan University of Medical Sciences, Isfahan 81746-73461, Iran*
[2]*Department of Immunology, School of Medicine, Isfahan University of Medical Sciences, Isfahan 81746-73461, Iran*

Correspondence should be addressed to Mansoor Salehi; m_salehi@med.mui.ac.ir

Academic Editor: Francine Durocher

The immune system consists of cells, proteins, and other molecules that beside each other have a protective function for the host against foreign pathogens. One of the most essential features of the immune system is distinguishability between self- and non-self-cells. This function has an important role in limiting development and progression of cancer cells. In this case, the immune system can detect tumor cell as a foreign pathogen; so, it can be effective in elimination of tumors in their early phases of development. This ability of the immune system resulted in the development of a novel therapeutic field for cancer treatment using host immune components which is called cancer immunotherapy. The main purpose of cancer immunotherapy is stimulation of a strong immune response against the tumor cells that can result from expressing either the immune activator cytokines in the tumor area or gene-modified immune cells. Because of the problems of culturing and manipulating immune cells *ex vivo*, in recent years, embryonic stem cell (ESC) and induced pluripotent stem cell (iPSC) have been used as new sources for generation of modified immune stimulatory cells. In this paper, we reviewed some of the progressions in iPSC technology for cancer immunotherapy.

1. Introduction

The immune system consists of cells, proteins, and other molecules that beside each other have a protective function for the host against foreign pathogens. This system has two major types called innate and adaptive immunity. The innate immune system acts as the first response against a pathogen that is a rapid and nonspecific response and has the ability to activate adaptive immune response [1, 2]. Some of the essential components of this system are macrophages, NK cells, dendritic cells (DCs), mast cells, neutrophils, and complement proteins. The adaptive immune system consists of B- and T-cells that can recognize antigens in a highly specific manner. Antibodies released by plasma cells make up the noncellular portion of the adaptive immune system [2].

One of the most essential features of the immune system is distinguishability between self- and non-self-cells. This function especially has an important role in limiting development and progression of cancer cells. In this case, the immune system can detect tumor cell as a foreign pathogen; so, it can be effective in inhibition and elimination of tumors in

their early phases of development [1]. Tumor cells have some mechanisms for escaping from an immune response, for example, reduction or absence of surface MHCI expression in tumor cells [3], defective or altered apoptotic signaling pathways [4], reduced expression of adhesion molecules in blood vessels of tumor mass for reducing the ability of immune cells to migrate into tumor area [5], and secretion of immune suppressor cytokines [6]. The detection of these escaping mechanisms and the different responses of the immune system to cancer cells resulted in the development and progression of a novel therapeutic field for cancer treatment using immune components which is called cancer immunotherapy [5, 7].

The main purpose of cancer immunotherapy is stimulation of a strong immune response against the tumor cells using the components of the host immune system. This strong response can result from expressing either the immune activator cytokines and antibodies in the tumor area or gene-modified immune cells [8, 9]. The immunological checkpoint blockade is also a new strategy for cancer immunotherapy whose main purpose is enhancing tumor-specific activity

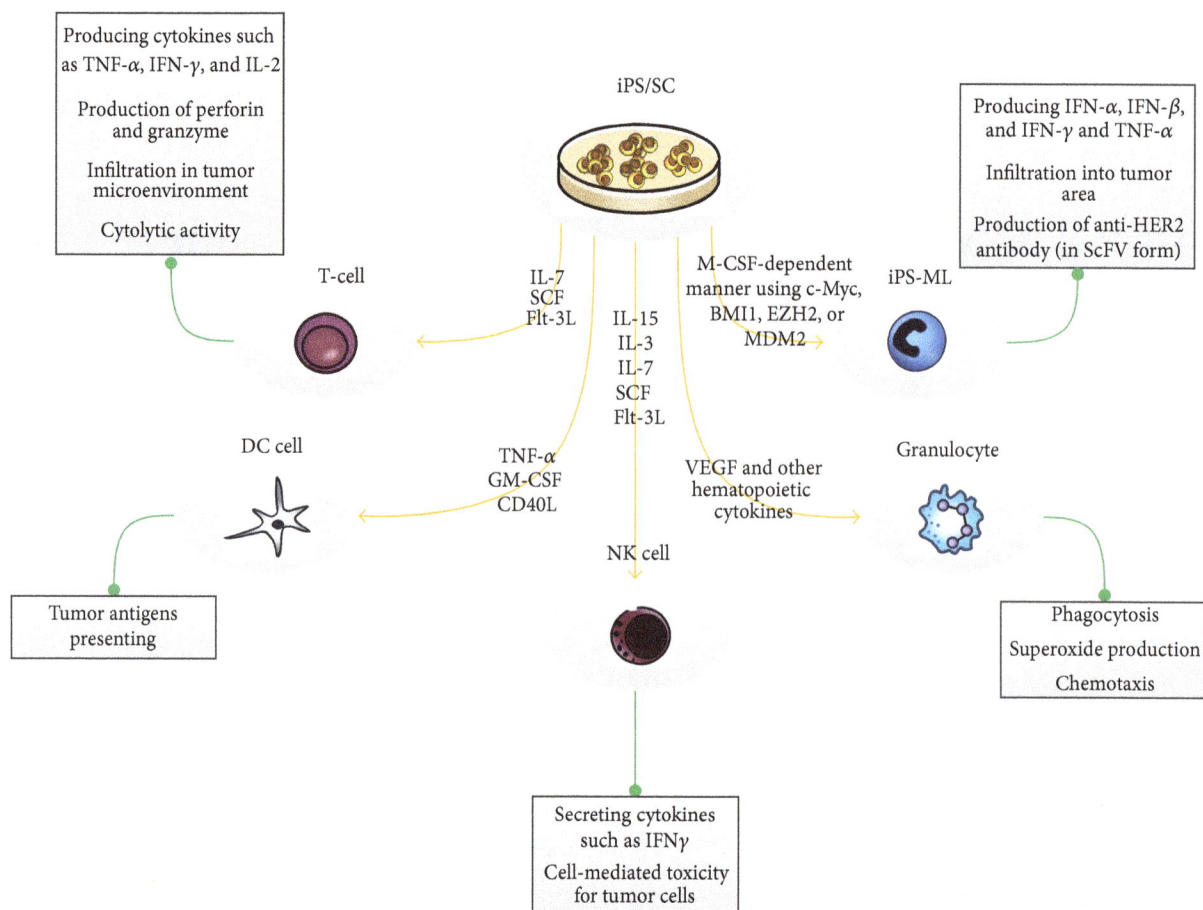

FIGURE 1: iPSC can differentiate into the immune system cells using some factors. These differentiated immune cells were indicated to have the ability to activate an immune response in different manner. Some of the factors for generating immune cells from iPSC and the function related to these cells are summarized in this schematic figure [14–18].

of immune system. The central components for immunological checkpoint blockade strategy are immunoglobulin superfamily and cell surface receptors such as CTLA4 that can act either as activators for initiation of an immune costimulatory signal or as inhibitors for initiation of an immune coinhibitory signal in the targeted tumor cells. This strategy can be used alone or in combination with other cancer immunotherapy methods [10].

As declared, the immune cells are the most important and central components for most of the immunotherapy methods; because of the problems of culturing and manipulating immune cells *ex vivo*, in recent years, embryonic stem cell (ESC) has been used as a new source for generation of modified immune stimulatory cells [11]. The ESC is a pluripotent cell with the ability to differentiate to most tissue cell types including most of the immune system cells [11, 12].

iPSC (induced pluripotent stem cell) is one of the other types of pluripotent cells with the same properties of ESC which made it a suitable choice for generation of immune cells. iPSC can be generated from a patient's somatic cells, such as blood cells and fibroblasts; then, it has the same genetic and histological structure of the patients cells. Using

this kind of cell, it might be possible to generate a personal immune cell with antitumor activity that can be effective in personal cancer treatment. Moreover, it does not have the ethical concerns related to ESC and the problem of undesired immune reaction against a foreign tissue [13]. Figure 1 summarized some of the factors for generating immune cells from iPSC and the function related to these generated cells.

In this paper, we reviewed some of the progressions in iPSC technology for cancer immunotherapy.

2. iPSC Applications in Stimulation of Antitumor Immune Response

Today, using the immune system for cancer therapy has progressed in different areas such as cancer vaccines, T- and NK cell therapy, antitumor antibodies, immune regulatory cytokines, and DC cancer therapy [5, 7]. These therapeutic methods are regarded as new hopes for cancer treatment, although the precise assessment of their therapeutic ability needs more studies [5]. Here, we introduce some of the iPSC generated cells applications in cancer immunotherapy.

2.1. Using iPSC for Dendritic Cells Generation. Dendritic cells (DCs) are types of hematopoietic cells with potent antigen presenting activity that localize in different tissues of the human body. They have the ability to activate naïve T lymphocytes in an immune response and also have a key role in proliferation of regulatory T-cells or anergy of autoreactive T-cells; they are central components of immune system regulation. When an antigen penetrates into a tissue, site localized DCs capture it by phagocytosis or pinocytosis; peptides that are products of digesting antigens in these cells, then, will be represented to lymphocytes using MHC molecules that can activate them [19].

Because of the large range of antigens that can be presented by DCs, they had been used in most of the cancer immunotherapy as an APC (antigen presenting cell) [20, 21]. Today, it has been revealed that tumor cells express some proteins that are different from normal ones. These antigens are detectable by the immune system but are not sufficient for stimulating an immune response [22]. Then, the basis of using DCs in cancer immunotherapy is presenting sufficient antigens for activating host T-cells against the tumor. In this case, DCs sensitized by tumor cells lysate, synthetic peptides, and complete proteins have been used for stimulation of T-cell response [21, 23, 24].

The first use of DCs for cancer therapy was in 1996 on a patient with follicular B-cell lymphoma. In this study, few numbers of DCs were directly isolated from the patient's blood and underwent spontaneous maturation [25]. In the latter studies, DCs were produced from monocytes isolated from patients peripheral blood [26]. However, this method had some problems such as uneasy proliferation of monocytes *in vitro* [27], limitation in the number of the obtained monocytes, and variable potential of differentiation based on blood donors [13].

In 2000, the first studies on using ESC for DC generation were performed [28]. These ESC-derived DCs could activate a more powerful immune response in comparison to previous studies [20, 28]. However, the unavailability of ESC genetically identical for each patient and the ethical issues in using human ESC create limitations for generating DC from ESC. Both of these problems have been solved using iPS cells [29].

The iPS cell-derived DCs have the characteristics of original DCs including the capability of T-cell stimulation, processing and presenting antigens, and the capability of producing cytokines. While using the OP9 culture system is the main method for generating DCs from iPSC, the xeno-free culture systems also are available to generate iPSC-DCs for clinical use [13, 29]. One of these reports belongs to Choi et al. that generate myelomonocytic cells, including DC, from human iPS cells [30]. Similar results are also indicated in the study of Senju et al. [29] and Zhang et al. [31] on the iPSCs derived from mouse cell lines.

iPS cells can generate hematopoietic cells similar to those derived from ES cells that are specific for each person and can be differentiated from a small number of available somatic cells such as fibroblast, but with a low efficiency [32]. Enhancement of iPSC-derived DCs apoptosis, limitation in cell growth and reduction in colony formation ability of these cells [33], and the problems of cost and time related to iPSC

also exist [32]. Because of these limitations, iPSC-derived DCs have not been used in trial studies, yet.

Most of the studies on cancer immunotherapy using DCs have been done for melanoma antigen presentation [9, 20, 34, 35]. The other studied cancers are prostate cancer [36], renal cell carcinoma [37], breast cancer [2, 38], hepatocellular carcinoma [39], multiple myeloma [40], leukemia [20], colorectal cancer [41], gastric cancer [42], and glioblastoma [22, 43]. Cells used in these researches for DC generation were mature and immature monocytes, CD34+ progenitors, ESC, and iPSC, while most of the trial studies were performed using mature monocyte-derived DCs and also CD34+ progenitors-derived DCs that differentiated using cytokines such as TNF-α, GM-CSF, and CD40L [9, 11, 34, 35]. These factors in addition to PGE2, IL-6, IL-12, IL-15, and IFN-γ were also used for stimulating differentiated DC [20, 40]. Some of the antigens that successfully have been presented by DC cells in these studies include oncogenes (such as RAS), epidermal growth factor receptor (HER-2/neu), embryonic genes (such as MAGE, BAGE, and GACE), normal development genes (such as tyrosinase, gp100, and MART-1/Melan-A), viral genes (such as HPV), and other tumor-associated proteins (such as PSMA and MUCI) [23].

2.2. Using iPS for T-Cell Generation. The principal mechanism of tumor immunity is killing of tumor cells by CD8+ CTLs. CTLs have a critical function by recognizing and killing potentially malignant cells. The malignant cells express peptides derived from mutant cellular proteins or oncogenic viral proteins and present them in association with class I MHC molecules. The activation of tumor-specific T-cells depends on DCs, which endocytose tumor cell debris and apoptotic vesicles. After intracellular processing, DCs present peptides derived from tumor-associated antigens in complex with MHC class I molecules to naive CD8+ T-cells. As soon as effector CTLs are generated, they are able to recognize and kill the tumor cells [44–47].

Then, the CD8+ T-cell response is specific for tumor antigens and requires cross-presentation of the tumor antigens by professional APCs, such as dendritic cells. The APCs express costimulator proteins that provide the signals needed for differentiation of CD8+ T-cells into antitumor CTLs. The APCs also express class II MHC molecules that present internalized tumor antigens and activate CD4+ helper T-cells as well [48].

CD4+ cells play their role in antitumor immune responses by providing cytokines such as interleukin-2 (IL-2) (for effective CTL development and clonal expansion of activated CTLs) [49], TNF, and IFN-γ (that can boost cellular components of the innate immunity (macrophages and NK cells), increasing tumor cell class I MHC expression and sensitivity to lysis by CTLs) [50, 51]. Furthermore, activated CD4+ T-cells can enhance the function of DCs to induce CTLs [52, 53]. Another subtype of CD4+ T-cell that is often present in tumor tissue is regulatory T-cell (Treg) that negatively regulates the immune system. It differentiates from CD4+ T-cell when recognizing antigens in a noninflamed condition and in the presence of

TGF-beta and IL-10. Existence of Treg cells in tumor tissue can decrease the expansion of CTLs and suppress the antitumor immune responses, so they are considered as targets for cancer immunotherapy [1, 53]. The ability of CTLs to provide effective antitumor immunity *in vivo* is most clearly seen in animal experiments using carcinogen and DNA virus-induced tumors. In addition, researches showed that tumor-infiltrating CD8+ cytotoxic T-cells can predict clinical outcome in colon, lung, and breast cancers [54].

Declared activation of tumor-specific CTLs is the main goal of cancer immunotherapy; so, adoptive transfer of tumor-specific T-cells is one of the effective therapeutic approaches for fighting against many types of malignancies [55–57]. The isolation of tumor-specific T-cells from a cancer patient, *in vitro* preparation (activation and expansion), and transfusion of these T-cells to the patient are basic steps of adaptive immunotherapy with T-cell [55], although there are some problems with this approach, for example, the low number of antigen-specific T-cells and senescence of these activated cells [55, 56, 58]. Then, iPSC technology can be used to improve the efficacy of adoptive cell transfer immunotherapy (ACT). The main idea of using this kind of cell is according to the capability of iPSC generation in patient or disease specific noninvasive manner without ethical concerns. The difficulty of obtaining ESCs or HSCs from cancer patients also makes iPSC cells a good option for cancer ACT compared to ESCs or HSCs [45, 56].

Previous studies showed that HSC and ESC can differentiate into lymphocyte lineage using the *in vitro* OP9 coculture system which included OP9 cells expressing a Notch ligand, delta-like 1 [59, 60]. Lei et al. differentiated mouse iPS-MEF-Ng-20D-17 cell line. The iPSC in this study was obtained from mouse embryonic fibroblasts induced through retroviral transfection of Oct3/4, Sox2, Klf4, and c-Myc (Yamanaka factors) into T-cell lineages by culturing it on monolayer OP9-DL1 cell system in addition to Flt-3 ligand and IL-7. Adaptive transfer of these iPS cell-derived T lymphocytes to Rag1-deficient mice (mice lacking mature T-cells) enabled them to reconstitute T-cell pool by generation of CD4+ and CD8+ T lymphocytes in lymph nodes and spleen [61].

An important advance in iPSC research was successful iPSC generation from reprogrammed primary CD34+ hematopoietic progenitor cells obtained from peripheral blood [62, 63]. However, due to the low number of these progenitor cells in nonmobilized adult peripheral blood, various studies tried to generate iPSC from peripheral blood mononuclear cells (PBMCs) [64, 65]. Molecular analysis of PBMC derived iPSC for T-cell receptor and immunoglobulin showed that they are derivations of cells from T lineage and nonlymphoid lineage [65].

A potentially efficient approach for generating antigen-specific CTLs is to generate iPSC from immune T-cells and, after their expansion, redifferentiate into T-cells. Brown and colleagues indicated that human T lymphocyte can act as cell source for iPSC generation. Peripheral blood mononuclear cells (PBMCs) were separated from whole blood by leukapheresis or venipuncture and then CD3+ T-cells were expanded by stimulation with IL-2 and anti-CD3

antibody. T-cell-derived iPSCs (TiPS) were generated from activated T-cell when exposed to retroviral transduction of the reprogramming factors [64]. These T-iPSCs preserve their original T-cell receptor (TCR) gene rearrangements, so they can be used as an unlimited source of hematopoietic stem cells bearing endogenous tumor-specific TCR gene for cancer ACT therapy. These T-iPS cells may bypass key step in the thymic development sequence by differentiating *in vitro* in a thymus-independent manner [64].

Some studies have demonstrated the successful differentiation of antigen specific T-cells from an iPSC that itself was generated from CTL specific for particular epitope [57, 66]. CTLs were transduced with Sendai virus bearing Yamanaka factors (Klf4, Sox2, Oct4, c-Myc, and miR-302 target sequence) and SV40 (large T antigen). Experiments on iPSCs generation from mature CTLs specific for the MART-1 (melanoma epitope) [57] and pp65 antigen (cytomegalovirus) [66] indicate that iPSC-derived CTLs (iPSC-CTLs) retain their original antigen specificity. Stimulation of CTL-iPSC-CTL cells with their specific antigens led to IFN-γ secretion and degranulation in a normal manner represents their normal and specific cytolytic reactivity [57, 66]. CTL-iPSC-CTL cells have some differences to parent CD8+ T-cells with elongated telomeres and excellent potential for proliferation and survival. Additionally, some of them display central memory T-cell (T_{CM}) and stem-cell memory T-cells (T_{SCM}) phenotypes which was associated with increasing expression of CCR7, CD27, and CD28 markers [57, 58, 66]. Several lines of evidence show that T_{SCM} and T_{CM} have superior antitumor immunity for ACT-based immunotherapy (due to the resistance to apoptosis, potent response to homeostatic cytokines, self-renewal, and efficient generation of other T-cells' population) [66–72].

Also, generation of iPSCs from murine splenic B-cell and redifferentiation into T-cell lineage have been reported. Isolated B-cells (CD19+, CD24+, and CD45R+ (B220+) and IgM+) were activated with IL-4 and LPS and then transduced with four retroviruses encoding reprogramming factors. Using OP9 coculture system, these B-iPS cells have been differentiated to T-cells that keep their original BCR rearrangement. iPS cell-derived T-cells contained both CD4/CD8 double-positive and CD8+ cells that have surface expression of TCR$\alpha\beta$ and TCR$\gamma\delta$ with normal function following TCR stimulation [73]. Further studies indicated that B-cell-derived iPSCs (B-iPS) and T-cell-derived iPSCs (TiPS) have the same characteristics as human embryonic stem cells (hESCs) [64, 73].

Combination of iPS generation technology with transduction of tumor antigen-specific T-cell receptors (TCRs) or chimeric antigen receptors (CARs) showed successful generation of tumor-specific transgene T-cells. This approach can solve the problems related to low number of tumor-specific T-cells in peripheral blood of patients, their recognition and separation, and invasive nature of biopsy [45, 46, 74]. Lei and colleagues used murine iPS cells for introducing a retrovirus vector encoding MHCI restricted ovalbumin- (OVA-) specific TCR (OT-I). OT-I/iPS cells developed to CD8 CTL following adoptive transfer into recipient mice, produced IL-2 and interferon (IFN) after stimulation, and penetrated

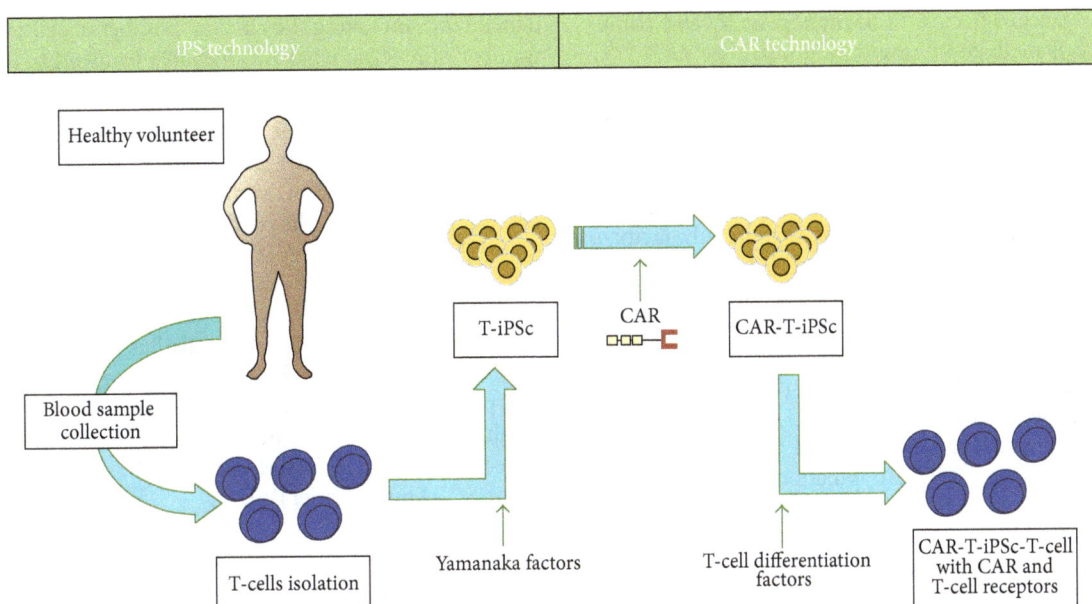

FIGURE 2: Differentiation of 1928z CAR engineered T-iPSCs into CD19-specific functional T lymphocytes.

into tumor tissue after adoptive transfer [45, 46]. This study showed that number of specific CTLs increased in lymph node and spleen after ACT in mice [46]. Also, cells could infiltrate into tumor tissue and 90-fold greater target cell lysis has been seen in these mice compared to control mice [45, 46]. Comparison of survival after ACT in two groups of tumor bearing mice receiving TCR gene-transduced iPS cells showed 100% survival of iPSCs receiving mice in comparison to CD8+ T-cells receiving mice [46].

In just one study, genetic modification was performed using the CAR technology. T-iPSCs are generated by retrovirus reprogrammed T-cells isolated from peripheral blood of healthy volunteers. In the next step, CAR sequence specific for CD19 has been added to human T-iPSCs colonies (Figure 2). iPSC-derived CAR specific T-cells were phenotypically similar to innate $T\gamma\delta$ cells and after ACT to mice showed potential ability to inhibit tumor progression in xenograft model [74]. Although the combination of iPS and TCRs/CARs techniques is efficient and may remove necessity for the detection of antigen-specific T-cells, this approach is costly and there are insertional mutagenesis risks [56].

OP-9 coculture system was not able to generate iPSCs-CD4+ T-cells in vitro [57, 61, 66, 73] because of the limitation in MHCII expression by OP9 cells [75, 76]. Normal development of CD4+ CD8 T-cells occurs by interaction with MHCII of thymic epithelium and expression of ThPOK, TOX, GATA-3, and RUNX factors essential for CD4+ lineage generation in vivo [61]. Only one study has shown the detection of few mature CD4+ T-cells in culture [16]. According to importance of CD4+ helper T-cells in antitumor immunity by promoting the permanence of memory CD8 T-cells [53], isolation of regulatory T-cells (Treg) based on CD4+ CD25+ CD127^{low} CD45RA+ from tumor microenvironment and reprogramming them into iPSCs and then generation of CD4+ helper T-cells may be an effective strategy for ACT [56].

Despite the great advantages of ACT with iPSC-derived T-cells in cancerous mice model [45, 46, 74], there are some limitations when applied in vivo; for example, differentiation of iPSC-derived T-cells takes a long time (at least six weeks) and because of their origin there is a teratoma genesis risk [45, 46]. However, the risk of tumorigenesis of iPSC-derived T-cells is just reported in one study. In the study of Lei and colleagues, when they used in vivo induction system for generation of antigen-specific T-cell differentiation from iPS cells, they did not observe any extrathymic mass in C57BL/6 mice although they observed an extrathymic mass in only one of the Rag1−/− mice. This finding can clear the importance of iPS genetic background for in vivo differentiation [45]. Other limitations are osteoporosis, hair loss, and autoimmune manifestation without any clear reasons (one explanation for it may be in vivo differentiation of other immune cells from iPSCs) [45, 66].

The immunogenicity of iPSC-derived cells is very complicated. Zhao et al. found that some but not all cells derived from mouse iPSC can be immunogenic and this immune rejection response was T-cell dependent. They reported that the inbred C57BL/6 (B6) iPSCs and their derived teratomas can induce T-cell-dependent immune responses after transplantation into the syngeneic B6 mice, although based on their report the immunogenicity of iPS was lower than ESC in vivo [77]. Abe's group studied the immunogenicity of different cell types derived from iPSCs including skin cells, bone marrow cells, and cardiomyocytes. This study indicated that iPSC-derived cardiomyocytes are highly immunogenic, although iPSC-derived skin and bone marrow cells have lower immunogenic effect [78].

It is widely accepted that reprogramming process can induce both genetic and epigenetic defects in produced iPSCs [79–82]. Abnormal overexpression of Hormad1, Zg16, Cyp3a11, Lce1f, Spt1, Lce3a, Chi3L4, Olr1, and Retn genes was also shown in iPSC-derived cells by gene expression analysis.

These genes can be effective in immunogenicity and stimulation of T-cell-mediated immune response after ACT [77]; so, the assessment of iPSC-derived T-cells immunogenicity should be considered before their clinical applications for cancer immunotherapy [45].

Researches showed that changing culture condition and adding multiple soluble proteins influence iPSC lineage differentiation. Presence of transforming growth factor-β (TGF-β) along with TCR stimulation led to differentiation of suppressor T-cells (Foxp3+ population) in iPSC-derived T-cells culture [73]. Also, it has been shown that stimulation of T-cells in the presence of IL-7, IL-15, and IL-21 results in memory phenotype with enhanced persistence in comparison to IL-2 primed T-cells before ACT [83–87] and inhibition of GSK3b (glycogen synthase kinase 3 beta) led to more efficient production of TSCM population *in vitro* [66]. So, combination of iPSC technology with CAR/TCR transgene technique [88] and optimization of culture media [66] may improve the iPSC-derived T-cells that are suitable for clinical applications in cancer ACT.

2.3. Using iPS for Cytokine Producing Cell Generation. One of the mechanisms used by tumor cells for escaping immune response is the suppression of immune response in the tumor area with secretion of immune suppressor cytokines [7] (such as PGE2, IDO, and TGF-β [14]). Then, the basis of using cytokine producing cells in cancers is generating cells with the capability of migration into tumor tissue and secretion of cytokines with the immune activation ability in this area [14–16, 89, 90]. Cytokine producing cells can be from myeloid or lymphoid lineages obtained from coculturing of iPSC or ESC with a mouse bone marrow stromal cell line (OP9) [16] and using of different growth factor (regarding cell type that it must be differentiated into).

One of the most effective cells in defense against tumor development with cytokine secretion activity is natural killer (NK) cells. NK cells have been used in some clinical studies including AML and some other hematological malignancies with low toxicity for patients [14]. However, a significant factor in success of treatment is obtaining a pure and functional NK cell population [91]. NK cell has been differentiated from both ESC and iPSC using two different sets of factors (IL-15, IL-3, IL-7, Flt-3L, SCF or BMP4, VEGF, SCF, FGF, TPO, and Flt-3L). These cells were reported to have the ability to secrete cytokines such as IFN-γ, in addition to their ability for cell-mediated toxicity [92].

The other cytokine producing cell is T lymphocyte that has been reported to have the ability to produce cytokines such as TNF-α, IFN-γ, and IL-2 and cytolytic proteins (perforin and granzyme B) when differentiated from iPSC [16]. T-cells with the ability to secrete CSF also have been reported to be effective in patients with myelodysplastic syndromes [93].

Macrophages are cells whose infiltration is frequently observed in cancer area [94]. This kind of cell has two different functions related to cancer: (1) tumor-associated macrophages (TAM) that cause cancer progression and (2) other macrophages with antitumor activity. Cancer immunotherapy using macrophages just like using NK cells has great dependence on taking the efficient number of cells; this large number of macrophages is achievable by using ESC or iPSC for differentiation into it [15].

Genetically modified iPSCs that differentiated into myeloid lineage (iPS-ML) with the capability of cytokine secretion are also reported to be effective in cancer cells elimination. iPS-ML cells producing IFN-α, IFN-β, and IFN-γ and TNF-α have been studied in gastric cancer cell line (NUGC-4) and human pancreatic cell line (MIAPaCa-2). In this study also, the production of anti-HER2 antibody (in ScFV form) using iPS-ML cells has been reported. The iPS-ML cells had the ability to infiltrate into tumor area and produce antibody and cytokines in the tumor site. The result of this study also indicated that IFN-β has greater local concentration and remains longer than IFN-α in cancer tissue in the SCID mice model [15].

3. Conclusion

iPS cells have been studied in different fields of cancer immunotherapy including tumor Ag presentation, T-cell activity regulation, and cytokine or Ab producing cells, many of which had a successful result for elimination of cancer cell lines. There are many hopes for the future of this technique, although it cannot be used in clinical treatment because of some obstacles that already exist such as generating hiPSC (human iPSC) in a safe manner, enhancing reprogramming and differentiation process efficiency [14], reducing the time and cost needed for the process [13], and proving iPSC safety for clinical use. These problems must be solved before any use of iPSC in patients treatment [14].

Taken together, cancer immunotherapy with iPSC can be considered as a new hope for cancer treatment but still on the early stages that need more studies before its real use in clinic [40].

References

[1] R. R. Raval, A. B. Sharabi, A. J. Walker, C. G. Drake, and P. Sharma, "Tumor immunology and cancer immunotherapy: summary of the 2013 SITC primer," *Journal for ImmunoTherapy of Cancer*, vol. 2, article 14, 2014.

[2] G. K. Koski, U. Koldovsky, S. Xu et al., "A novel dendritic cell-based immunization approach for the induction of durable Th1-polarized Anti-HER-2/neu responses in women with early breast cancer," *Journal of Immunotherapy*, vol. 35, no. 1, pp. 54–65, 2012.

[3] M. Meissner, T. E. Reichert, M. Kunkel et al., "Defects in the human leukocyte antigen class I antigen-processing machinery in head and neck squamous cell carcinoma: association with clinical outcome," *Clinical Cancer Research*, vol. 11, no. 7, pp. 2552–2560, 2005.

[4] J. P. Medema, J. De Jong, L. T. C. Peltenburg et al., "Blockade of the granzyme B/perforin pathway through overexpression of

the serine protease inhibitor PI-9/SPI-6 constitutes a mechanism for immune escape by tumors," *Proceedings of the National Academy of Sciences of the United States of America*, vol. 98, no. 20, pp. 11515–11520, 2001.

[5] K. Töpfer, S. Kempe, N. Müller et al., "Tumor evasion from T cell surveillance," *Journal of Biomedicine and Biotechnology*, vol. 2011, Article ID 918471, 19 pages, 2011.

[6] U. Bogdahn, P. Hau, G. Stockhammer et al., "Targeted therapy for high-grade glioma with the TGF-β2 inhibitor trabedersen: results of a randomized and controlled phase IIb study," *Neuro-Oncology*, vol. 13, no. 1, pp. 132–142, 2011.

[7] S. P. Kerkar and N. P. Restifo, "Cellular constituents of immune escape within the tumor microenvironment," *Cancer Research*, vol. 72, no. 13, pp. 3125–3130, 2012.

[8] K. Dietrich and M. Theobald, "Immunologische tumor therapie," *Der Internist*, vol. 56, no. 8, pp. 907–916, 2015.

[9] C. L.-L. Chiang, D. A. Maier, L. E. Kandalaft et al., "Optimizing parameters for clinical-scale production of high IL-12 secreting dendritic cells pulsed with oxidized whole tumor cell lysate," *Journal of Translational Medicine*, vol. 9, no. 1, article 198, 2011.

[10] A. J. Korman, K. S. Peggs, and J. P. Allison, "Checkpoint blockade in cancer immunotherapy," *Advances in Immunology*, vol. 90, pp. 297–339, 2006.

[11] S. Senju, M. Haruta, Y. Matsunaga et al., "Characterization of dendritic cells and macrophages generated by directed differentiation from mouse induced pluripotent stem cells," *STEM CELLS*, vol. 27, no. 5, pp. 1021–1031, 2009.

[12] M. Patel and S. Yang, "Advances in reprogramming somatic cells to induced pluripotent stem cells," *Stem Cell Reviews and Reports*, vol. 6, no. 3, pp. 367–380, 2010.

[13] S. Senju, Y. Matsunaga, S. Fukushima et al., "Immunotherapy with pluripotent stem cell-derived dendritic cells," *Seminars in Immunopathology*, vol. 33, no. 6, pp. 603–612, 2011.

[14] C. Eguizabal, O. Zenarruzabeitia, J. Monge et al., "Natural killer cells for cancer immunotherapy: pluripotent stem cells-derived NK cells as an immunotherapeutic perspective," *Frontiers in Immunology*, vol. 5, article 439, 2014.

[15] C. Koba, M. Haruta, Y. Matsunaga et al., "Therapeutic effect of human iPS-cell-derived myeloid cells expressing IFN-β against peritoneally disseminated cancer in xenograft models," *PLoS ONE*, vol. 8, no. 6, Article ID e67567, 2013.

[16] C.-W. Chang, Y.-S. Lai, L. S. Lamb Jr., and T. M. Townes, "Broad T-cell receptor repertoire in T-lymphocytes derived from human induced pluripotent stem cells," *PLoS ONE*, vol. 9, no. 5, Article ID e97335, 2014.

[17] T. Morishima, K. Watanabe, A. Niwa et al., "Neutrophil differentiation from human-induced pluripotent stem cells," *Journal of Cellular Physiology*, vol. 226, no. 5, pp. 1283–1291, 2011.

[18] M. Haruta, Y. Tomita, A. Yuno et al., "TAP-deficient human iPS cell-derived myeloid cell lines as unlimited cell source for dendritic cell-like antigen-presenting cells," *Gene Therapy*, vol. 20, no. 5, pp. 504–513, 2013.

[19] F. Antignano, M. Ibaraki, C. Kim et al., "SHIP is required for dendritic cell maturation," *Journal of Immunology*, vol. 184, no. 6, pp. 2805–2813, 2010.

[20] A. Ballestrero, D. Boy, E. Moran, G. Cirmena, P. Brossart, and A. Nencioni, "Immunotherapy with dendritic cells for cancer," *Advanced Drug Delivery Reviews*, vol. 60, no. 2, pp. 173–183, 2008.

[21] Z. Su, C. Frye, K.-M. Bae, V. Kelley, and J. Vieweg, "Differentiation of human embryonic stem cells into immunostimulatory dendritic cells under feeder-free culture conditions," *Clinical Cancer Research*, vol. 14, no. 19, pp. 6207–6217, 2008.

[22] L. M. Liau, K. L. Black, N. A. Martin et al., "Treatment of a patient by vaccination with autologous dendritic cells pulsed with allogeneic major histocompatibility complex class I-matched tumor peptides. Case Report," *Neurosurgical Focus*, vol. 9, no. 6, article e8, 2000.

[23] M. Tewari, S. Sahai, R. R. Mishra, S. K. Shukla, and H. S. Shukla, "Dendritic cell therapy in advanced gastric cancer: a promising new hope?" *Surgical Oncology*, vol. 21, no. 3, pp. 164–171, 2012.

[24] H. Iwamoto, T. Ojima, M. Nakamori et al., "Cancer vaccine therapy using genetically modified induced pluripotent stem cell-derived dendritic cells expressing the TAA gene," *Gan to Kagaku Ryōhōsha*, vol. 40, no. 12, pp. 1575–1577, 2013.

[25] F. J. Hsu, C. Benike, F. Fagnoni et al., "Vaccination of patients with B-cell lymphoma using autologous antigen-pulsed dendritic cells," *Nature Medicine*, vol. 2, no. 1, pp. 52–58, 1996.

[26] M. A. Jakobsen, B. K. Møller, and S. T. Lillevang, "Serum concentration of the growth medium markedly affects monocyte-derived dendritic cells' phenotype, cytokine production profile and capacities to stimulate in MLR," *Scandinavian Journal of Immunology*, vol. 60, no. 6, pp. 584–591, 2004.

[27] Y. Matsunaga, D. Fukuma, S. Hirata et al., "Activation of antigen-specific cytotoxic T lymphocytes by β_2-microglobulin or TAP1 gene disruption and the introduction of recipient-matched MHC class I gene in allogeneic embryonic stem cell-derived dendritic cells," *The Journal of Immunology*, vol. 181, no. 9, pp. 6635–6643, 2008.

[28] P. J. Fairchild, F. A. Brook, R. L. Gardner et al., "Directed differentiation of dendritic cells from mouse embryonic stem cells," *Current Biology*, vol. 10, no. 23, pp. 1515–1518, 2000.

[29] S. Senju, M. Haruta, K. Matsumura et al., "Generation of dendritic cells and macrophages from human induced pluripotent stem cells aiming at cell therapy," *Gene Therapy*, vol. 18, no. 9, pp. 874–883, 2011.

[30] K.-D. Choi, M. A. Vodyanik, and I. I. Slukvin, "Generation of mature human myelomonocytic cells through expansion and differentiation of pluripotent stem cell–derived lin^{--}CD34$^+$CD43$^+$CD45$^+$ progenitors," *The Journal of Clinical Investigation*, vol. 119, no. 9, pp. 2818–2829, 2009.

[31] Q. Zhang, M. Fujino, S. Iwasaki et al., "Generation and characterization of regulatory dendritic cells derived from murine induced pluripotent stem cells," *Scientific Reports*, vol. 4, article 3979, 2014.

[32] S. Senju, S. Hirata, Y. Motomura et al., "Pluripotent stem cells as source of dendritic cells for immune therapy," *International Journal of Hematology*, vol. 91, no. 3, pp. 392–400, 2010.

[33] G. T.-J. Huang, "Induced pluripotent stem cells—a new foundation in medicine," *Journal of Experimental and Clinical Medicine*, vol. 2, no. 5, pp. 202–217, 2010.

[34] B. Thurner, I. Haendle, C. Röder et al., "Vaccination with Mage-3A1 peptide-pulsed nature, monocyte-derived dendritic cells expands specific cytotoxic T cells and induces regression of some metastases in advanced stage IV melanoma," *The Journal of Experimental Medicine*, vol. 190, no. 11, pp. 1669–1678, 1999.

[35] J. Banchereau, A. K. Palucka, M. Dhodapkar et al., "Immune and clinical responses in patients with metastatic melanoma to CD34$^+$ progenitor-derived dendritic cell vaccine," *Cancer Research*, vol. 61, no. 17, pp. 6451–6458, 2001.

[36] L. Fong, D. Brockstedt, C. Benike et al., "Dendritic cell-based xenoantigen vaccination for prostate cancer Immunotherapy," *Journal of Immunology*, vol. 167, no. 12, pp. 7150–7156, 2001.

[37] L. Höltl, C. Zelle-Rieser, H. Gander et al., "Immunotherapy of metastatic renal cell carcinoma with tumor lysate-pulsed autologous dendritic cells," *Clinical Cancer Research*, vol. 8, no. 11, pp. 3369–3376, 2002.

[38] S. Baek, C.-S. Kim, S.-B. Kim et al., "Combination therapy of renal cell carcinoma or breast cancer patients with dendritic cell vaccine and IL-2: results from a phase I/II trial," *Journal of Translational Medicine*, vol. 9, article 178, 2011.

[39] Y. Motomura, S. S. Senju, T. Nakatsura et al., "Embryonic stem cell-derived dendritic cells expressing glypican-3, a recently identified oncofetal antigen, induce protective immunity against highly metastatic mouse melanoma, B16-F10," *Cancer Research*, vol. 66, no. 4, pp. 2414–2422, 2006.

[40] G. Schuler, B. Schuler-Thurner, and R. M. Steinman, "The use of dendritic cells in cancer immunotherapy," *Current Opinion in Immunology*, vol. 15, no. 2, pp. 138–147, 2003.

[41] C.-Q. Bao, C. Jin, B.-H. Xu, Y.-L. Gu, J.-P. Li, and Xiao-Lu, "Vaccination with apoptosis colorectal cancer cell pulsed autologous dendritic cells in advanced colorectal cancer patients: report from a clinical observation," *African Journal of Biotechnology*, vol. 10, no. 12, pp. 2319–2327, 2011.

[42] J. Ananiev, M. V. Gulubova, and I. Manolova, "Prognostic significance of CD83 positive tumor-infiltrating dendritic cells and expression of TGF-beta 1 in human gastric cancer," *Hepato-Gastroenterology*, vol. 58, no. 110-111, pp. 1834–1840, 2011.

[43] I. Nakano and E. A. Chiocca, "Hope and challenges for dendritic cell-based vaccine therapy for glioblastoma," *World Neurosurgery*, vol. 77, no. 5-6, pp. 633–635, 2012.

[44] N. Kayagaki, N. Yamaguchi, M. Nakayama, E. Hiroshi, K. Okumura, and H. Yagita, "Type I interferons (IFNs) regulate tumor necrosis factor-related apoptosis-inducing ligand (TRAIL) expression on human T cells: a novel mechanism for the antitumor effects of type I IFNs," *Journal of Experimental Medicine*, vol. 189, no. 9, pp. 1451–1460, 1999.

[45] F. Lei, R. Haque, X. Xiong, and J. Song, "Directed differentiation of induced pluripotent stem cells towards T lymphocytes," *Journal of Visualized Experiments*, vol. 63, Article ID e3986, 2012.

[46] F. Lei, B. Zhao, R. Haque et al., "In vivo programming of tumor antigen-specific T lymphocytes from pluripotent stem cells to promote cancer immunosurveillance," *Cancer Research*, vol. 71, no. 14, pp. 4742–4747, 2011.

[47] I. Rousalova and E. Krepela, "Granzyme B-induced apoptosis in cancer cells and its regulation (review)," *International Journal of Oncology*, vol. 37, no. 6, pp. 1361–1378, 2010.

[48] B. Platzer, M. Stout, and E. Fiebiger, "Antigen cross-presentation of immune complexes," *Frontiers in Immunology*, vol. 5, article 140, 2014.

[49] S. R. Clarke, "The critical role of CD40/CD40L in the CD4-dependent generation of CD8+ T cell immunity," *Journal of Leukocyte Biology*, vol. 67, no. 5, pp. 607–614, 2000.

[50] C. Le Page, P. Génin, M. G. Baines, and J. Hiscott, "Interferon activation and innate immunity," *Reviews in Immunogenetics*, vol. 2, no. 3, pp. 374–386, 2000.

[51] B. Seliger, F. Ruiz-Cabello, and F. Garrido, "IFN inducibility of major histocompatibility antigens in tumors," *Advances in Cancer Research*, vol. 101, pp. 249–276, 2008.

[52] M. Cella, D. Scheidegger, K. Palmer-Lehmann, P. Lane, A. Lanzavecchia, and G. Alber, "Ligation of CD40 on dendritic cells triggers production of high levels of interleukin-12 and

enhances T cell stimulatory capacity: T-T help via APC activation," *The Journal of Experimental Medicine*, vol. 184, no. 2, pp. 747–752, 1996.

[53] J. M. Kirkwood, L. H. Butterfield, A. A. Tarhini, H. Zarour, P. Kalinski, and S. Ferrone, "Immunotherapy of cancer in 2012," *CA Cancer Journal for Clinicians*, vol. 62, no. 5, pp. 309–335, 2012.

[54] A. Corthay, "Does the immune system naturally protect against cancer?" *Frontiers in Immunology*, vol. 5, article 197, 2014.

[55] T. D. S. Fernandez, C. de Souza Fernandez, and A. L. Mencalha, "Human induced pluripotent stem cells from basic research to potential clinical applications in cancer," *BioMed Research International*, vol. 2013, Article ID 430290, 11 pages, 2013.

[56] P. Sachamitr, S. Hackett, and P. J. Fairchild, "Induced pluripotent stem cells: challenges and opportunities for cancer immunotherapy," *Frontiers in Immunology*, vol. 5, article 176, 2014.

[57] R. Vizcardo, K. Masuda, D. Yamada et al., "Regeneration of human tumor antigen-specific T cells from iPSCs derived from mature CD8+ T cells," *Cell Stem Cell*, vol. 12, no. 1, pp. 31–36, 2013.

[58] J. G. Crompton, M. Rao, and N. P. Restifo, "Memoirs of a reincarnated T cell," *Cell Stem Cell*, vol. 12, no. 1, pp. 6–8, 2013.

[59] R. N. La Motte-Mohs, E. Herer, and J. C. Zúñiga-Pflücker, "Induction of T-cell development from human cord blood hematopoietic stem cells by Delta-like 1 in vitro," *Blood*, vol. 105, no. 4, pp. 1431–1439, 2005.

[60] T. M. Schmitt, R. F. de Pooter, M. A. Gronski, S. K. Cho, P. S. Ohashi, and J. C. Zúñiga-Pflücker, "Induction of T cell development and establishment of T cell competence from embryonic stem cells differentiated in vitro," *Nature Immunology*, vol. 5, no. 4, pp. 410–417, 2004.

[61] F. Lei, R. Haque, L. Weiler, K. E. Vrana, and J. Song, "T lineage differentiation from induced pluripotent stem cells," *Cellular Immunology*, vol. 260, no. 1, pp. 1–5, 2009.

[62] Y.-H. Loh, S. Agarwal, I.-H. Park et al., "Generation of induced pluripotent stem cells from human blood," *Blood*, vol. 113, no. 22, pp. 5476–5479, 2009.

[63] Z. Ye, H. Zhan, P. Mali et al., "Human-induced pluripotent stem cells from blood cells of healthy donors and patients with acquired blood disorders," *Blood*, vol. 114, no. 27, pp. 5473–5480, 2009.

[64] M. E. Brown, E. Rondon, D. Rajesh et al., "Derivation of induced pluripotent stem cells from human peripheral blood T lymphocytes," *PLoS ONE*, vol. 5, no. 6, Article ID e11373, 2010.

[65] Y.-H. Loh, O. Hartung, H. Li et al., "Reprogramming of T cells from human peripheral blood," *Cell Stem Cell*, vol. 7, no. 1, pp. 15–19, 2010.

[66] T. Nishimura, S. Kaneko, A. Kawana-Tachikawa et al., "Generation of rejuvenated antigen-specific T cells by reprogramming to pluripotency and redifferentiation," *Cell Stem Cell*, vol. 12, no. 1, pp. 114–126, 2013.

[67] J. Hataye, J. J. Moon, A. Khoruts, C. Reilly, and M. K. Jenkins, "Naïve and memory CD4+ T cell survival controlled by clonal abundance," *Science*, vol. 312, no. 5770, pp. 114–116, 2006.

[68] C. S. Hinrichs, Z. A. Borman, L. Gattinoni et al., "Human effector CD8+ T cells derived from naive rather than memory subsets possess superior traits for adoptive immunotherapy," *Blood*, vol. 117, no. 3, pp. 808–814, 2011.

[69] Y.-I. Seki, J. Yang, M. Okamoto et al., "IL-7/STAT5 cytokine signaling pathway is essential but insufficient for maintenance

of naive CD4 T cell survival in peripheral lymphoid organs," *Journal of Immunology*, vol. 178, no. 1, pp. 262–270, 2007.

[70] C. Siewert, U. Lauer, S. Cording et al., "Experience-driven development: effector/memory-like $\alpha E^+ Foxp3^+$ regulatory T cells originate from both naive T cells and naturally occurring naive-like regulatory T cells," *Journal of Immunology*, vol. 180, no. 1, pp. 146–155, 2008.

[71] C. Stemberger, K. M. Huster, M. Koffler et al., "A single naive $CD8^+$ T cell precursor can develop into diverse effector and memory subsets," *Immunity*, vol. 27, no. 6, pp. 985–997, 2007.

[72] L.-X. Wang and G. E. Plautz, "Tumor-primed, in vitro-activated $CD4^+$ effector T cells establish long-term memory without exogenous cytokine support or ongoing antigen exposure," *The Journal of Immunology*, vol. 184, no. 10, pp. 5612–5618, 2010.

[73] H. Wada, S. Kojo, C. Kusama et al., "Successful differentiation to T cells, but unsuccessful B-cell generation, from B-cell-derived induced pluripotent stem cells," *International Immunology*, vol. 23, no. 1, pp. 65–74, 2011.

[74] M. Themeli, C. C. Kloss, G. Ciriello et al., "Generation of tumor-targeted human T lymphocytes from induced pluripotent stem cells for cancer therapy," *Nature Biotechnology*, vol. 31, no. 10, pp. 928–933, 2013.

[75] M. Ciofani, G. C. Knowles, D. L. Wiest, H. von Boehmer, and J. C. Zúñiga-Pflücker, "Stage-specific and differential notch dependency at the $\alpha\beta$ and $\gamma\delta$ T lineage bifurcation," *Immunity*, vol. 25, no. 1, pp. 105–116, 2006.

[76] T. M. Schmitt, M. Ciofani, H. T. Petrie, and J. C. Zúñiga-Pflücker, "Maintenance of T cell specification and differentiation requires recurrent notch receptor-ligand interactions," *Journal of Experimental Medicine*, vol. 200, no. 4, pp. 469–479, 2004.

[77] T. Zhao, Z.-N. Zhang, Z. Rong, and Y. Xu, "Immunogenicity of induced pluripotent stem cells," *Nature*, vol. 474, no. 7350, pp. 212–215, 2011.

[78] R. Araki, M. Uda, Y. Hoki et al., "Negligible immunogenicity of terminally differentiated cells derived from induced pluripotent or embryonic stem cells," *Nature*, vol. 494, no. 7435, pp. 100–104, 2013.

[79] K. Kim, A. Doi, B. Wen et al., "Epigenetic memory in induced pluripotent stem cells," *Nature*, vol. 467, no. 7313, pp. 285–290, 2010.

[80] A. Doi, I.-H. Park, B. Wen et al., "Differential methylation of tissue- and cancer-specific CpG island shores distinguishes human induced pluripotent stem cells, embryonic stem cells and fibroblasts," *Nature Genetics*, vol. 41, no. 12, pp. 1350–1353, 2009.

[81] J. M. Polo, S. Liu, M. E. Figueroa et al., "Cell type of origin influences the molecular and functional properties of mouse induced pluripotent stem cells," *Nature Biotechnology*, vol. 28, no. 8, pp. 848–855, 2010.

[82] R. Lister, M. Pelizzola, Y. S. Kida et al., "Hotspots of aberrant epigenomic reprogramming in human induced pluripotent stem cells," *Nature*, vol. 471, no. 7336, pp. 68–73, 2011.

[83] C. S. Hinrichs, R. Spolski, C. M. Paulos et al., "IL-2 and IL-21 confer opposing differentiation programs to $CD8^+$ T cells for adoptive immunotherapy," *Blood*, vol. 111, no. 11, pp. 5326–5333, 2008.

[84] A. Moroz, C. Eppolito, Q. Li, J. Tao, C. H. Clegg, and P. A. Shrikant, "IL-21 enhances and sustains $CD8^+$ T cell responses to achieve durable tumor immunity: comparative evaluation of IL-2, IL-15, and IL-21," *The Journal of Immunology*, vol. 173, no. 2, pp. 900–909, 2004.

[85] P. Neeson, A. Shin, K. M. Tainton et al., "Ex vivo culture of chimeric antigen receptor T cells generates functional $CD8^+$ T cells with effector and central memory-like phenotype," *Gene Therapy*, vol. 17, no. 9, pp. 1105–1116, 2010.

[86] K. S. Schluns and L. Lefrançois, "Cytokine control of memory T-cell development and survival," *Nature Reviews Immunology*, vol. 3, no. 4, pp. 269–279, 2003.

[87] R. Zeng, R. Spolski, S. E. Finkelstein et al., "Synergy of IL-21 and IL-15 in regulating $CD8^+$ T cell expansion and function," *The Journal of Experimental Medicine*, vol. 201, no. 1, pp. 139–148, 2005.

[88] H. Singh, M. J. Figliola, M. J. Dawson et al., "Reprogramming CD19-specific T cells with IL-21 signaling can improve adoptive immunotherapy of B-lineage malignancies," *Cancer Research*, vol. 71, no. 10, pp. 3516–3527, 2011.

[89] L. Zhang, X. Song, Y. Mohri, and L. Qiao, "Role of inflammation and tumor microenvironment in the development of gastrointestinal cancers: what induced pluripotent stem cells can do?" *Current Stem Cell Research & Therapy*, vol. 10, no. 3, pp. 245–250, 2015.

[90] E. Haga, Y. Endo, M. Haruta et al., "Therapy of peritoneally disseminated colon cancer by TAP-Deficient embryonic stem cell-derived macrophages in allogeneic recipients," *Journal of Immunology*, vol. 193, no. 4, pp. 2024–2033, 2014.

[91] A. M. Bock, D. Knorr, and D. S. Kaufman, "Development, expansion, and in vivo monitoring of human NK cells from human embryonic stem cells (hESCs) and and induced pluripotent stem cells (iPSCs)," *Journal of Visualized Experiments*, no. 74, Article ID e50337, 2013.

[92] A. Larbi, J.-M. Gombert, C. Auvray et al., "The HOXB4 homeoprotein promotes the ex vivo enrichment of functional human embryonic stem cell-derived NK cells," *PLoS ONE*, vol. 7, no. 6, Article ID e39514, 2012.

[93] S. Merchav, A. Nagler, E. Sahar, and I. Tatarsky, "Production of human pluripotent progenitor cell colony stimulating activity (CFU-GEMMCSA) in patients with myelodysplastic syndromes," *Leukemia Research*, vol. 11, no. 3, pp. 273–279, 1987.

[94] C. E. Lewis and J. W. Pollard, "Distinct role of macrophages in different tumor microenvironments," *Cancer Research*, vol. 66, no. 2, pp. 605–612, 2006.

17

Procaine Induces Epigenetic Changes in HCT116 Colon Cancer Cells

Hussein Sabit,[1] **Mariam B. Samy,**[1] **Osama A. M. Said,**[1] **and Mokhtar M. El-Zawahri**[1,2]

[1]*College of Biotechnology, Misr University for Science and Technology, Giza, Egypt*
[2]*Center of Research and Development, Misr University for Science and Technology, Giza, Egypt*

Correspondence should be addressed to Hussein Sabit; hussein.sabit@must.edu.eg

Academic Editor: Norman A. Doggett

Colon cancer is the third most commonly diagnosed cancer in the world, and it is the major cause of morbidity and mortality throughout the world. The present study aimed at treating colon cancer cell line (HCT116) with different chemotherapeutic drug/drug combinations (procaine, vorinostat "SAHA," sodium phenylbutyrate, erlotinib, and carboplatin). Two different final concentrations were applied: $3 \mu M$ and $5 \mu M$. Trypan blue test was performed to assess the viability of the cell before and after being treated with the drugs. The data obtained showed that there was a significant decrease in the viability of cells after applying the chemotherapeutic drugs/drug combinations. Also, DNA fragmentation assay was carried out to study the effect of these drugs on the activation of apoptosis-mediated DNA degradation process. The results indicated that all the drugs/drug combinations had a severe effect on inducing DNA fragmentation. Global DNA methylation quantification was performed to identify the role of these drugs individually or in combination in hypo- or hypermethylating the CpG dinucleotide all over the genome of the HCT116 colon cancer cell line. Data obtained indicated that different combinations had different effects in reducing or increasing the level of methylation, which might indicate the effectiveness of combining drugs in treating colon cancer cells.

1. Introduction

Cancer, the uncontrolled cell growth, is one of the most fatal diseases worldwide [1]. One of the most widespread and common types of cancer is colorectal cancer (CRC), which represents the third most common cancer after lung and breast cancers, and it is considered the second most common cause of cancer death [2–4].

It is well known that epigenetic alterations, particularly in the disease-related genes, are associated with various disorders including many cancer types [5]. Colorectal carcinoma is one of those diseases in which epigenetic inactivation of multiple tumor suppressor genes plays a crucial role in the tumorigenesis process [6].

The role of DNA methylation in the organization of the cancer epigenetic profile is still unclear. However, several studies have been conducted on HCT116 colon cancer cells to elucidate the landscape of DNA methylation [7, 8].

Chemotherapy is considered one of the effective therapeutic ways to control several types of cancer, although the standard chemotherapy plans often have limited survival benefits due to its severe cellular toxicity and the inability to target only the malignant cells [9]. Unfortunately, this off-targeting highlights the main concern of using chemotherapy [10].

Several literatures have focused on the application of a combination of chemotherapeutic drugs to treat cancer [11–14].

Procaine is one of the conventional chemotherapeutic agents, which was used also as a local anesthetic drug in surgeries. It showed an epigenetic mode of action, as a demethylating agent for the hypermethylated CpG island of DNA, and hence, it became one of the choices in treating different types of cancers [15].

Here, the main aim of the present study was to identify the role of procaine (as a representative of DNMT inhibitor drugs) combined with other chemotherapeutic drugs such as carboplatin (as a representative of DNMT inhibitor drugs), erlotinib, sodium phenylbutyrate, and vorinostat (as representatives of HDAC inhibitor drugs) in demethylating the whole genome of the HCT116 colon cancer cells.

2. Materials and Methods

2.1. Cell Line Maintenance. Colon cancer cell line (HCT116) was purchased from the Holding Company for Vaccines and Biological Products (VACSERA), Cairo, Egypt. Cells were cultured in RPMI-1640 media supplemented with 10% fetal bovine serum (FBS) and 1% antibiotic mix (ampicillin/streptomycin). Cells were maintained under the normal laboratory conditions, that is, 5% CO_2 at 37°C.

2.2. Cell Viability Test. Trypan blue is a vital stain that is used to selectively color dead cells blue, while leaving live cells with intact cell membranes not colored. It was conducted to assess the number of cells before and after treatment with the chemotherapeutic drugs. The test is straightforward. Briefly, cell suspension was diluted with equal volume (1:1) of the dye and left for 3 min and then loaded to the hemocytometer slide. Cells were counted under inverted microscope as the bright cells were considered viable while the blue ones were considered dead. The total number of viable cells was calculated using the following equations:

The total number of viable cells

$$= \text{Average number of viable cells} \times \text{dilution factor} \quad (1)$$
$$\times 10^4.$$

2.3. Chemotherapy Drugs. Five chemotherapeutic drugs, procaine, carboplatin, vorinostat, sodium phenylbutyrate, and erlotinib, were used. All the drugs were purchased from Santa Cruz Biotechnology, USA. These drugs represent two groups of chemotherapeutic drugs: HDAC inhibitor and DNMT inhibitor.

2.4. Drug Preparation and Application. Five micrograms of each drug was dissolved into 5 mL of injection water to prepare the stock solutions 1 μg/mL. Final concentration of 3 μM and 5 μM was prepared and applied to the HCT116 cells cultured in a 6-well plate. Two plates were used, one for each concentration. The 6-well plate layouts are presented in Tables 1 and 2.

2.5. Cell Harvesting. Cells were harvested for the downstream analysis after 3 days of incubation with drugs. Briefly, old media were decanted and the cells were trypsinized for 2 min and then collected *via* low speed centrifugation (200 rpm for 10 min). Cell viability was assessed also after treatment.

2.6. DNA Extraction. Total DNA from all samples was extracted using G-Spin™ Total DNA Extraction Kit (Boca Scientific, USA). The extracted DNA was used in both DNA degradation assay and methylation quantification.

TABLE 1: The 6-well plate layout for the concentration of 3 μM.

Control	2 μL P + 2498 μL cell suspension*	2 μL P + 2 μL V + 2496 μL cell suspension*
2 μL P + 2 μL S + 2496 μL cell suspension*	2 μL P + 3 μL E + 2495 μL cell suspension*	2 μL P + 5 μL C + 2493 μL cell suspension*

P: procaine, V: vorinostat, S: sodium phenylbutyrate, E: erlotinib, and C: carboplatin.
*Cell suspension was 10^6 cells per mL.

TABLE 2: The 6-well plate layout for the concentration of 5 μM.

Control	3 μL P + 2497 μL cell suspension*	3 μL P + 4 μL V + 2493 μL cell suspension*
3 μL P + 3 μL S + 2494 μL cell suspension*	3 μL P + 5 μL E + 2492 μL cell suspension*	3 μL P + 8 μL C + 2489 μL cell suspension*

P: procaine, V: vorinostat, S: sodium phenylbutyrate, E: erlotinib, and C: carboplatin.
*Cell suspension was 10^6 cells per mL.

2.7. DNA Degradation Assay. The extracted DNA from all samples was subjected to electrophoresis by loading a suitable volume on 1.2% agarose gel. Initial voltage (15 volts) for 5 min was applied and then the run was continued at 120 volts for 30 min. Gels were visualized and photographed under UV transilluminator after being stained with ethidium bromide.

2.8. Quantification of DNA Methylation. The extracted DNA from each sample was used to quantify the global DNA methylation using Global DNA Methylation ELISA Kit (Cell Biolabs Inc., USA). Briefly, a standard curve was initially generated and the kit's instruction was followed. The OD was read at 450 nm using plate reader.

2.9. Statistical Analysis. All statistical analyses were performed with SAS statistical software (SAS Institute, Cary, NC). Data were analyzed using the 2-factor repeated measures with interaction of analysis of variance (ANOVA) general linear models (GLM) procedure. Values were given as mean ± SD and differences among means were separated by Duncan's multiple range tests. A *P* value of 0.01 was considered significant. Correlation between variables was performed using Pearson correlation coefficient analysis.

3. Results

3.1. Cell Viability after Treatment. Trypan blue assay was performed to assess the cell viability after treatment with the chemotherapeutic drug/drugs combinations as it can stain the dead cells while leaving the viable cells unstained. Results obtained showed that there was a significant decrease in the cell viability after being treated with different concentrations/combinations of the chemotherapeutics under study (Figures 1 and 2 and Table 3) ($P < 0.01$). Meanwhile, the most effective drug/combination was procaine combined with

TABLE 3: The mean values of Duncan's multiple range test for cell viability of control and treated cells.

Treat./Conc.	N	Mean	Duncan grouping			
Control	4	350875	A			
P + S/3 μM	4	200000		B		
P + E/5 μM	4	124875			C	
P + V/3 μM	4	112500			C	D
P + V/5 μM	4	112500			C	D
P + E/3 μM	4	75000			C	D	E
P/5 μM	4	75000			C	D	E
P + C/3 μM	4	62575				D	E
P/3 μM	4	62500				D	E
P + C/5 μM	4	62500				D	E
P + S/5 μM	4	37500					E

P: procaine, V: vorinostat, S: sodium phenylbutyrate, E: erlotinib, and C: carboplatin.

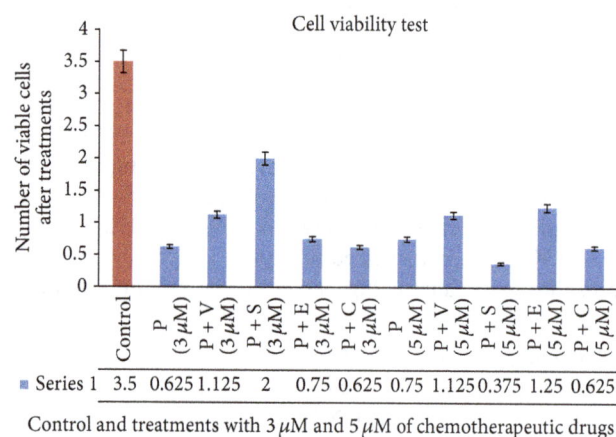

FIGURE 1: The effect of different drug combinations/concentration on the viability of HCT116 colon cancer cells compared to control. P: procaine, V: vorinostat, S: sodium phenylbutyrate, E: erlotinib, and C: carboplatin. Numbers are multiplied with 10^5.

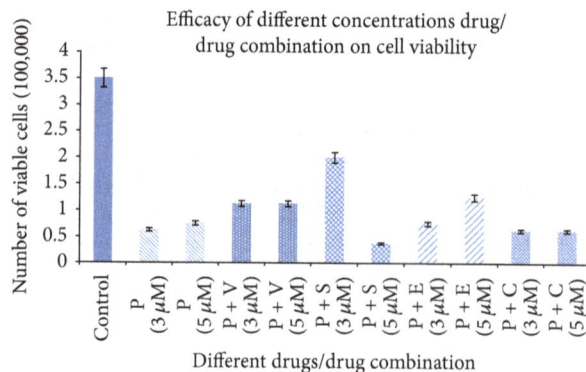

FIGURE 2: Drug combinations/concentration efficacy on the viability of HCT116 colon cancer cells, organized in drug-wise. P: procaine, V: vorinostat, S: sodium phenylbutyrate, E: erlotinib, and C: carboplatin. Numbers are multiplied with 10^5.

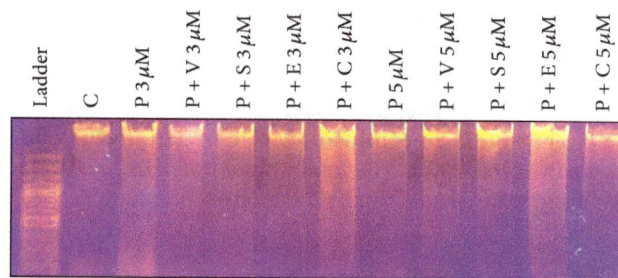

FIGURE 3: DNA degradation assay of treated and nontreated HCT116 colon cancer cells. P: procaine, V: vorinostat, S: sodium phenylbutyrate, E: erlotinib, and C: carboplatin.

sodium phenylbutyrate at a concentration of 5 μM (37,500 viable cells). While the same combination had less effect on malignant cell viability (200,000 viable cells) when applied at a lower concentration (3 μM). This might indicate the efficacy of the higher doses of the combined chemotherapeutic drugs, despite the profile obtained with procaine combined with erlotinib as the low concentration of this combination gave better efficacy. Figure 2 highlights a noticeable pattern as there was no significant variation of the number of viable cells when changing the drug/drug combination concentration. This pattern has been shown when using procaine, procaine combined with vorinostat, and procaine combined with carboplatin.

3.2. DNA Degradation Assay.
Chemotherapeutic drugs' effect can be studied by studying DNA degradation [16]. In the present study, DNA fragmentation was assessed in the HCT116 colon cancer cells after being treated with different drugs/drug combinations. Data obtained (Figure 3) indicated the severe damage in cellular DNA of all cells regardless of the drug/drug combination. However, the most effective drug/drug combination in inducing DNA fragmentation was procaine alone (3 μM) followed by procaine combined with both carboplatin (3 μM) and erlotinib (5 μM).

3.3. Quantification of DNA Methylation.
In colon cancer, epigenetic changes, such as promoter CpG island hypermethylation, resulted from the expression of DNMT. Promoter methylation occurs more frequently than genetic mutations [17, 18], so quantification of the global methylome in colon cancer cells might provide insights about the epigenetic changes, particularly after being treated with chemotherapy. In the present study the global methylation pattern was identified to evaluate the role of procaine associated with other drugs in changing the methylation profile in colon cancer cells (Figures 4 and 5 and Table 4). Results obtained indicated that procaine alone at higher concentration (5 μM) and procaine combined with carboplatin in the lower concentration (3 μM) were the most efficient drug/drug combination in demethylating the whole genome of colon cancer cells compared to control ($P < 0.01$). However, the same combination in the higher concentration (5 μM) was able to hypermethylate the whole genome of the cells compared to control. Meanwhile procaine combined with the other

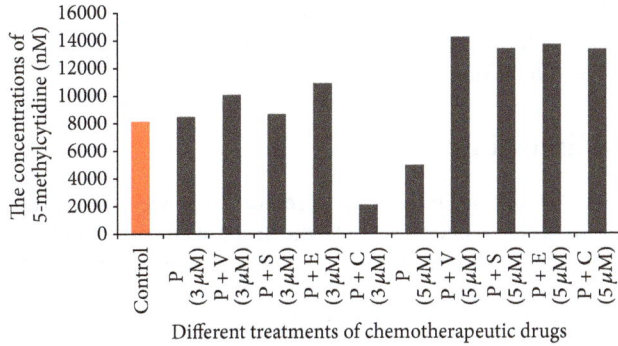

FIGURE 4: The concentration of 5-methylcytidine in all cells after being treated with different chemotherapeutic drugs compared to control. P: procaine, V: vorinostat, S: sodium phenylbutyrate, E: erlotinib, and C: carboplatin.

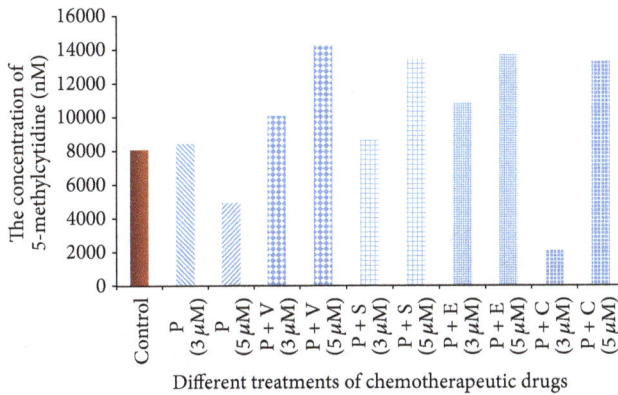

FIGURE 5: The concentration of 5-methylcytidine of treated and nontreated HCT116 colon cancer cells, organized drug-wise. P: procaine, V: vorinostat, S: sodium phenylbutyrate, E: erlotinib, and C: carboplatin.

TABLE 4: Mean values of Duncan's multiple range test for quantification of global DNA methylation.

Treat./Conc.	N	Mean	Duncan grouping		
P + V/5 μM	4	14100	A		
P + E/5 μM	4	14000	A		
P + S/5 μM	4	13900	A		
P + C/5 μM	4	13900	A		
P + E/3 μM	4	10400	A	B	
P + V/3 μM	4	10000	A	B	
P + S/3 μM	4	8300		B	
P/3 μM	4	8200		B	
Control	4	8000		B	
P/5 μM	4	5100		B	C
P + C/3 μM	4	2200			C

P: procaine, V: vorinostat, S: sodium phenylbutyrate, E: erlotinib, and C: carboplatin.

4. Discussion

4.1. Cell Viability. One of the most direct tools to identify the effect of any treatment on malignant cells is assessing the cell viability [19]. The significant decrease of the number of viable cells compared to control could be attributed to the activation of extrinsic apoptotic pathway through which the procaine combined with sodium phenylbutyrate demethylated some apoptosis related gene, which, in turn, activated the apoptosis machinery and then cell death [20, 21].

Several studies have indicated the efficacy of chemotherapeutic drugs/drug combinations in reducing malignant cell counts [22, 23]. Procaine was the main drug applied to the HCT116 colon cancer cells. When it was applied solely (either in 3 or 5 μM), a significant reduction in the cell count was obtained. This might indicate that procaine was able to enforce the cells to commit apoptosis. Other research groups have indicated the same profile [24–26]. A reduction in the cell count has also been obtained when procaine was combined with vorinostat in both concentrations, and similar profiles were obtained by other research groups [25, 27]. Meanwhile, in this combination, the dose has no significant effect, which might indicate the differences in the role of both procaine and vorinostat. When combined with sodium phenylbutyrate, procaine in both concentrations exhibited an ability to significantly reduce the cell count which was lower than using procaine solely. This might highlight the antagonistic effect of sodium phenylbutyrate when combined with procaine, in which the former drug inhibited procaine from performing its action [28–31].

However, when procaine was combined with either erlotinib or carboplatin, a significant reduction in the cells count was obtained. This data might also show that erlotinib and carboplatin have also a differed mode of action compared to procaine. Several studies have identified similar outcomes [32–35].

drugs, that is, vorinostat (HDAC inhibitor), sodium phenylbutyrate (HDAC inhibitor), and erlotinib (HDAC inhibitor), in higher concentration (5 μM) was also able to increase the methylation level significantly compared to control. This might indicate that combining procaine with these drugs might antagonize its DNMT inhibitory effect and lead to promotion of the cell proliferation and hence increase the total methylation amount.

However, data showed (Figure 5) a general trend of a correlation between the methylation level and the drug doses. The lower doses (3 μM) of procaine combined with vorinostat, sodium phenylbutyrate, erlotinib, and carboplatin were more effective in reducing the methylation level of the HCT116 colon cancer cells compared to the high doses (5 μM) of the same combinations. The only exception appears when procaine was used solely as the high dose was more effective than the low dose in reducing the methylation level. The present study, therefore, indicates the effectiveness of using lower doses of the specified chemotherapy.

4.2. DNA Degradation Assay. Several mechanisms could explain the DNA degradation in cells treated with chemotherapeutic drugs. One of these mechanisms postulates that demethylating agent such as procaine could activate tumor suppressor gene(s), which, in turn, enforce the cells to commit apoptosis [36, 37]. Other mechanisms suggest the activation of caspase-activated DNase (CAD), which degrades DNA after a cascade of activation processes [38]. In the present investigation all drug combinations applied to HCT116 colon cancer cells have resulted in a degradation pattern that might indicate the occurrence of drugs-induced apoptosis [27, 39, 40].

4.3. Quantification of DNA Methylation. DNA methylation quantification is considered one of the most widespread tools to assess the nonsequence dependent gene regulation [41]. In the present study, global methylation was quantified in the treated and untreated HCT116 colon cancer cells. Results obtained indicated that procaine applied solely ($3\,\mu$M) has increased the methylation level compared to the control untreated cells, while the same compound in the higher dose ($5\,\mu$M) has resulted in decreased methylation level compared to control. This data might indicate that procaine applied solely activated DNMT and hence increased the methylation level. This data was also noticed by several research groups [42–44]. Meanwhile, when procaine was combined with vorinostat or sodium phenylbutyrate or erlotinib, a hypermethylation pattern was obtained in the lower concentration ($3\,\mu$M), while the higher concentration ($5\,\mu$M) gave a significant increase in the methylation level compared either to control or to the lower concentration. This data might indicate that combining procaine with either vorinostat or sodium phenylbutyrate or erlotinib has a synergistic effect on activating DNMT. This profile was observed in several previously published researches [17, 39, 45, 46]. However, when procaine was combined with carboplatin, the lower dose ($3\,\mu$M) has resulted in hypomethylating the whole genome of HCT116 colon cancer cells, and this might indicate the activity of this combination in inactivating DNMT [32, 42], while the higher dose has resulted in hypermethylation of the whole genome compared to control [47, 48].

5. Conclusion

Colon cancer is one of the leading causes of death worldwide. Therefore, the seeking for effective chemotherapeutic drugs or drug combinations is still a big demand. In the present study, HCT116 colon cancer cell line was treated with two concentrations ($3\,\mu$M and $5\,\mu$M) of procaine solely or combined with other drugs aiming to control the disease. Global DNA methylation and the cell viability were assessed. Data showed that using procaine combined with carboplatin in low dose ($3\,\mu$M) was the most effective treatment that was capable of reducing the level of global methylation. On the other hand, data indicated that applying higher doses of the drugs under study has resulted in promoting the cell proliferation and hence the methylation amount. Meanwhile, using the lower doses of the specified drugs was more effective in controlling

colon cancer cells. However, further analysis is required to elucidate the mechanism by which the higher doses promoted the proliferation of HCT116 colon cancer cells.

Additional Points

Study Limitation. Here in the present study, we assessed the role of different chemotherapeutic drugs in altering the global methylation pattern of HCT116 colon cancer cells. Despite the obtained data, this study needs more conformational investigation to deeply assess the mode of action of each combination applied. Meanwhile, the drug combinations used in this study are not intended for clinical uses.

Competing Interests

The authors declare no conflict of interests.

References

[1] M. López-Gómez, E. Malmierca, M. de Górgolas, and E. Casado, "Cancer in developing countries: the next most preventable pandemic. The global problem of cancer," *Critical Reviews in Oncology/Hematology*, vol. 88, no. 1, pp. 117–122, 2013.

[2] K. Esposito, P. Chiodini, A. Capuano et al., "Colorectal cancer association with metabolic syndrome and its components: a systematic review with meta-analysis," *Endocrine*, vol. 44, no. 3, pp. 634–647, 2012.

[3] A. Gado, B. Ebeid, A. Abdelmohsen, and A. Axon, "Colorectal cancer in Egypt is commoner in young people: is this cause for alarm?" *Alexandria Journal of Medicine*, vol. 50, no. 3, pp. 197–201, 2014.

[4] W. Zheng, L. Zhao, G. Wang et al., "Promoter methylation and expression of RASSF1A genes as predictors of disease progression in colorectal cancer," *International Journal of Clinical and Experimental Medicine*, vol. 9, no. 2, pp. 2027–2036, 2016.

[5] R. Cacabelos, "Epigenetic biomarkers in cancer," *Clinical & Medical Biochemistry*, vol. 1, no. 1, article e101, 2015.

[6] E. Sakai, A. Nakajima, and A. Kaneda, "Accumulation of aberrant DNA methylation during colorectal cancer development," *World Journal of Gastroenterology*, vol. 20, no. 4, pp. 978–987, 2014.

[7] C. Michailidi, S. Theocharis, G. Tsourouflis et al., "Expression and promoter methylation status of hMLH1, MGMT, APC, and CDH1 genes in patients with colon adenocarcinoma," *Experimental Biology and Medicine*, vol. 240, no. 12, pp. 1599–1605, 2015.

[8] P. A. Jones, "How DNA methylation organizes the cancer epigenome," in *Proceedings of the AACR Special Conference on Chromatin and Epigenetics in Cancer*, Atlanta, Ga, USA, September 2015.

[9] S. Yin, W. Wei, F. Jian, and N. Yang, "Therapeutic applications of herbal medicines for cancer patients," *Evidence-Based Complementary and Alternative Medicine*, vol. 2013, 15 pages, 2013.

[10] S. P. Kumar and V. Sisodia, "Chemotherapy-induced or chemotherapy-associated? Does physical therapy play a role in prevention and/or management of peripheral neurotoxicity and neuropathy?" *Indian Journal of Palliative Care*, vol. 19, no. 1, pp. 77–78, 2013.

[11] J. Durant, P. Clevenbergh, P. Halfon et al., "Drug-resistance genotyping in HIV-1 therapy: the VIRADAPT randomised controlled trial," *Lancet*, vol. 353, no. 9171, pp. 2195–2199, 1999.

[12] P. E. A. da Silva and J. C. Palomino, "Molecular basis and mechanisms of drug resistance in *Mycobacterium tuberculosis*: classical and new drugs," *Journal of Antimicrobial Chemotherapy*, vol. 66, no. 7, Article ID dkr173, pp. 1417–1430, 2011.

[13] P. Borst, J. Jonkers, and S. Rottenberg, "What makes tumors multidrug resistant?" *Cell Cycle*, vol. 6, no. 22, pp. 2782–2787, 2007.

[14] L. Xie, L. Xie, S. L. Kinnings, and P. E. Bourne, "Novel computational approaches to polypharmacology as a means to define responses to individual drugs," *Annual Review of Pharmacology and Toxicology*, vol. 52, pp. 361–379, 2012.

[15] S. Dhivya, N. Khandelwal, S. K. Abraham, and K. Premkumar, "Impact of anthocyanidins on mitoxantrone-induced cytotoxicity and genotoxicity: an in vitro and in vivo analysis," *Integrative Cancer Therapies*, 2016.

[16] Y. R. Saadat, N. Saeidi, S. Z. Vahed, A. Barzegari, and J. Barar, "An update to DNA ladder assay for apoptosis detection," *BioImpacts*, vol. 5, no. 1, pp. 25–28, 2015.

[17] S. Flis, A. Gnyszka, and K. Flis, "DNA methyltransferase inhibitors improve the effect of chemotherapeutic agents in SW48 and HT-29 colorectal cancer cells," *PLoS ONE*, vol. 9, no. 3, Article ID e92305, 2014.

[18] J. M.-K. Ng and J. Yu, "Promoter hypermethylation of tumour suppressor genes as potential biomarkers in colorectal cancer," *International Journal of Molecular Sciences*, vol. 16, no. 2, pp. 2472–2496, 2015.

[19] G. Rivaa, S. Baronchellia, L. Paolettaa et al., "In vitro anticancer drug test: a new method emerges from the model of glioma stem cells," *Toxicology Reports 1*, 2014.

[20] M. H. Lee, J. Y. Yang, Y. Cho et al., "Menadione induces apoptosis in a gastric cancer cell line mediated by down-regulation of X-linked inhibitor of apoptosis," *International Journal of Clinical and Experimental Medicine*, vol. 9, no. 2, pp. 2437–2443, 2016.

[21] J. Lin, H.-J. Yao, and R.-Y. Li, "Bakuchiol inhibits cell proliferation and induces apoptosis and cell cycle arrest in SGC-7901 human gastric cancer cells," *Biomedical Research*, vol. 27, no. 1, pp. 181–185, 2016.

[22] N. Andre and W. Schmiegel, "Chemoradiotherapy for colorectal cancer," *Gut*, vol. 54, no. 8, pp. 1194–1202, 2005.

[23] J. Nautiyal, S. S. Kanwar, Y. Yu, and A. P. N. Majumdar, "Combination of dasatinib and curcumin eliminates chemo-resistant colon cancer cells," *Journal of Molecular Signaling*, vol. 6, article 7, 2011.

[24] A. Villar-Garea, M. F. Fraga, J. Espada, and M. Esteller, "Procaine is a DNA-demethylating agent with growth-inhibitory effects in human cancer cells," *Cancer Research*, vol. 63, no. 16, pp. 4984–4989, 2003.

[25] B. Brueckner, R. G. Boy, P. Siedlecki et al., "Epigenetic reactivation of tumor suppressor genes by a novel small-molecule inhibitor of human DNA methyltransferases," *Cancer Research*, vol. 65, no. 14, pp. 6305–6311, 2005.

[26] C. Stresemann, B. Brueckner, T. Musch, H. Stopper, and F. Lyko, "Functional diversity of DNA methyltransferase inhibitors in human cancer cell lines," *Cancer Research*, vol. 66, no. 5, pp. 2794–2800, 2006.

[27] M. Sachan and M. Kaur, "Epigenetic modifications: therapeutic potential in cancer," *Brazilian Archives of Biology and Technology*, vol. 58, no. 4, pp. 526–539, 2015.

[28] M. Rodríguez-Paredes and M. Esteller, "Cancer epigenetics reaches mainstream oncology," *Nature Medicine*, vol. 17, no. 3, pp. 330–339, 2011.

[29] Y. K. Walia and V. Sharma, "Role of HDACs and DNMTs in cancer therapy: a review," *Asian Journal of Advanced Basic Sciences*, vol. 1, no. 1, pp. 62–78, 2013.

[30] M. Ushijima, Y. Ogata, H. Tsuda, Y. Akagi, K. Matono, and K. Shirouzu, "Demethylation effect of the antineoplaston AS2-1 on genes in colon cancer cells," *Oncology Reports*, vol. 31, no. 1, pp. 19–26, 2014.

[31] S. R. Burzynski, T. J. Janicki, G. S. Burzynski, and S. Brookman, "Preliminary findings on the use of targeted therapy in combination with sodium phenylbutyrate in colorectal cancer after failure of second-line therapy—a potential strategy for improved survival," *Journal of Cancer Therapy*, vol. 5, no. 13, pp. 1270–1288, 2014.

[32] S. Amatori, I. Bagaloni, B. Donati, and M. Fanelli, "DNA demethylating antineoplastic strategies: a comparative point of view," *Genes and Cancer*, vol. 1, no. 3, pp. 197–209, 2010.

[33] M. S. Kim, J. Lee, and D. Sidransky, "DNA methylation markers in colorectal cancer," *Cancer and Metastasis Reviews*, vol. 29, no. 1, pp. 181–206, 2010.

[34] C. A. Townsley, P. Major, L. L. Siu et al., "Phase II study of erlotinib (OSI-774) in patients with metastatic colorectal cancer," *British Journal of Cancer*, vol. 94, no. 8, pp. 1136–1143, 2006.

[35] S. M. Mitchell, J. P. Ross, H. R. Drew et al., "A panel of genes methylated with high frequency in colorectal cancer," *BMC Cancer*, vol. 14, no. 1, article 54, 2014.

[36] D. Lai, S. Visser-Grieve, and X. Yang, "Tumour suppressor genes in chemotherapeutic drug response," *Bioscience Reports*, vol. 32, no. 4, pp. 361–374, 2012.

[37] J.-H. Xu, S.-L. Hu, G.-D. Shen, and G. Shen, "Tumor suppressor genes and their underlying interactions in paclitaxel resistance in cancer therapy," *Cancer Cell International*, vol. 16, article 13, 2016.

[38] L. Gao, K. Huang, D.-S. Jiang et al., "Novel role for caspase-activated DNase in the regulation of pathological cardiac hypertrophy," *Hypertension*, vol. 65, no. 4, pp. 871–881, 2015.

[39] A. Nebbiosoa, V. Carafaa, R. Benedettia, and L. Altuccia, "Trials with 'epigenetic' drugs: an update," *Molecular Oncology*, vol. 6, no. 6, pp. 657–682, 2012.

[40] S. Witta, "Histone deacetylase inhibitors in non-small-cell lung cancer," *Journal of Thoracic Oncology*, vol. 7, no. 12, supplement 5, pp. S404–S406, 2012.

[41] S. D. Fouse, R. P. Nagarajan, and J. F. Costello, "Genome-scale DNA methylation analysis," *Epigenomics*, vol. 2, no. 1, pp. 105–117, 2010.

[42] F. Lyko and R. Brown, "DNA methyltransferase inhibitors and the development of epigenetic cancer therapies," *Journal of the National Cancer Institute*, vol. 97, no. 20, pp. 1498–1506, 2005.

[43] A. A. Johnson, K. Akman, S. R. G. Calimport, D. Wuttke, A. Stolzing, and J. P. de Magalhães, "The role of DNA methylation in aging, rejuvenation, and age-related disease," *Rejuvenation Research*, vol. 15, no. 5, pp. 483–494, 2012.

[44] Y. Delpu, P. Cordelier, W. C. Cho, and J. Torrisani, "DNA methylation and cancer diagnosis," *International Journal of Molecular Sciences*, vol. 14, no. 7, pp. 15029–15058, 2013.

[45] K. A. Kang, M. J. Piao, K. C. Kim et al., "Epigenetic modification of Nrf2 in 5-fluorouracil-resistant colon cancer cells: involvement of TET-dependent DNA demethylation," *Cell Death and Disease*, vol. 5, no. 4, Article ID e1183, 2014.

[46] P. Wongtrakoongate, "Epigenetic therapy of cancer stem and progenitor cells by targeting DNA methylation machineries," *World Journal of Stem Cells*, vol. 7, no. 1, pp. 137–148, 2015.

[47] J. Ren, B. N. Singh, Q. Huang et al., "DNA hypermethylation as a chemotherapy target," *Cellular Signalling*, vol. 23, no. 7, pp. 1082–1093, 2011.

[48] D. Subramaniam, R. Thombre, A. Dhar, and S. Anant, "DNA methyltransferases: a novel target for prevention and therapy," *Frontiers in Oncology*, vol. 4, article 80, 2014.

18

Molecular Characterization of a Novel Germline *VHL* Mutation by Extensive *In Silico* Analysis in an Indian Family with Von Hippel-Lindau Disease

Gautham Arunachal,[1] **Divya Pachat,**[1] **C. George Priya Doss,**[2] **Sumita Danda,**[1] **Rekha Pai,**[3] **and Andrew Ebenazer**[3]

[1]*Department of Clinical Genetics, Christian Medical College, Vellore, Tamil Nadu 632004, India*
[2]*School of Biosciences and Technology, VIT University, Vellore, Tamil Nadu 632014, India*
[3]*Department of Pathology, Christian Medical College, Vellore, Tamil Nadu 632004, India*

Correspondence should be addressed to Gautham Arunachal; gautham@cmcvellore.ac.in

Academic Editor: Bernard Weissman

Von Hippel-Lindau [VHL] disease, an autosomal dominant hereditary cancer syndrome, is well known for its complex genotype-phenotype correlations. We looked for germline mutations in the *VHL* gene in an affected multiplex family with Type 1 VHL disease. Real-Time quantitative PCR for deletions and Sanger sequencing of coding regions along with flanking intronic regions were performed in two affected individuals and one related individual. Direct sequencing identified a novel heterozygous single nucleotide base substitution in both the affected members tested, segregating with VHL phenotype in this family. This variant in exon 3, c.473T>A, results in substitution of leucine, a highly conserved acid, to glutamine at position 158 [p.L158Q] and has not been reported thus far as a variant associated with disease causation. Further, this variant was not observed in 50 age and ethnicity matched healthy individuals. Extensive *in silico* prediction analysis along with molecular dynamics simulation revealed significant deleterious nature of the substitution L158Q on pVHL. The results of this study when collated support the view that the missense variation p.L158Q in the Elongin C binding domain of pVHL may be disease causing.

1. Introduction

Von Hippel-Lindau [VHL] syndrome is an autosomal dominantly inherited cancer syndrome with an incidence of 1 in 36,000 live births [1]. VHL syndrome is characterized by multiple tumor types affecting different organ systems with an age dependent penetrance pattern. Among the most frequent tumors are hemangioblastomas [HB] of the CNS and retina, clear cell carcinoma of kidneys [RCC], pheochromocytoma, endolymphatic sac tumors [ELSTs], pancreatic neuroendocrine tumors [NETs], and paragangliomas. Also, cysts of the pancreas, renal, epididymal, and broad ligament are seen in varying proportions.

The disease not only is known for its phenotypic heterogeneity, but also demonstrates significant interfamilial and intrafamilial variations which are partly explained by well known complex genotype-phenotype correlations. *VHL*,

a tumor suppressor, is the only gene in which heterozygous germline mutations are known to cause VHL disease [2]. The evidence of biallelic inactivation of *VHL* found in tumor samples is consistent with Knudson's "two-hit model" where a mutation in one of the alleles is inherited, and the wild-type allele is somatically inactivated leading to tumorigenesis.

All types of mutations including deletions have been described in the literature. Over 900 nonrecurrent mutations have been described so far, although mutational hot spots and some founder mutations in certain ethnic groups have also been observed [2]. With the exception of deletions which account for 30–40% of cases, most mutations cluster in two regions of high functional importance in VHL protein [pVHL], the Elongin C binding site [amino acid residues 157–171] and HIF1 alpha [amino acid residues 91–113], which are highly evolutionarily conserved regions [3, 4].

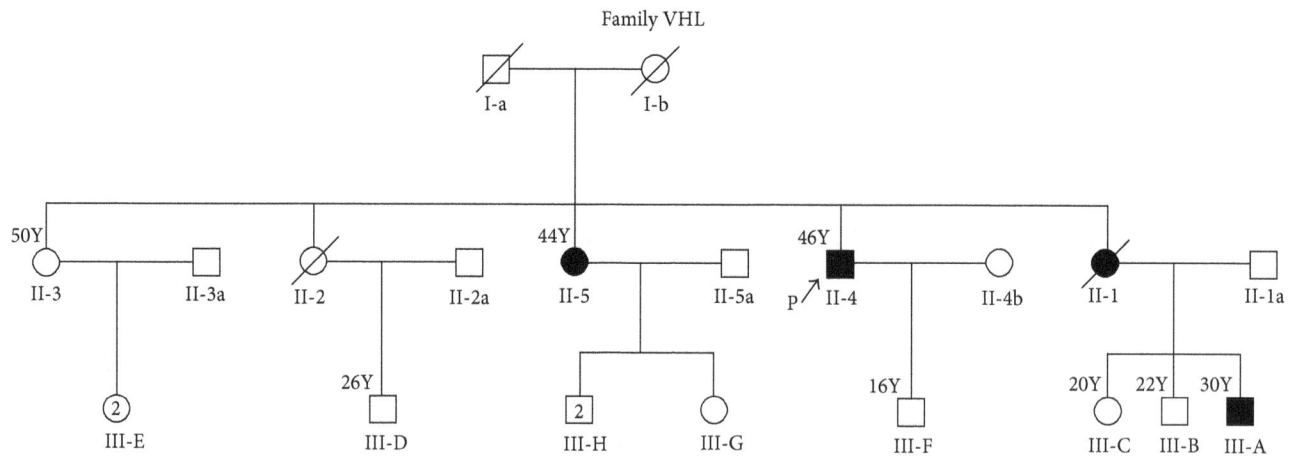

FIGURE 1: Pedigree of the affected family.

Considering the occurrence of phenotypic heterogeneity and interfamilial variations in VHL, historically it has been classified into five disease phenotypes based on the probability of consistent occurrence of either pheochromocytoma or RCC in the kindred [5]. VHL Type 1 families are characterized by HB with RCC and are thought to have a low risk of pheochromocytoma. It is further subtyped as Type 1a in those families with predisposition to both HB and RCC and Type 1b in those families with predisposition predominantly to HB with low risk of RCC. The loss of function mutations due to either truncating mutations or exonic deletions is often associated with Type 1 VHL phenotype [5]. VHL Type IIa is associated with high risk of pheochromocytoma and HB as well as low risk of renal cell carcinoma while Type IIb is associated with high risk of pheochromocytoma, renal cell carcinoma along with HB. Type IIc VHL disease is characterized by occurrence of only pheochromocytoma in the particular kindred. Often missense mutations are associated with Type 2 VHL phenotype which accounts for around 10–20% of VHL kindreds [6, 7]. There are exceptions to these generalizations and in recent periods enough evidence exists to suggest that the families move from one class to another making the classification less meaningful clinically.

Genetic testing is indicated in all individuals suspected to have VHL syndrome along with the family members at risk [3]. Predictive testing of asymptomatic individuals and prenatal diagnosis have become possible because of increasing availability of molecular genetic testing. Here we describe the genetic workup of a family with multiple affected members with VHL and its role in genetic counseling. In addition, we employed a set of *in silico* prediction methods along with molecular dynamics simulation analysis to investigate whether the novel germline variant is disease causing or neutral.

2. Methods

2.1. Case Summary. Proband, a 45-year-old man, born of nonconsanguineous parentage, presented to the Medical Genetics Clinic with history of bilateral retinal angiomas, multifocal cerebellar hemangioblastomas, and bilateral RCC. Most of the tumors had been promptly treated and at the time of presentation, he was on a recommended surveillance program. Upon further intensive investigations, there were no evidences for pheochromocytoma or ELSTs and pancreatic NETs. Family history revealed multiple first- and second-degree relatives affected with similar illness. Detailed pedigree is as shown in Figure 1. Of the four siblings, two were deceased. Among the two surviving, one is affected and another yet unaffected. Parents of the proband were deceased; the cause for their demise is unknown. With this history a diagnosis of Type 1 VHL was made and genetic testing was carried out for proband [II-4 in the pedigree], another affected relative [III-A, in the pedigree], and one yet unaffected relative [III-F, in the pedigree] after appropriate counseling and consenting process.

2.2. Molecular Genetic Testing. Peripheral blood samples were collected from two individuals affected with VHL, proband (II-4 in the pedigree) and his nephew (III-A in the pedigree), after appropriate consenting process (Figure 1). Genomic DNA was extracted from blood samples using NucleoSpin® Kit method (Macherey-Nagel, Germany). DNA extracted was quantified using Nano-Drop ND-1000 spectrophotometer [Thermo Fisher Scientific Commercial Services] and was subjected to polymerase chain reaction [PCR]. PCR was performed using appropriate primer sets and PCR conditions for exons 1, 2, and 3 covering the flanking intronic sequences as described previously [6]. Sequencing was performed using ABI 3130 genetic analyzer. Real-Time quantitative PCR [Applied 97500] was done as described previously [8], to check for large deletions/duplications which has been used reliably across many studies [9].

The above described method was employed after an appropriate consenting process to detect the presence or absence of the variants in the proband's son who is yet unaffected at the age of 18 years. Direct sequencing of 50 age and ethnicity matched controls was performed to detect the presence or absence of the novel variant found in the proband after receiving approval by the Institutional Review Board

[IRB], Christian Medical College, Vellore. For the variants found, appropriate literature search, database search like dbSNP, HGMD, and extensive *in silico* prediction analysis were performed as necessary.

2.3. In Silico Prediction Analysis. To discriminate the impact of the new variant as disease causing or neutral, we employed various *in silico* prediction tools such as Sorting Intolerant from Tolerant (SIFT) [10], Polymorphism Phenotyping 2 (PolyPhen 2) [11], Screening for Non-Acceptable Polymorphisms (SNAP) [12], PROVEAN (Protein Variation Effect Analyzer) [13], MUpro [14], iStable [15] MuStab [16], and I-Mutant Suite [17]. These methods utilize sequence homology (conservation), amino acid physicochemical and structural properties to predict variant's effect on protein function and protein stability and pathogenicity. We submitted NCBI GI number, wild-type protein FASTA sequence, and wild and new residue after mutation (single-letter amino acid code) as input data for making their predictions.

In addition, we used Project Have yOur Protein Explained (HOPE) [18], which is an automatic mutant web server, to analyze the effect of a certain mutation on the protein structure.

2.4. ConSurf. To investigate the extent of conservation of amino acids, we used MUSCLE (Multiple Sequence Comparison by Log-Expectation), a web based tool to align multiple sequences from several vertebrate species including humans [19]. ConSurf utilizes Bayesian analysis to evaluate the evolutionary significance of the missense variant involved in a protein sequence [20]. We submitted the 3D structure with PDB ID as input to measure the functionally conserved residues in pVHL. ConSurf combines evolutionary and structural attributes such as solvent accessibility to make predictions. The scores represent extent of conservation with three categories: variable (1–4), intermediate (5-6), and conserved (7–9), respectively.

2.5. Molecular Dynamics. To investigate the structural and dynamical information about the native and mutant proteins at the atomic level, molecular dynamics simulation was performed. It was carried out for 25 ns using GROMACS 4.6 package [21]. To create the mutant model of pVHL, we downloaded X-ray crystallized structure 1LM8 with 1.85 Å [22] from the Protein Data Bank [23]. Mutation analysis was performed using SWISSPDB viewer at position 158 [24]. Human pVHL contains two domains, alpha and beta domain. In our analysis we used "V" chain of PDB ID 1LM8 (Figure 2). The UCSF Chimera (http://www.cgl.ucsf.edu/chimera/) was used to visualize the protein models. Then the entire molecular system was subjected to energy minimization by steepest-descent followed by conjugated gradient method implementing GROMOS96 43a1 force field [25]. Energy minimized structures of native and mutant protein molecules were used as a starting point for MD simulations. We solvated native and mutant proteins in a cubic box with simple point charge (SPC) water molecules [26]. Periodic boundary conditions

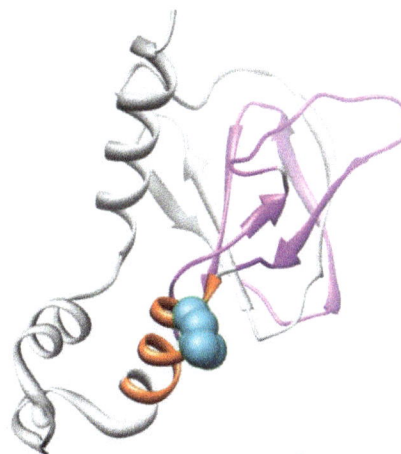

FIGURE 2: Structural visualization of VHL protein with PDB ID: 1LM8 shown in ribbon representation. The protein is colored in orange [157–166: interaction with Elongin BC complex] and violet [100–155: involved in binding to CCT complex] color; the side chain of the native residue is colored cyan and shown as small balls.

were applied on the system to maintain the number of particles and constant pressure and temperature. Berendsen temperature coupling method [27] was used to regulate the temperature inside the box. Isotropic pressure coupling was performed using Parrinello-Rahman method. In order to obtain electrically neutralized system, we utilized GENION from the GROMACS package to replace random water molecule with Na^+ or Cl^- ions. In our analysis, we added six chloride ions [Cl^-] in the simulation box to neutralize the system. Position restrained molecular dynamics simulation was performed at 300 K for 25000 ps to equilibrate the system in an aqueous environment. The temperature was kept constant by using Berendsen algorithm with a coupling time of 0.2. To constrain bond lengths involving hydrogen bond formations, the SHAKE algorithm [28] was used at a time step of 2 fs. The coordinates were saved at regular time intervals of 1 ps. The coulomb interactions were truncated at 0.9 nm, and the Van der Waals force was maintained at 1.4 nm.

3. Results

Sequence analysis showed a novel heterozygous single nucleotide base substitution in the two affected individuals, which was not seen in the unaffected individual. The variant was seen in *exon 3, c.473T>A [TRANSCRIPT ID-ENST00000256474, NM_000551]* (Figure 3), with genomic coordinate chr3: 10,149,796 [*hg38 build*] leading to substitution of amino acid leucine to glutamine at position 158 [p.L158Q] in the Elongin C binding domain of pVHL. After a thorough search of databases and literature, this variation was previously unreported as a causative variant of VHL. However, according to the dbSNP build 142, this variant has been flagged as clinically associated, the details of which will be scrutinized further under Section 4. Deletion analysis did not reveal any major loss of genetic material in all the three samples. This variant was also not observed in 50 age

TABLE 1: Summary of prediction scores of *in silico* methods.

In silico methods	Prediction
SIFT	Deleterious
PROVEAN	Deleterious
PolyPhen 2	Probably damaging
SNAP	Nonneutral
iStable	Decrease protein stability
MUpro (DDG)	Decrease protein stability
MuStab (DDG)	Decrease protein stability
I-Mutant Suite (DDG)	Decrease protein stability

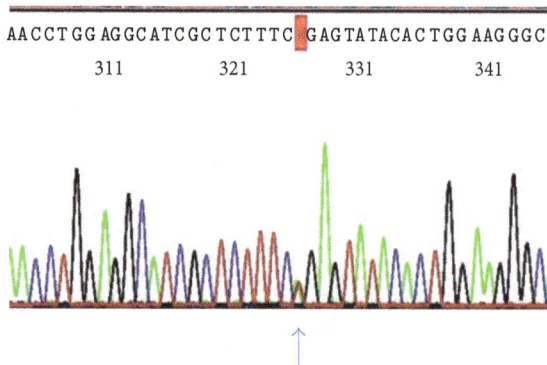

AACCTGG AGGCATCGC TCTTTC GAGTA TACAC TGG AAG GGC

311   321   331   341

FIGURE 3: Electropherogram showing the heterozygous mutation *c.473T>A* in exon 3.

and ethnicity matched controls. Further in the two affected individuals a variation in *intron 1, c.340+5G>C*, with genomic coordinate 10,42,192 [*hg38 build*] was also observed. This has been previously reported as a benign variant [29].

3.1. In Silico Analysis. So far among VHL patients the amino acid change at position 158 from leucine to glutamine has not been reported previously. In view of this being a new variant, we considered doing an extensive *in silico* analysis. The results of the eight different *in silico* methods SIFT, PolyPhen 2, SNAP, PROVEAN, MUpro, iStable, MuStab, and I-Mutant Suite to predict the impact of p.L158Q are shown in Table 1. Sequence based methods SIFT and PROVEAN classified the variant as deleterious and structure based methods PolyPhen 2 and SNAP classified the variant as possibly damaging and nonneutral, respectively. In order to quantify the destabilization effects, the protein stability change upon mutation was evaluated by calculating the difference in folding free energy change between native and mutant proteins (DDG or ddG). The results from protein stability predictors, namely, iStable, MUpro, MuStab, and I-Mutant Suite, indicated that the p.L158Q is less stable and deleterious (Table 1). The mutation analysis of amino acid at their corresponding position 158 was performed by SWISS-PDB viewer independently to achieve modeled structures.

3.2. Effect of Mutation on Protein Structure. Mutation leads to change in amino acid properties which in turn lead to loss or gain of interactions and hydrogen bonds. This may disrupt

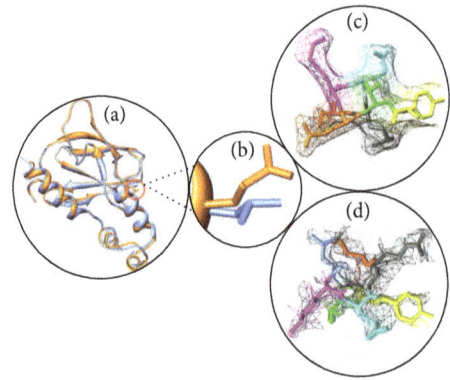

FIGURE 4: (a) Superimposed structure of native and mutant structure of VHL. (b) Close-up view of native amino acid leucine (orange) and mutant amino acid glutamine (blue) in dot. (c) Local environment change visualized for native amino acid [leucine] in dot model. (d) Local environment change visualized for mutant amino acid [glutamine] in dot model. These figures were drawn using Chimera.

the local environment based on the structural interactions in the nearby sites of substitution and manifest their deleterious effects by bringing in changes in structural characteristics such as change in size, surface charge distribution, and hydrophobic contacts [19]. Local environment change upon mutation was defined using PyMol within 4A and visualized using UCSF Chimera (http://www.cgl.ucsf.edu/chimera/). No local environment change was observed upon substitution of leucine by glutamine at the 158th position in VHL (Figure 4). We defined the local environment of native and mutant amino acid by HOPE server to portray the distinct pattern of amino acid substitutions on protein structure based on size, charge, and hydrophobicity. The mutant residue is much bigger than the wild-type residue. The hydrophobicity of the wild-type and mutant residue differed; mutant residue is more hydrophobic than the wild type. The mutation might cause loss of hydrophobic interactions with other molecules on the surface of the protein [18].

3.3. Conservation Analysis. Primary sequence of a protein provides the most direct information regarding the clues for functional mutation sites. Multiple sequence alignments of amino acid residues provide valuable information about the conservation pattern and represent localized evolution. The substitutions of conserved residues are mostly deleterious in nature. In this context, we applied ConSurf to predict the evolutionary significance of the missense variant p.L158Q. Bayesian analyzer ConSurf predicted the p.L158Q as highly conserved (Figure 5). The obtained results are in agreement with the *in silico* prediction tools that predicted p.L158Q as highly deleterious and less stable.

3.4. Molecular Dynamics Simulation Analysis. In order to measure the effect and conformational change upon mutation, we conducted molecular dynamics simulation for 25 ns. Structural properties of the native and mutant protein were calculated from the trajectory files with the built-in functions

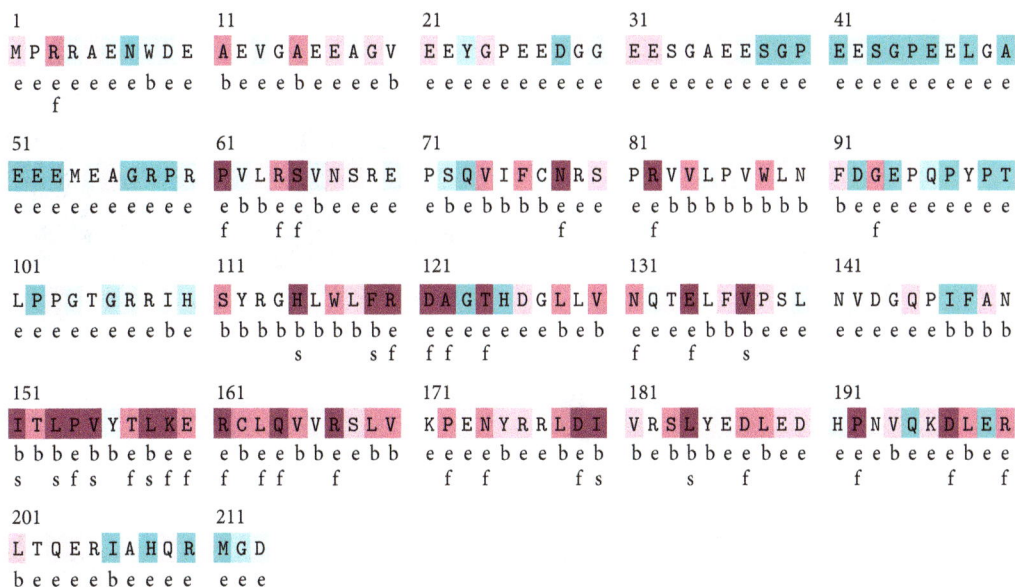

```
1              11             21             31             41
M P R R A E N W D E  A E V G A E E A G V  E E Y G P E E D G G  E E S G A E E S G P  E E S G P E E L G A
e e e e e e b e e    b e e e b e e e e b  e e e e e e e e e e  e e e e e e e e e e  e e e e e e e e e e
    f

51             61             71             81             91
E E E M E A G R P R  P V L R S V N S R E  P S Q V I F C N R S  P R V V L P V W L N  F D G E P Q P Y P T
e e e e e e e e e e  e b b e e b e e e e  e b e b b b b e e e  e e b b b b b b b b  b e e e e e e e e e
                       f     f f              f f                f                        f

101            111            121            131            141
L P P G T G R R I H  S Y R G H L W L F R  D A G T H D G L L V  N Q T E L F V P S L  N V D G Q P I F A N
e e e e e e e e b e  b b b b b b b b b e  e e e e e e b e b    e e e b b b e e e    e e e e e b b b b
                         s         s f    f f f                f     f   s

151            161            171            181            191
I T L P V Y T L K E  R C L Q V V R S L V  K P E N Y R R L D I  V R S L Y E D L E D  H P N V Q K D L E R
b b b e b b e b e e  e b e e b b e e b b  e e e b e e e b e b  b e b b b e e b e e  e e e b e e e b e e
s   s f s   f s f f  f   f f         f    f   f           f s      s       f              f       f f

201            211
L T Q E R I A H Q R  M G D
b e e e e b e e e e  e e e
```

The conservation scale

| 1 | 2 | 3 | 4 | 5 | 6 | 7 | 8 | 9 |

Variable   Average   Conserved

e: an exposed residue according to the neural-network algorithm

b: a buried residue according to the neural-network algorithm

f : a predicted functional residue (highly conserved and exposed)

s : a predicted structural residue (highly conserved and buried)

x: insufficient data: the calculation for this site was performed
    on less than 10% of the sequences

FIGURE 5: The conservation pattern of amino acid sequence in VHL. The location of amino acid residues in VHL based on the evolutionary conservation pattern. Color intensity increases with a degree of conservation (variable (1–4), intermediate (5-6), and conserved (7–9)).

of GROMACS software. The trajectory files were analyzed through the verity of GROMACS utility. To generate the 3D backbone of the native and mutant protein, RMSD, RMSF, hydrogen bonding, and SASA analysis were plotted for the simulations using Graphing, Advanced Computation and Exploration (GRACE) program.

RMSD for all the backbone atoms from the initial structure as a function of time was calculated for the trajectories of the native and mutant protein. In 25 ns trajectory files of RMSD (Figure 6), we can visualize the change in deviation between the native and mutant protein which illustrates the mild or no stabilizing effect on the structure. Mutant protein trajectory showed slightly distinct fashion of deviation after ~12 ns when compared to native, followed by smaller deviations for the rest of the simulations. The RMSDs trajectories of both the simulations reach a stable value after the relaxation period of ~5 ns led to the conclusion that the simulation produced stable trajectories, which provides a suitable basis for further analysis.

The change in conformational flexibility of the native and mutant protein was calculated using RMSF values of Cα

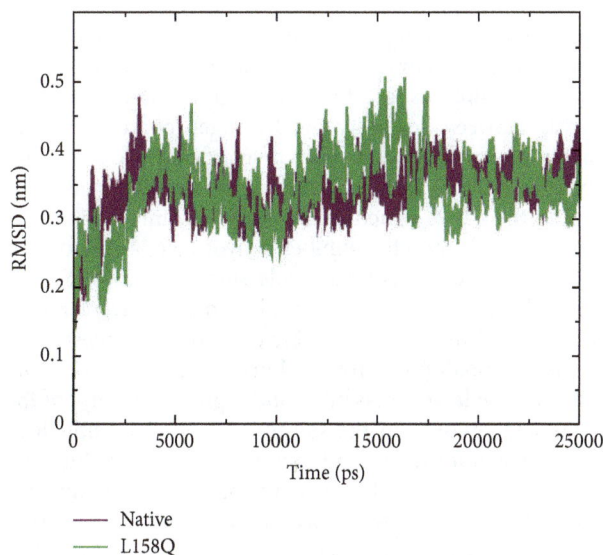

FIGURE 6: Backbone Root Mean Square Deviation (RMSD) of native and mutant (L158Q) VHL protein. The ordinate is RMSD (nm) and the abscissa is time (ps).

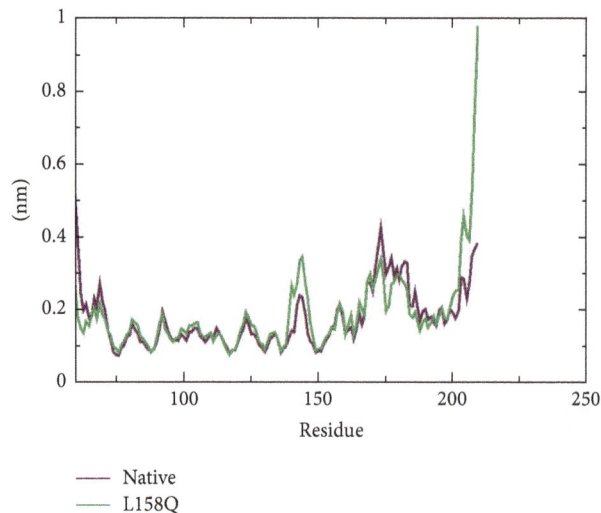

FIGURE 7: C-alpha Root Mean Square Fluctuation (RMSF) of native and mutant (L158Q) VHL protein. The ordinate is RMSF (nm) and the abscissa is amino acid residues.

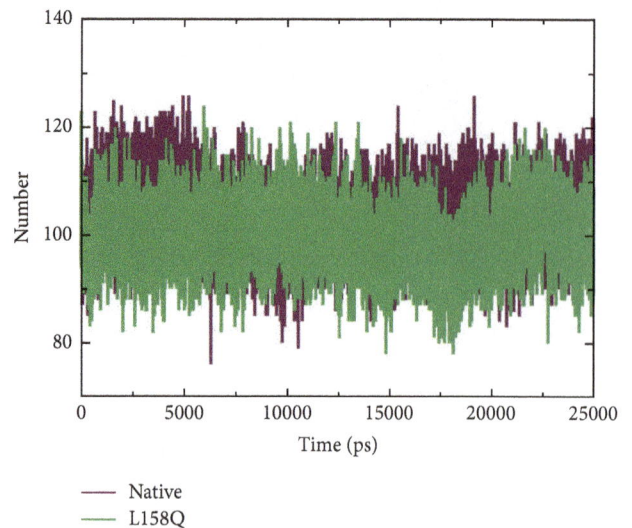

FIGURE 8: Total number of hydrogen bonds formed between native and mutant (L158Q) VHL protein.

atom (Figure 7). Analysis of fluctuation value revealed the presence of higher degree of flexibility in the mutant structure as compared to the native protein. The two proteins display a different pattern of fluctuation in few regions depicting that the mutation affected the flexibility of protein. Moreover, it was very interesting to observe that the amino acid residues present in the region of Elongin C binding of pVHL showed very high fluctuation. This illustrates that the mutation has affected the conformation of protein in a different fashion, which caused the rise in flexibility of residues whereas it induced constraints in the flexibility of the residues in other regions. We conclude from these findings that the mutation induces increased fluctuations in some parts of the structure while not decreasing the overall stability of this structure as judged by the RMSD. Hydrogen bond accounts for a major factor for maintaining the stable conformation of protein. Hydrogen bond analyses of native and mutant proteins were performed with respect to time to understand the relationship between flexibility and hydrogen bond formation. Notably mutant structure showed significantly less number of hydrogen bonds formation during the simulation as compared to the native structure (Figure 8). Atomic flexibility is greatly dependent on the number of hydrogen bonds formed by amino acids. Solvent accessible surface area (SASA) of a biomolecule is that it is accessible to the solvent and it can be related to the hydrophobic core. Solvent accessibility was divided predominantly into buried and exposed region, indicating the least accessibility and high accessibility of the amino acid residues to the solvent. Solvent accessible area was calculated for native and mutant trajectories values and depicted in Figure 9. It is evident that native and mutant exhibited similar fashion of solvent accessible surface area of ~48–56 nm^2 in the simulation period of ~0–7 ns and very low solvent accessible area of ~46 nm^2 in the simulation period of ~8-9 ns (Figure 9). Increase or decrease in the solvent accessible surface area indicates change in exposed amino

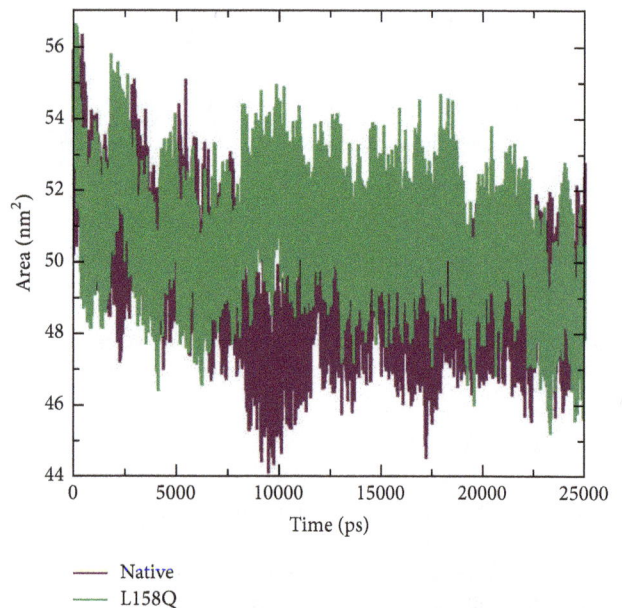

FIGURE 9: Solvent accessible surface area (SASA) analysis of native and mutant (L158Q) VHL protein.

acid residues and it could affect the tertiary structure of the proteins.

4. Discussion

The variant observed in our study is heterozygous, in keeping with the autosomal dominant pattern of inheritance of VHL. The variant in exon 3 *c.473T>A* identified in our patients results in a leucine-to-glutamine substitution at amino acid 158 in the Elongin C binding domain of pVHL. For the regulation of hypoxic gene response pathway, it is essential that pVHL forms a complex with Elongin C and Elongin B. Elongin C directly binds to pVHL whereas Elongin B binds to

```
tr|R4GLJ6|R4GLJ6_CHICK        DAGTHDGLLVNRQELFVAAPDVNK----ADITLPVFTLKERCLQVVRSLVRPGDYRKLDI
sp|Q5Q9Z2|VHL_CANFA           DAGTYDGLLVNQTELFVPSLNVDGQPIFANITLPVYTLKERCLQVVRSLVKPENYRRLDI
sp|Q5Q9Z2|VHL_CANFA           DAGTYDGLLVNQTELFVPSLNVDGQPIFANITLPVYTLKERCLQVVRSLVKPENYRRLDI
tr|G7MKR1|G7MKR1_MACMU        DAGTHDGLLVNQTELFVPSLNVDGQPIFANITLPVYTLKERCLQVVRSLVKPENYRRLDI
sp|P40337|VHL_HUMAN           DAGTHDGLLVNQTELFVPSLNVDGQPIFANITLPVYTLKERCLQVVRSLVKPENYRRLDI
tr|K7BID8|K7BID8_PANTR        DAGTHDGLLVNQTELFVPSLNVDGQPIFANITLPVYTLKERCLQVVRSLVKPENYRRLDI
tr|G3R4I8|G3R4I8_GORGO        DAGTHDGLLVNQTELFVPSLNVDGQPIFANITLPVYTLKERCLQVVRSLVKPENYRRLDI
sp|P40338|VHL_MOUSE           DAGTHDGLLVNQTELFVPSLNVDGQPIFANITLPVYTLKERCLQVVRSLVKPENYRRLDI
sp|P40338|VHL_MOUSE           DAGTHDGLLVNQTELFVPSLNVDGQPIFANITLPVYTLKERCLQVVRSLVKPENYRRLDI
sp|Q64259|VHL_RAT             DAGTHDGLLVNQTELFVPSLNVDGQPIFANITLPVYTLKERCLQVVRSLVKPENYRRLDI
                              ****:******.  ****.: :*:       *:*****:**************.* :**.***
```

FIGURE 10: Evolutionary conserved pattern of leucine residue at the 158th position in VHL across cross species. Multiple sequence alignment was performed by MUSCLE.

Elongin C thereby stabilizing the complex [30]. This complex in turn recruits CUL2 and regulates ubiquitination of proteins involved in hypoxic gene response pathway. *In vitro* protein binding studies have demonstrated that residues from 157 to 171 on pVHL form the Elongin C binding site [31]. This region [157–171] forms a helix in the crystal structure that fits into the concave surface present on Elongin C and has been a hot spot for mutations causing VHL. The protein binding studies have also revealed that residues at T157, L158, R161, and C162 are crucial for Elongin C binding [32]. Of these residues L158 produces the strongest of contacts with Elongin C as per *in vitro* alanine scanning experiments [substitution of leucine at position 158 by alanine]. Along with C162 and R161, L158 forms the most significant Van der Waals contacts, with Elongin C residues 17–50. *In vitro* alanine scanning experiments have clearly demonstrated the complete inactivation of pVHL leading to absolute loss of Elongin C binding when leucine at position 158 is replaced by alanine [31]. In addition as Figure 10 illustrates leucine residue at this position is highly evolutionarily conserved making it functionally a crucial amino acid for pVHL. ConSurf analysis provided the highest score of 9 in the conservation scale for the residue L158 as well revealing buried nature of the residue. Previously leucine-to-proline [p.L158P] and leucine-to-valine [p.L158V] change at this position have been reported as causative variants of VHL disease [32]. These evidences suggest the importance of residue leucine at position 158.

To estimate whether the amino acid substitution L158Q in our proband can affect protein function we applied web based programs SIFT, PolyPhen 2, SNAP, PROVEAN, iStable, MUpro, MuStab, and I-Mutant Suite. *In silico* prediction analysis revealed the p.L158Q as deleterious and less stable. Analytical tool HOPE predicted protein structural alterations due to heavier and more hydrophobic mutant residue, also predicting possible loss of hydrophobic interactions at the surface of the protein. Molecular dynamics simulation analysis showed formation of less number of hydrogen bonds and altered flexibility of molecules in and around the region due to the mutation, affecting ultimately the final tertiary structure of pVHL. Summating the results of *in silico* analysis, it can be reliably concluded that p.L158Q is not a neutral variant.

As mentioned earlier in Section 3 according to the dbSNP build 142, *c.473T>A* with genomic coordinate 10,149,796 [*hg38 build*] has been considered a flagged SNP [single nucleotide polymorphism] with unknown minor allele frequency [MAF] and clinically associated. The SNP was deemed clinically associated based on the locus specific database by a single submitter in a somatic sample with germline allele being T. The frequency of allele A is unknown [allele A observed only once in a somatic sample], which was not validated and has never been reported in 1000 genomes or in any other source like HGMD.

Further, the change was not observed in any of our 50 age and ethnicity matched controls, making *c.473T>A* being a polymorphism highly unlikely. The two single nucleotide base substitution variants were shown to segregate with VHL phenotype in this family in which the variant in intron 1, c.340+5G>C, is previously reported and is a recognized neutral variant. Therefore it is possible that the variant in exon 3 *c.473T>A*, a missense variant [p.L158Q] with unknown MAF, never reported previously in any database of germline variants, which segregated with VHL phenotype only in the family and which was not observed in the ethnic population the proband belongs to, maybe is disease causing.

4.1. Genotype-Phenotype Correlation. In the past it was generally believed that protein truncating mutations are considered to be responsible for many of the Type 1 VHL cases; however growing evidence indicates that it is actually complete loss of function of pVHL irrespective of the type of the mutation that leads to Type 1 VHL. Missense mutations in Elongin C binding domain have been established as a cause of Type 1 VHL by involving key amino acids leading to complete loss of function of pVHL as described earlier. This is in keeping with the phenotype observed in the proband and his family. This also highlights the role of an independent pathway other than hypoxic gene response pathway for the genesis of pheochromocytoma [33]. Usually in pheochromocytoma the missense variations which account for 80–95% [34] of mutations in Type 2 VHL leave pVHL with some residual activity [2] which is in contrast to the possibility of complete abrogation of Elongin C binding in our case. This could explain the absence of pheochromocytoma in the proband's family. However transethnic differences in genotype-phenotype correlations are described. As an example p.L158V which was associated with Type 1 VHL in Caucasians was associated with Type 2 disease in Japanese [35]. Hence even with the entire available database on mutations and basic research, phenotype prediction is difficult.

4.2. Biological Relevance. It is clear from the alanine scanning experiments and previous literature that the mutations that affect L158 lead to complete abolishment of Elongin C binding and thereby complete inactivation of pVHL. The highly conserved amino acid leucine at position 158 forms one of the strongest bonding regions with Elongin C. Hence mutations of this amino acid affect interactions with Elongin C and in turn a host of other molecules. This paper illustrates in general the importance of Elongin C binding domain and L158 in particular in the functioning of pVHL.

5. Conclusions

In silico predictive analyses, available relevant literature, and results of genotype analysis of healthy controls provide strong evidence for pathogenicity of the previously unreported variant *c.473T>A [p.L158Q]*. Hence, this variant is highly likely to be causative of VHL. Our finding supports the functional importance of residue L158 in pVHL functioning and further illustrates the diversity of *VHL* gene defects underlying phenotypic heterogeneity in VHL. The results of this study can serve as a harbinger of functional studies to be performed in the future. The analysis also helped us in genetic counseling, as we could reassure the proband about his son's status.

Competing Interests

The authors declare that they have no competing interests.

Acknowledgments

The authors thank the proband and his family for the samples and cooperation. The authors take this opportunity to thank the management of VIT University for providing the facilities to carry out computational analysis.

References

[1] E. R. Maher, L. Iselius, J. R. W. Yates et al., "Von Hippel-Lindau disease: a genetic study," *Journal of Medical Genetics*, vol. 28, no. 7, pp. 443–447, 1991.

[2] M. Nordstrom-O'Brien, R. B. Van Der Luijt, E. Van Rooijen et al., "Genetic analysis of von Hippel-Lindau disease," *Human Mutation*, vol. 31, no. 5, pp. 521–537, 2010.

[3] P. H. Maxwell, M. S. Wlesener, G.-W. Chang et al., "The tumour suppressor protein VHL targets hypoxia-inducible factors for oxygen-dependent proteolysis," *Nature*, vol. 399, no. 6733, pp. 271–275, 1999.

[4] C. E. Stebbins, W. G. Kaelin Jr., and N. P. Pavletich, "Structure of the VHL-elonginC-elonginB complex: implications for VHL tumor suppressor function," *Science*, vol. 284, no. 5413, pp. 455–461, 1999.

[5] E. R. Maher, A. R. Webster, F. M. Richards et al., "Phenotypic expression in von Hippel-Lindau disease: correlations with germline VHL gene mutations," *Journal of Medical Genetics*, vol. 33, no. 4, pp. 328–332, 1996.

[6] P. A. Crossey, F. M. Richards, K. Foster et al., "Identification of intragenic mutations in the Von Hippel-Lindau disease tumour suppressor gene and correlation with disease phenotype," *Human Molecular Genetics*, vol. 3, no. 8, pp. 1303–1308, 1994.

[7] M.-A. Abbott, K. L. Nathanson, S. Nightingale, E. R. Maher, and R. M. Greenstein, "The von Hippel-Lindau (VHL) germline mutation V84L manifests as early-onset bilateral pheochromocytoma," *American Journal of Medical Genetics, Part A*, vol. 140, no. 7, pp. 685–690, 2006.

[8] J. Hoebeeck, R. van der Luijt, B. Poppe et al., "Rapid detection of VHL exon deletions using real-time quantitative PCR," *Laboratory Investigation*, vol. 85, no. 1, pp. 24–33, 2005.

[9] A. Ebenazer, S. Rajaratnam, and R. Pai, "Detection of large deletions in the VHL gene using a real-time PCR with SYBR Green," *Familial Cancer*, vol. 12, no. 3, pp. 519–524, 2013.

[10] P. Kumar, S. Henikoff, and P. C. Ng, "Predicting the effects of coding non-synonymous variants on protein function using the SIFT algorithm," *Nature Protocols*, vol. 4, no. 7, pp. 1073–1082, 2009.

[11] I. A. Adzhubei, S. Schmidt, L. Peshkin et al., "A method and server for predicting damaging missense mutations," *Nature Methods*, vol. 7, no. 4, pp. 248–249, 2010.

[12] Y. Bromberg and B. Rost, "SNAP: predict effect of nonsynonymous polymorphisms on function," *Nucleic Acids Research*, vol. 35, no. 11, pp. 3823–3835, 2007.

[13] Y. Choi, G. E. Sims, S. Murphy, J. R. Miller, and A. P. Chan, "Predicting the functional effect of amino acid substitutions and indels," *PLoS ONE*, vol. 7, no. 10, Article ID e46688, 2012.

[14] J. Cheng, A. Randall, and P. Baldi, "Prediction of protein stability changes for single-site mutations using support vector machines," *Proteins: Structure, Function and Genetics*, vol. 62, no. 4, pp. 1125–1132, 2006.

[15] C.-W. Chen, J. Lin, and Y.-W. Chu, "iStable: off-the-shelf predictor integration for predicting protein stability changes," *BMC Bioinformatics*, vol. 14, supplement 2, article S5, 2013.

[16] S. Teng, A. K. Srivastava, and L. Wang, "Sequence feature-based prediction of protein stability changes upon amino acid substitutions," *BMC Genomics*, vol. 11, supplement 2, article S5, 2010.

[17] E. Capriotti, P. Fariselli, I. Rossi, and R. Casadio, "A three-state prediction of single point mutations on protein stability changes," *BMC Bioinformatics*, vol. 9, supplement 2, article S6, 2008.

[18] H. Venselaar, T. A. H. te Beek, R. K. P. Kuipers, M. L. Hekkelman, and G. Vriend, "Protein structure analysis of mutations causing inheritable diseases. An e-Science approach with life scientist friendly interfaces," *BMC Bioinformatics*, vol. 11, article 548, 2010.

[19] R. C. Edgar, "MUSCLE: multiple sequence alignment with high accuracy and high throughput," *Nucleic Acids Research*, vol. 32, no. 5, pp. 1792–1797, 2004.

[20] H. Ashkenazy, E. Erez, E. Martz, T. Pupko, and N. Ben-Tal, "ConSurf 2010: calculating evolutionary conservation in sequence and structure of proteins and nucleic acids," *Nucleic Acids Research*, vol. 38, no. 2, pp. W529–W533, 2010.

[21] S. Pronk, S. Páll, R. Schulz et al., "GROMACS 4.5: a high-throughput and highly parallel open source molecular simulation toolkit," *Bioinformatics*, vol. 29, no. 7, pp. 845–854, 2013.

[22] J.-H. Min, H. Yang, M. Ivan, F. Gertler, W. G. Kaelin Jr.,

and N. P. Pavietich, "Structure of an HIF-1α-pVHL complex: hydroxyproline recognition in signaling," *Science*, vol. 296, no. 5574, pp. 1886–1889, 2002.

[23] A. Kouranov, L. Xie, J. de la Cruz et al., "The RCSB PDB information portal for structural genomics," *Nucleic Acids Research*, vol. 34, pp. D302–D305, 2006.

[24] W. Kaplan and T. G. Littlejohn, "Swiss-PDB viewer (Deep View)," *Briefings in Bioinformatics*, vol. 2, no. 2, pp. 195–197, 2001.

[25] W. F. Van Gunsteren, S. R. Billeter, A. A. Eising et al., *Biomolecular Simulation: The GROMOS96 Manual and User Guide*, vdf Hochschulverlag AG, an der ETH Zurich, Zürich, Switzerland; BIOMOS, Groningen, The Netherlands, 1996.

[26] W. L. Jorgensen, J. Chandrasekhar, J. D. Madura, R. W. Impey, and M. L. Klein, "Comparison of simple potential functions for simulating liquid water," *The Journal of Chemical Physics*, vol. 79, no. 2, pp. 926–935, 1983.

[27] H. J. C. Berendsen, J. P. M. Postma, W. F. Van Gunsteren, A. Dinola, and J. R. Haak, "Molecular dynamics with coupling to an external bath," *The Journal of Chemical Physics*, vol. 81, no. 8, pp. 3684–3690, 1984.

[28] R. Elber, A. P. Ruymgaart, and B. Hess, "SHAKE parallelization," *European Physical Journal: Special Topics*, vol. 200, no. 1, pp. 211–223, 2011.

[29] Z. Erlic, M. M. Hoffmann, M. Sullivan et al., "Pathogenicity of dna variants and double mutations in multiple endocrine neoplasia type 2 and Von Hippel-Lindau syndrome," *Journal of Clinical Endocrinology and Metabolism*, vol. 95, no. 1, pp. 308–313, 2010.

[30] K. M. Lonergan, O. Iliopoulos, M. Ohh et al., "Regulation of hypoxia-inducible mRNAs by the von Hippel-Lindau tumor suppressor protein requires binding to complexes containing elongins B/C and Cul2," *Molecular and Cellular Biology*, vol. 18, no. 2, pp. 732–741, 1998.

[31] M. Ohh, Y. Takagi, T. Aso et al., "Synthetic peptides define critical contacts between elongin C, elongin B, and the von Hippel-Lindau protein," *Journal of Clinical Investigation*, vol. 104, no. 11, pp. 1583–1591, 1999.

[32] V. Bangiyeva, A. Rosenbloom, A. E. Alexander, B. Isanova, T. Popko, and A. R. Schoenfeld, "Differences in regulation of tight junctions and cell morphology between VHL mutations from disease subtypes," *BMC Cancer*, vol. 9, article 229, 2009.

[33] S. Lee, E. Nakamura, H. Yang et al., "Neuronal apoptosis linked to EglN3 prolyl hydroxylase and familial pheochromocytoma genes: developmental culling and cancer," *Cancer Cell*, vol. 8, no. 2, pp. 155–167, 2005.

[34] B. Zbar, T. Kishida, F. Chen et al., "Germline mutations in the Von Hippel-Lindau disease (VHL) gene in families from North America, Europe and Japan," *Human Mutation*, vol. 8, no. 4, pp. 348–357, 1996.

[35] M. Yoshida, S. Ashida, K. Kondo et al., "Germ-line mutation analysis in patients with Von Hippel-Lindau disease in Japan: an extended study of 77 families," *Japanese Journal of Cancer Research*, vol. 91, no. 2, pp. 204–212, 2000.

19

Screening for Subtelomeric Rearrangements in Thai Patients with Intellectual Disabilities using FISH and Review of Literature on Subtelomeric FISH in 15,591 Cases with Intellectual Disabilities

Chariyawan Charalsawadi,[1] Jariya Khayman,[1] Verayuth Praphanphoj,[2] and Pornprot Limprasert[1]

[1]*Department of Pathology, Faculty of Medicine, Prince of Songkla University, Songkhla 90110, Thailand*
[2]*Medical Genetics Center, Bangkok 10220, Thailand*

Correspondence should be addressed to Pornprot Limprasert; lpornpro@yahoo.com

Academic Editor: Norman A. Doggett

We utilized fluorescence in situ hybridization (FISH) to screen for subtelomeric rearrangements in 82 Thai patients with unexplained intellectual disability (ID) and detected subtelomeric rearrangements in 5 patients. Here, we reported on a patient with der(20)t(X;20)(p22.3;q13.3) and a patient with der(3)t(X;3)(p22.3;p26.3). These rearrangements have never been described elsewhere. We also reported on a patient with der(10)t(7;10)(p22.3;q26.3), of which the same rearrangement had been reported in one literature. Well-recognized syndromes were detected in two separated patients, including 4p deletion syndrome and 1p36 deletion syndrome. All patients with subtelomeric rearrangements had both ID and multiple congenital anomalies (MCA) and/or dysmorphic features (DF), except the one with der(20)t(X;20), who had ID alone. By using FISH, the detection rate of subtelomeric rearrangements in patients with both ID and MCA/DF was 8.5%, compared to 2.9% of patients with only ID. Literature review found 28 studies on the detection of subtelomeric rearrangements by FISH in patients with ID. Combining data from these studies and our study, 15,591 patients were examined and 473 patients with subtelomeric rearrangements were determined. The frequency of subtelomeric rearrangements detected by FISH in patients with ID was 3%. Terminal deletions were found in 47.7%, while unbalanced derivative chromosomes were found in 47.9% of the rearrangements.

1. Introduction

An intellectual disability (ID) is a condition wherein the development of the mentality is arrested or incomplete, which contributes to the impairment of the overall level of intelligence, including cognitive, language, motor, and social abilities [1]. ID affects 1% to 3% of the global population. Besides environmental factors, genetic factors are a significant cause of ID. More than half of all patients with ID are categorized as having an unexplained ID, with subtelomeric rearrangements having been observed in a number of these patients, ranging between 0 and 29.4% [2].

The subtelomere is a region between chromosome-specific sequences and telomeric caps of each chromosome. This region is located in close proximity to a gene-rich area. Due to sequence homology between subtelomeres of different chromosomes, it can facilitate recombination and may subsequently result in detrimental effects, that is, promoting disease-causing chromosomal rearrangements [3].

Subtelomeric rearrangements can be detected by various methods. In this study, we utilized fluorescent in situ hybridization (FISH) method with probes specific to the subtelomere region of all chromosome ends to screen for subtelomeric rearrangements in patients with unexplained

ID. We identified new rearrangements and described clinical entities of patients with the rearrangements.

2. Subjects and Methods

Inclusion criteria for subject recruitment were patients with ID of unknown causes, with or without multiple congenital anomalies (MCA) and/or dysmorphic features (DF) and with normal G-banding chromosome analysis result at the 450–550 bands' levels. All patients who met these inclusion criteria were referred from clinicians at Rajanukul Institute. The institute is the governmental agency under the Department of Mental Health, Ministry of Public Health, providing medical care primarily for individuals with intellectual and developmental disabilities. Clinical features of some patients may be limited as we collected data from the laboratory order form. A total of 82 Thai patients with ID were recruited to the study. These patients included 50 males and 32 females aged between 1 year and 39 years (mean age being 4 years). ID without MCA/DF was present in 35 cases (22 males and 13 females), and ID with MCA/DF was present in 47 cases (28 males and 19 females). This study was a one-year prospective study that was conducted between the years of 2005 and 2006.

We performed subtelomeric FISH analysis on metaphase spreads obtained from lymphocyte cultures, which were initiated and harvested following a standard protocol. The FISH probes used in this study were constructed using bacterial artificial chromosome (BAC) as well as P1-derived artificial chromosome (PAC) clones. These clones contained 41 different subtelomeric-specific sequences for all human chromosome ends, located within 2 Mb distance from the telomere. For the FISH analysis of each patient, the p-arm and q-arm probes of each chromosome were denatured and hybridized onto the denatured metaphase spreads. FISH was performed following a standard protocol. We examined at least 10 informative metaphase spreads for each chromosome. For patients with detected subtelomeric rearrangements, G-banding karyotype and FISH analyses using the same probes were carried out in parental blood when available. In addition, because the subtelomeric probe for the short arms of both X and Y chromosomes is specific to the pseudoautosomal region, the other X- and Y-specific probes were used to distinguish the chromosomes. This study was approved by the ethical committee of the Ministry of Public Health.

3. Results

We detected 5 subtelomeric rearrangements in 5 patients. The frequency of subtelomeric rearrangements in our study was 6.1%. Two patients had deletions, including one with del(4)(p16.3) (Patient 1, Figure 1(a)) and one with del(1)(p36.3)dn (Patient 2, Figure 1(b)). The other three patients had derivative chromosomes, including one with der(10)t(7;10)(p22.3;q26.3) (Patient 3, Figure 1(c)), one with der(20)t(X;20)(p22.3;q13.3) (Patient 4, Figure 1(d)), and one with der(3)t(X;3)(p22.3;p26.3)dn (Patient 5, Figure 1(e)). Two patients had moderate degrees of ID, and 3 patients had severe degrees of ID, and those 3 patients included 2 patients with deletions. MCA/DF were observed in almost all of the

patients with subtelomeric rearrangements, except a patient with der(20)t(X;20), who had only minor DF (Table 1). The frequency of subtelomeric rearrangements in patients with ID and MCA/DF was 8.5%, while in patients with only ID it was 2.9%.

4. Discussion

4.1. Detection of Subtelomeric Rearrangements. Subtelomeric rearrangements can be detected using various methods, such as FISH [4, 26], multiplex ligation-dependent probe amplification (MLPA) [32–34], and a microarray-based method [35, 36]. The latter method allows the whole human genome to be scanned at high resolution in a single experiment. As a result, not only cryptic subtelomeric rearrangements but also other unbalanced chromosomal abnormalities can be detected, such as interstitial deletions, duplications, and unbalanced translocations. However, FISH has the advantage of providing an instant location of certain rearrangements, such as insertions, inversions, and balanced translocations, which cannot be achieved by the microarray-based method.

We reviewed literature on subtelomeric rearrangements in individuals with ID, which were detected by subtelomeric FISH (Table 2). We search for articles in PubMed using the following keywords: subtelomeric FISH, intellectual disability, developmental delay, and mental retardation. We excluded studies from authors that did not provide details regarding number of the patients with ID, studies when FISH was performed to confirm findings of other molecular methods, studies where a selected panel of subtelomeric probes was used in each case, and extended studies with additional cases from the previous populations. We found 28 publications on the detection of subtelomeric rearrangements by FISH in patients with ID. Combining data from these studies (15,509 patients) and our study (82 patients), 15,591 patients with ID were examined by FISH and 474 subtelomeric rearrangements were identified in 473 patients. There was one patient with 2 rearrangements. Frequencies of subtelomeric rearrangements ranged from 0 to 20%, with an average of 3% (473/15,591) (Table 2). Familial variants and possible variants were found in approximately 1.0% (149/15,591) (Table 3). One of the most common variants is the del(2)(qter), in which the detection of a deletion depends on the subtelomeric probe used. In this literature review, we found that frequency of the del(2)(qter) was 42% of all variants (63/149). It is important to do parental analysis with the same subtelomeric probe when an abnormality is detected in a patient to determine the clinical significance of the finding.

We divided subtelomeric rearrangements into 3 categories: (1) deletion, (2) derivative, and (3) others. The latter category included duplications, insertions, inversions, isochromosome, and balanced translocations. Deletions were found in approximately 47.7% (226/474), while derivative chromosomes were found in approximately 47.9% (227/474) (Figure 2). The most frequent deletions (>5% of all deletions) involved chromosomes 1p, 22q, 9q, and 4p (Figure 3(a)). The derivative category was composed of an unbalanced translocation with both deletion and duplication of subtelomere regions, an unbalanced translocation with duplication of

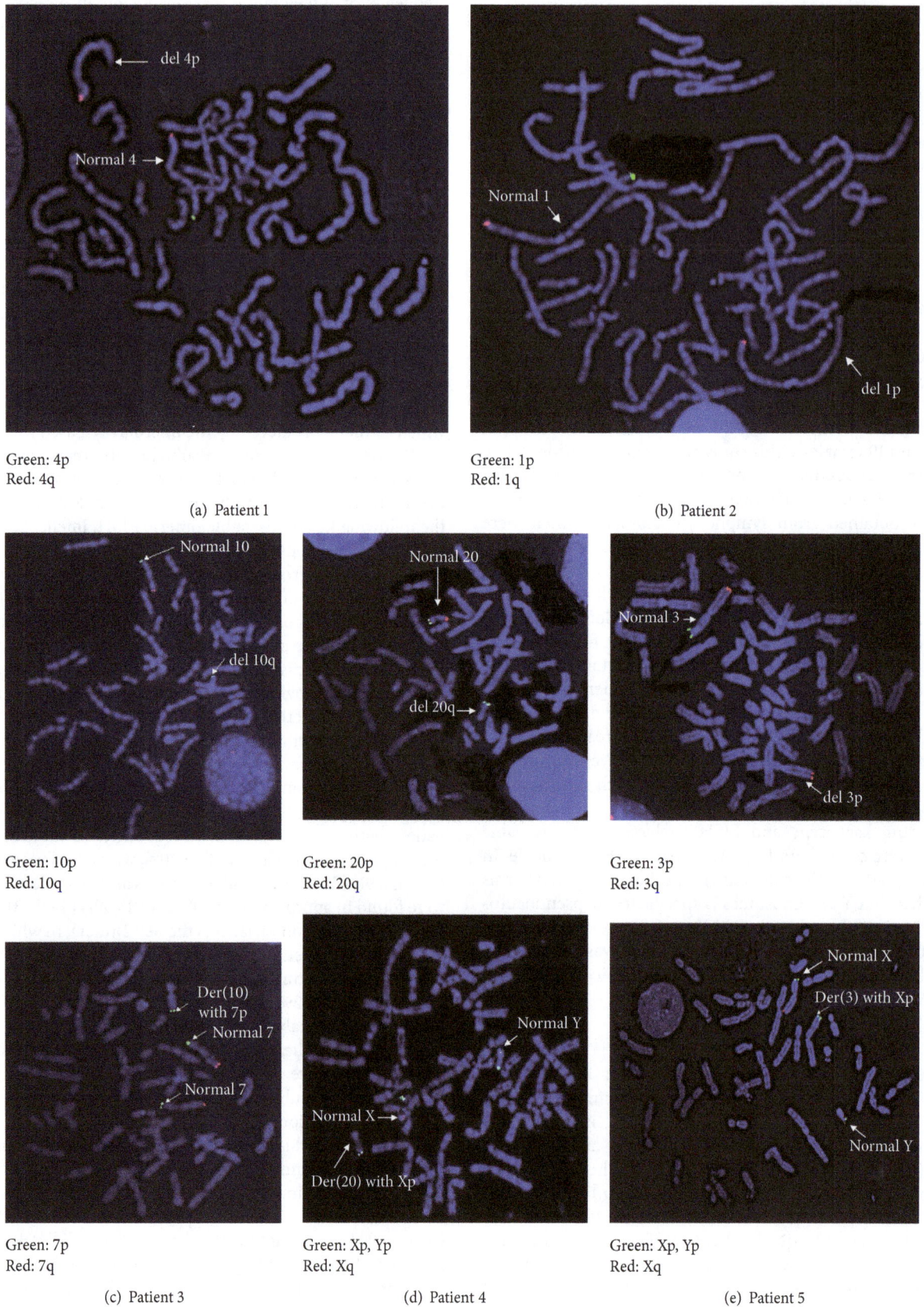

Green: 4p
Red: 4q

(a) Patient 1

Green: 1p
Red: 1q

(b) Patient 2

Green: 10p
Red: 10q

Green: 20p
Red: 20q

Green: 3p
Red: 3q

Green: 7p
Red: 7q

(c) Patient 3

Green: Xp, Yp
Red: Xq

(d) Patient 4

Green: Xp, Yp
Red: Xq

(e) Patient 5

FIGURE 1: FISH results of 5 patients with subtelomeric rearrangements.

TABLE 1: Clinical features of 5 patients with subtelomeric rearrangements.

Patient number	Patient 1	Patient 2	Patient 3	Patient 4	Patient 5
Age (mo)	60	96	180	36	72
Gender	Male	Female	Female	Male	Male
Karyotype	46,XY.ish del(4)(p16.3)(GS-36-P21−)	46,XX.ish del(1)(p36.32)(RPI1-465B22−)dn	46,XX.ish der(10)t(7;10)(p22.3;q26.3)(GS-164-D18+;RPI1-108K14−)	46,XY.ish der(20)t(X;20)(p22.3;q13.3)(RPI1-465BI7+;RPI1-11M20−)	46,XY.ish der(3)t(X;3)(p22.3;p26.3)(RPI3-465BI7+;RPI1-306H5−)dn
Interpretation	4p16.3 deletion	1p36.3 deletion	7p22.3 duplication and 10q26.3 deletion	Xp22.3 duplication and 20q13.3 deletion	Xp22.3 duplication and 3p26.3 deletion
Growth		Height and weight below the 3rd percentile	Height and weight below the 3rd percentile		
Head, face	Dolichocephaly, tall and prominent forehead	Triangular face with flat midface, prominent forehead	Microcephaly, low posterior hair line, webbed-neck		Dolichocephaly, prominent forehead, fair hair, low posterior hair line
Ocular region			Left esotropia	Mild hypertelorism	Upslanting eyebrows, mild hypertelorism, downslanting palpebral fissure
Nose	Broad nasal bridge				
Ears	Prominent ears with thick helix, bilaterally	Simple ears	Low set ears		
Mouth region		Cleft palate	Downturned corner of the mouth		Downturned corner of the mouth
Thorax, abdomen, extremities	Normal	Congenital heart defects including ventricular septal defect, patent ductus arteriosus, coarctation of the aorta	Cubitus valgus, simian crease on the right hand	Simian crease on both hands	Normal
Other anomalies	Sacral dimple				Shawl scrotum
Level of ID	IQ ~ 28–30 (severe)	IQ < 24 (severe)	IQ ~ 39–41 (moderate)	IQ = 42 (moderate)	IQ < 30 (severe)
Other tests	Abnormal EEG compatible with seizure disorder; normal hearing and thyroid function tests	Normal brain MRI; normal vision and hearing tests		Normal	Normal EEG and skull X-ray; normal hearing test

TABLE 2: Frequencies and details of subtelomeric rearrangements detected by FISH in patients with ID.

Reference	Total cases	Positive cases	Frequency (%)	De novo	Familial	Unknown inheritance
Knight et al., 1999 [4]	466†	22	4.7	Deletion: del(1)(p36.3)dn [n = 2], del(1)(q44)dn, del(6)(p25)dn, del(9)(p24)dn, del(13)(q34)dn, del(22)(q13.3)dn [n = 3]; Derivative: der(9)t(9;16)(p24;q24)dn, der(13)t(Y;13)(p11.3;q34)dn, der(18)t(X;18)(q28;q23)dn	Derivative: der(12)t(6;12)(q27;p13.3)mat, der(4)t(4;22)(p16;q13.3)pat, der(9)t(9;13)(q34;p11.1)mat, der(1)t(1;13)(q44;q34)mat, der(4)t(4;11)(p16;p15.5)pat, der(4)t(4;6)(q35;q27)pat, der(1)t(1;19)(p36.3;q13.4)pat, der(4)t(4;20)(q35;p13)pat, der(8)t(8;20)(p23;p13)mat, der(7)t(2;7)(q37;q36)pat	
Joyce et al., 2001 [5]	213^{a,†}	11	5.2	Deletion: del(1)(p36.2)dn, del(15)(q26.2)dn; Derivative: der(13)t(13;19)(q34;q13.43)dn, der(4)t(4;8)(p16.1;p23.1)dn, der(4)t(4;11)(p16.3;p15.5)dn, mos der(18)t(4;18)(p16.3;q23)dn	Deletion: del(8)(p23.3)pat*; Derivative: der(2)t(2;14)(q27.3;p11)pat, der(2)t(2;7)(q37.2;q36.3)mat, der(17)t(11;17)(p15.5;p13.3)mat, der(22)t(14;22)(q32.33;q13.31)pat	
Fan et al., 2001 [6]	150‡	6	4.0	Deletion: del(1)(ptel)dn [n = 2]	Derivative: der(1)t(1;3)(qtel;qtel)pat, der(18)t(7;18)(ptel;qtel)mat	Derivative: der(4)t(4;12)(ptel;qtel), der(5)t(5;20)(ptel;ptel)
Sismani et al., 2001 [7]	70†	1	1.4	Deletion: del(8)(pter)dn		
Anderlid et al., 2002 [8]	111†	11	9.9	Deletion: del(2)(qter)dn, del(4)(pter)dn, del(6)(qter)dn, del(9)(qter)dn; Derivative: der(22)t(20;22)(qter;qter)dn	Derivative: der(4)t(2;4)(qter;qter)pat, der(21)t(9;21)(qter;qter)mat, der(17)t(12;17)(qter;qter)pat; Others: rec(6)dup(6p)inv(6)(p23q27)pat [n = 2 siblings]	Deletion: del(22)(qter)
Baker et al., 2002 [9]	250‡	8	3.2	Deletion: 22q– dn; Derivative: der(5)t(5;16)(q35.3;q24.3)dn	Deletion: 20p– mat*; Derivative: der(5)t(5;16)(q35.3;q24.3)pat, der(14)t(9;14)(q34.3;q32.33)mat; Others: insertion (1p– and 9p+)pat	
Popp et al., 2002 [10]	30‡	6	20.0	Derivative: der(3)t(3;16)(p25;p13.2)dn, der(8)t(8;11)(p23.1;p15.5)dn, der(7)t(7;7)(p22;q36)dn	Deletion: del(8)(p23.1p23.1)pat^b; Derivative: der(5)t(3;5)(q27;p15.3)pat, der(16)t(16;19)(p13.3;p13.3)mat	
Dawson et al., 2002 [11]	40†	3	7.5	Deletion: del(9)(q34.3)dn	Derivative: der(9)t(5;9)(q35.3;q34.3)pat	Derivative: der(3)t(3;11)(p36.3;p15.5)
Clarkson et al., 2002 [12]	50†	2	4.0		Derivative: der(5)t(5;21)(p13;q21.1)mat; Others: rec(11)dup(11p)inv(11)(p15.5q24.3)pat	
van Karnebeek et al., 2002 [13]	184†	1	0.5	Deletion: del(12)(q24.33)dn		

TABLE 2: Continued.

Reference	Total cases	Positive cases	Frequency (%)	De novo	Familial	Unknown inheritance
Kirchhoff et al., 2004 [14]	94‡	3	3.2	*Deletion:* del(22)(qtel)dn [n = 2] *Derivative:* del(4)(qtel) and dup(7)(ptel) dn		
Font-Montgomery et al., 2004 [15]	43‡	6	14.0	*Deletion:* del(6)(q27)dn, del(9)(q34.3)dn *Derivative:* der(6)t(1;6)(q44;q27)dn, der(Y)dn (trisomy for PAR of Xp/Yp)	*Derivative:* der(4)t(4;5)(q35;p15.3)pat, der(2)t(2;17)(q37.3;p13)pat	
Walter et al., 2004 [16]	50‡	10	20.0	*Deletion:* del(22)(q13.32)dn, del(4)(p16.3)dn *Derivative:* der(7)t(7;9)(q36.3p24.1)dn, der(12)t(12;20)(p13p13.3)dn *Others:* t(17;20)(p13.3q13.33)dn	*Derivative:* der(4)t(4;7)(p16.3p22)pat, der(9)t(9;18)(p23q22.3)pat, der(7)t(7;10)(q36.3q26.3)mat, der(18)t(18;21)(q22.1q21.3)mat *Others:* der(7)inv(7)(p22q36.3)mat	
Rodriguez-Revenga et al., 2004 [17]	30‡	2	6.7	*Deletion:* del(1)(p36)dn	*Derivative:* der(13)t(1;13)(qter;qter)mat	
Bocian et al., 2004 [18]	83ᶜ⁺	9	10.8	*Deletion:* del(4)(p16.1p16.3)dn *Derivative:* der(13)t(X;13)(q28;q34)dn	*Derivative:* der(13)t(4;13)(p16;q34)pat, der(2)t(2;7)(q37;q36)pat, der(4)t(4;21)(p16;q22)pat, der(6)t(4;6)(q35;q27)pat, der(4)t(4;7)(q33;q34)mat, der(10)t(10;19)(q26;p13.3)mat *Others:* rec(5)dup(5)(q35.3)inv(5)(p15.33q35.3)pat	
Velagaleti et al., 2005 [19]	18†	2	11.1			*Deletion:* del(2)(q27.3) *Others:* t(3;7)(q27;p21.2)
Baroncini et al., 2005 [20]	219†	12	5.5	*Deletion:* del(1pter)dn, del(7pter)dn, del(9pter)dn, del(9qter)dn, del(20pter)dn, del(22qter)dn *Derivative:* der(6)t(6;18)(qter;pter)dn, der(18)t(8;18)(pter;qter)dn		*Derivative:* der(2)t(2;17)(qter;qter) *Others:* t(4;18)(pter;qter), t(1;16)(pter;pter)
Sogard et al., 2005 [21]	132‡	9	6.8	*Deletion:* del(1)(p36.3)dn, del(4)(p16.1)dn, del(5)(q35)dn *Derivative:* mos der(22)t(12;22)(p13;p?)dn	*Derivative:* der(13)t(5;13)(q35.2;q34)mat [n = 2 siblings], der(9)t(9;22)(q34.2;q13.3)pat	*Deletion:* del(2)(q37.2) *Derivative:* der(2)t(2;22)(q37.2;q1?)
Yu et al. 2005 [22]	534‡	7	1.3	*Deletion:* del(9)(q34)dn *Derivative:* der(4)t(4;10)(q35.2;p15.3)dn	*Derivative:* der(22)t(4;22)(p16q14;q13)pat, der(10)t(10;11)(q26.1;q23.3)mat, der(X)t(X;4)(p22.3;q35.2)mat, der(10)t(7;10)(p22.3;q26.3)pat	*Deletion:* del(4)(q35)
Erjavec-Škerget et al., 2006 [23]	100‡	6	6.0	*Deletion:* del(X)(pter−)dn	*Derivative:* der(21)t(21;8)(qter−;qter+)pat, der(11)t(11;10)(qter−;qter+)pat	*Deletion:* del(9)(pter−)(23→pter−) *Derivative:* der(13)t(13;10)(qter−;q23.3→qter+) *Others:* rec(X)(qter+, pter−)

TABLE 2: Continued.

Reference	Total cases	Positive cases	Frequency (%)	De novo	Familial	Unknown inheritance
Palomares et al., 2006 [24]	50‡	5	10.0	*Deletion:* del(9)(q34.3)dn, del(2)(q37.3)dn *Derivative:* der(1)t(1;22)(p36.3;q13.3)dn, der(15)t(15;17)(q26.3;p25.3)dn	*Derivative:* der(2)t(2;10)(q37.3;q26.1)pat	
Rauch et al., 2006 [25]	500†	9	1.8	*Deletion:* del(1)(pter)dn, del(5)(qter)dn, del(9)(qter)dn, del(16)(pter)dn, del(22)(qter)dn *Others:* t(11;20)(pter;qter)dn	*Derivative:* der(14)t(14;20)(qter;qter)pat, der(10)t(9;10)(q34.1;q26.1)familial, der(10)t(10;22)(qter;qter)pat	
Ravnan et al., 2006 [26]	11688†	291d	2.5	*Deletion:* del(1)(pter)dn [n = 10], del(1)(qter)dn, del(2)(qter)dn, del(3)(qter)dn [n = 4], del(4)(pter)dn, del(4)(qter)dn, del(6)(qter)dn [n = 2], del(7)(pter)dn, del(8)(pter)dn, del(9)(pter)dn [n = 4], del(9)(qter)dn [n = 6], del(10)(pter)dn, del(10)(qter)dn [n = 2], del(13)(qter)dn, del(16)(pter)dn [n = 4], del(18)(qter)dn, del(20)(pter)dn [n = 3], del(20)(qter)dn [n = 2], del(22)(qter)dn [n = 2] *Derivative:* der(19)t(19;19)(pter;qter)dn, der(13)t(9;13)(qter;qter)dn, der(13)t(13;17)(pter;qter)dn, der(14)t(14;14)(pter;qter)dn, der(14)t(14;16)(pter;pter)dn, der(15)t(X;15)(qter;pter)dn, der(22)t(22;22)(pter;qter)dn, der(1)t(1;4)(pter;qter)dn, der(1)t(1;10)(pter;pter)dn, der(4)t(X;4)(qter;qter)dn, der(6)t(6;16)(qter;qter)dn, der(7)t(7;22)(qter;qter)dn, der(8)t(8;9)(pter;pter)dn, der(9)t(3;9)(pter;pter)dn, der(9)t(X;9)(qter;pter)dn, der(10)t(10;16)(pter;pter)dn, der(13)t(3;13)(qter;qter)dn, der(15)t(X;15)(qter;qter)dn, der(16)t(16;22)(qter;qter)dn, der(20)t(9;20)(pter;pter)dn, der(22)t(6;22)(pter;qter)dn, der(Y)t(Y;X or Y)(qter;pter)dn	*Deletion:* del(4)(qter)mat*, del(11)(qter)mat* *Derivative:* der(7)t(7;16)(pter;pter)pat, der(10)t(10;16)(qter;pter)mat, der(15)t(15;19)(pter;qter)mat, der(2)t(2;17)(qter;pter)pat, der(2)t(2;20)(qter;qter)pat, der(2)t(2;22)(qter;qter)mat, der(4)t(3;4)(qter;pter)pat, der(4)t(4;5)(qter;pter)pat, der(5)t(2;5)(pter;pter)mat, der(5)t(5;7)(pter;pter)mat, der(5)t(5;20)(pter;pter)pat, der(6)t(6;21)(qter;qter)mat, der(7)t(7;19)(pter;qter)mat, der(7)t(5;7)(pter;qter)mat, der(7)t(7;8)(qter;pter)mat, der(7)t(7;11)(qter;pter)mat, der(8)t(2;8)(pter;pter)mat, der(9)t(9;16)(qter;pter)mat, der(10)t(1;10)(pter;qter)mat, der(10)t(8;10)(qter;qter)mat, der(10)t(10;17)(qter;qter)mat, der(11)t(2;11)(qter;qter)mat, der(12)t(12;17)(pter;pter)pat, der(12)t(12;19)(qter;qter)pat, der(12)t(12;19)(qter;pter)pat, der(17)t(9;17)(qter;qter)mat, der(18)t(4;18)(qter;qter)mat, der(21)t(5;21)(pter;qter)pat, der(22)t(16;22)(qter;qter)pat, der(X)t(X;X or Y)(pter;qter)mat*	*Deletion:* del(1)(pter) [n = 18], del(2)(pter) [n = 2], del(2)(qter) [n = 6], del(3)(pter) [n = 4], del(3)(qter) [n = 4], del(4)(pter) [n = 7], del(4)(qter) [n = 4], del(4)(qter) and der(8)t(4;8)e, del(5)(pter), del(6)(pter), del(6)(qter) [n = 2], del(7)(pter), del(7)(qter), del(8)(pter) [n = 3], del(9)(pter), del(9)(qter) [n = 5], del(10)(pter) [n = 2], del(10)(qter) [n = 6], del(11)(qter) [n = 3], del(12)(pter), del(13)(qter) [n = 2], del(14)(qter) [n = 3], del(15)(qter), del(16)(pter), del(17)(pter) [n = 2], del(18)(qter) [n = 2], del(20)(pter) [n = 5], del(21)(qter), del(22)(qter) [n = 13], del(X)(pter), del(X)(qter), del(Y)(pter) [n = 2] *Derivative:* der(1)t(1;17)(pter;pter), der(10)t(10;16)(qter;pter), der(15)t(7;15)(qter;qter), der(19 or 20)t(X or Y;19 or 20)(qter;pter or qter), der(13)t(11;13)(qter;pter), der(14)t(7;14)(qter;pter), der(15)t(15;16)(pter;qter), der(15)t(X or Y;15)(qter;pter), der(21)(17;21)(qter;pter), der(21)(21;22)(pter;qter), der(22)t(1;22)(qter;pter), der(1)t(1;1)(pter;pter), der(1)t(1;10)(pter;pter), der(1)t(X or Y;1)(qter;pter), der(1)t(1;4)(qter;qter), der(1)t(1;3)(qter;qter), der(1)t(1;15)(qter;qter), der(1)t(1;18)(qter;qter), der(1)t(1;22)(qter;qter), der(2)t(1;2)(qter;qter), der(2)t(2;12)(qter;qter), der(4)t(4;6)(pter;qter) [n = 2], der(4)t(4;8)(pter;pter) [n = 2], der(4)t(4;11)(pter;pter), der(4)t(X or Y;4)(qter;qter), der(5)t(2;5)(pter;pter), der(5)t(5;9)(pter;pter), der(5)t(5;10)(pter;qter), der(5)t(5;14)(pter;qter), der(6)t(6;10)(pter;qter), der(6)t(6;12)(pter;qter), der(6)t(3;6)(qter;qter), der(6)t(6;7)(qter;pter), der(7)t(7;7)(pter;qter), der(7)t(7;16)(pter;pter), der(7)t(2;7)(qter;qter), der(7)t(7;9)(qter;qter), del(4)(qter) and der(8)t(4;8)e, der(8)t(8;8)(pter;qter), der(8)t(8;10)(pter;qter), der(8)t(8;12)(pter;pter) [n = 2], der(8)t(8;18)(pter;qter), der(9)t(7;9)(pter;qter), der(9)t(9;17)(qter;qter), der(10)t(7;10)(qter;pter), der(10)t(4;10)(pter;qter), der(10)t(4;10)(qter;qter), der(10)t(8;10)(qter;qter), der(10)t(10;10)(qter;qter), der(10)t(10;21)(qter;qter), der(11)t(11;12)(qter;qter) [n = 2], der(12)t(12;20)(pter;qter), der(13)t(2;13)(qter;qter), der(13)t(3;13)(qter;qter), der(14)t(X or Y;14)(pter;qter), der(15)t(3;15)(qter;qter), der(15)t(9;15)(qter;qter), der(16)t(16;16)(qter;qter), der(18)t(18;18)(pter;qter), der(18)t(2;18)(pter;qter), der(18)t(4;18)(qter;qter) [n = 2], der(18)t(10;18)(pter;qter), der(18)t(14;18)(qter;qter), der(20)t(5;20)(qter;qter), der(21)t(14;21)(qter;qter), der(22)t(12;22)(qter;qter), der(X)t(X;14)(pter;qter), der(X)t(X;X or Y)(pter;qter), der(X)t(X;3)(qter;pter), der(X)(pter)/der(X)t(X;15)(pter;qter) *Others:* dup(4)(qter), idic(Y)(q11.2), t(1;5)(qter;qter), t(6;12)(qter;qter), t(19;21)(pter;qter), t(5;6)(qter;qter), der(22)ins(22;X or Y)(q11.2;pter)

TABLE 2: Continued.

Reference	Total cases	Positive cases	Frequency (%)	De novo	Familial	Unknown inheritance
Utine et al., 2009 [27]	130†	3	2.3	*Deletion:* del(22)(qter)dn; *Derivative:* der(4)t(4;8)(pter;pter)dn	*Derivative:* der(9)t(4;9)(qter;pter)mat	
Mihçi et al., 2009 [28]	107‡	9	8.4	*Deletion:* del(1)(pter–)dn [n = 2], del(3)(qter–)dn, del(9)(pter–)dn, del(9)(qter–)dn; *Derivative:* der(18)t(18;22)(qter–;qter+)dn	*Derivative:* der(5)t(5;15)(pter–;qter+)pat [n = 2 siblings]; *Others:* rec(10)dup(10p)inv(10)(p13q26)mat	
Belligni et al., 2009 [29]	76‡	10	13.2	*Deletion:* del(1)(p36)dn [n = 2], del(9)(q34qter)dn; *Derivative:* der(6)(ptel–;qtel++)dn, der(5)t(5pter;10qter)dn, t(1;13)(p32.2;q31.1)dn^f	*Derivative:* der(9)t(9;i6)(9pter–9q34.3;:16q24.3–qter)pat, der(20)t(16;20)(q24;q13.3)pat, der(6)t(6;1)(p22.3;q44)mat, der(7)t(7;12)(q34;q24.32)mat	
Bogdanowicz et al., 2010 [30]	76‡	4	5.3	*Deletion:* del(1)(p36.3–)dn [n = 2]; *Derivative:* der(1)t(1;12)(p36.3;q24.3+)dn; *Others:* t(19;22)(q13.4–;q13.3+;q13.3–;q13.4+)dn		
dos Santos and Freire-Maia, 2012 [31]	15‡	0	0			
Our study	82†	5	6.1	*Deletion:* del(1)(p36.32)dn; *Derivative:* der(3)t(X;3)(p22.3;p26.3)dn		*Deletion:* del(4)(p16.3); *Derivative:* der(10)t(7;10)(p22.3;q26.3), der(20)t(X;20)(p22.3;q13.3)
Summary	*15,591*	*473*	*3.0*			

aWe included 13 patients, who had subtle subtelomeric rearrangements detected by high resolution G-banding. Of these, 10 cases had subtelomeric rearrangements detected by FISH. We excluded del(4)(q35.2)dn and der(Y)t(Y;17)(pter;q25.3)dn that were detected in 150 controls. bFather was balanced translocation with deletion and duplication of 8p23. cWe excluded one case as FISH was performed in a normal father of 3 deceased patients with severe ID. The FISH result in the father revealed t(7;10)(q36;q26). dWe excluded 9 patients with interstitial deletions that were detected with control probes. eThe same individual. fKaryotype nomenclature here is as shown in the original article but it was actually an unbalanced rearrangement (trisomy for 1p32.2 and monosomy for 13q31.1). Note that 1p32.1 and 13q31.1 are actually not subtelomeric regions. *The same abnormality also found in parent but not a familial variant. †ID with or without dysmorphism. ‡ID with dysmorphism. +: gain, –: loss, dn: de novo, del: deletion, der: derivative, dim: partial deletion or diminished signal, dup: duplication, ins: insertion, inv: inversion, idic: isodicentric, mat: maternal, mos: mosaic, pat: paternal, rec: recombinant, tel: telomere, ter: terminal, and t: translocation.

TABLE 3: Familial variants and possible variants detected by FISH in patients with ID.

Reference	Number of cases	Variant
Fan et al., 2001 [6]	8	del(2)(qter) [n = 8]
Anderlid et al., 2002 [8]	2	del(2)(qter) [n = 2]
Baker et al., 2002 [9]	5	del(12)(pter) [n = 1], del(2)(qter) [n = 4]
Dawson et al., 2002 [11]	1	del(Y)(qter) [n = 1]
Clarkson et al., 2002 [12]	1	del(2)(qter) [n = 1]
van Karnebeek et al., 2002 [13]	11	del(2)(qter) [n = 7], del(X)(pter) [n = 3], del(Y)(pter) [n = 1]
Kirchhoff et al., 2004 [14]	5	del(2)(qter) [n = 5]
Ravnan et al., 2006 [26]	56	del(3)(pter) [n = 1], del(4)(qter) [n = 2], del(10)(qter) [n = 1], del(17)(pter) [n = 1], del(20)(pter) [n = 1], del(21)(qter) [n = 2], del(Y)(qter) [n = 10], dim(4)(pter) [n = 1], dim(10)(qter) [n = 1], dim(14)(qter) [n = 14*], dim(Y)(pter) [n = 3], dup(8)(pter) [n = 1], ?dup(10)(qter) [n = 1], mos dup(10)(qter) [n = 1], der(3)t(3;14)(pter;qter) [n = 1], der(18)t(16;18)(qter;pter) [n = 5], der(18)t(17;18)(pter;pter) [n = 1], der(20)t(X or Y;20)(qter;pter) [n = 1], der(X)t(X;X or Y)(qter;pter) [n = 5], der(22)t(4;22)(pter;pter) [n = 3]
Erjavec-Škerget et al., 2006 [23]	5	del(2)(qter) [n = 5**]
Rauch et al., 2006 [25]	52	del(9)(pter) [n = 1], del(Y)(qter) [n = 1], del(2)(qter) [n = 5], dim(2)(qter) [n = 24], dim(4)(qter) [n = 18], dim(2qter and 4qter) [n = 1], dim(15)(qter) [n = 2]
Bogdanowicz et al., 2010 [30]	3	del(2)(qter) [n = 2***], dup(7)(qter) [n = 1]
Total	*149*	

*One patient had dim(10)(qter) familial variant and a clinically significant del(16)(pter). **One patient had del(2)(qter) familial variant and a clinically significant del(X)(pter). ***One patient had del(2)(qter) familial variant and a clinically significant del(1)(pter).

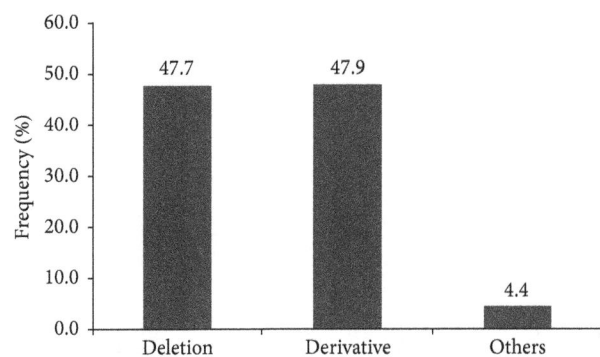

FIGURE 2: Frequency (%) of each category of subtelomeric rearrangements.

a subtelomere region onto the short arm of acrocentric chromosome, and an unbalanced translocation with only duplication or deletion of a subtelomere region detected. Monosomies associated with unbalanced derivative chromosomes frequently involved chromosomes 4p, 10q, 2q, 5p, 13q, 18q, and 7q (frequency >5% of all the monosomies) (Figure 3(b)). In this study, we detected subtelomeric rearrangements in 6.1% of the patients. Deletions of chromosomes 1p and 4p, which were among the most frequently detected deletions, were also detected in our study.

Attempts have been made to increase the diagnostic yield of the subtelomeric FISH method. A clinical checklist was developed to preselect patients for investigation. The selection criteria included prenatal/postnatal growth retardation, the presence of dysmorphism and/or congenital anomalies, and a family history of ID [37]. We found that frequency of subtelomeric rearrangements in patients with ID and MCA/DF was higher than that of those patients with ID alone (8.5% versus 2.9%). Our finding supported the suggestion that preselection of patients using the above-mentioned checklist, or a similar one, for subtelomeric FISH may increase the positive findings in patients with ID.

4.2. Clinical Features of Patients with Subtelomeric Rearrangements. In this study, we detected two well-recognized syndromes, namely, 4p deletion syndrome or Wolf-Hirschhorn syndrome and 1p36 deletion syndrome, in Patient 1 and Patient 2, respectively. Both syndromes have overlapping clinical features, including growth retardation, variable degree of ID/developmental delay, structural brain abnormalities, hypotonia, seizures, skeletal abnormalities, congenital heart defects, and hearing loss [38–40]. However, they are clinically recognized by distinct facial features, such as a Greek warrior helmet appearance of the nose, along with prominent glabella in the 4p deletion syndrome, and straight eyebrows, deeply set eyes, and midface retrusion in the 1p36 deletion syndrome. Clinical features of these syndromes may be variable depending on the extent of the deletions, in addition to the number and significance of the deleted genes. The deletions can be detected by using standard karyotype analysis in 50–60% of patients with 4p deletion syndrome and 25% of patients with 1p36 deletion syndrome, while FISH and chromosomal microarray (CMA) can detect chromosomal

rearrangements in over 95% of the patients [38, 39]. In this study, our patients with 4p deletion and 1p36 deletion syndromes had subtle facial features, most likely because they had cryptic subtelomeric deletions.

Approximately half of the subtelomeric rearrangements detected in patients with ID were unbalanced derivative chromosomes. The presence of derivative chromosome results in deletion (i.e., partial monosomy) along with duplication (i.e., partial trisomy) of distinct subtelomere regions. Clinical features that presented in our patients with derivative chromosomes were influenced by the coexistence of two genomic imbalances, from which phenotypic consequence resulting from one type of genomic imbalance confounds the phenotypic consequence resulting from the other genomic imbalance. In addition, variable expressivity, incomplete penetrance, and the degree of skewed X-inactivation when X chromosome involved in the derivative chromosomes could have an influence on the phenotypes of the patients.

Patient 3 possessed der(10)t(7;10)(p22.3;q26.3), representing three 7p subtelomeres and only one 10q subtelomere. We found only one previous report of a patient with der(10)t(7;10)(p22.3;q26.3). She was a 17-year-old woman with short stature and moderate ID. Unlike our patient, she showed no dysmorphic facies and microcephaly [22]. Difference in clinical features may be due to the extent of the deletion and duplication. Duplication of 7p22.3 was reported on a patient with Asperger syndrome [41] and a patient with DF and skeletal abnormalities, including abnormal distal humeri [42]. A deformity of the elbow was present in our patient. In addition, developmental delay and minor DF were reported in the other patient with interstitial deletion of 10q26.3 [43].

Duplication of Xp22.3 that was detected in Patient 4 and Patient 5 was also detected in 2 out of 129 Thai patients with unexplained ID from a different cohort, who had subtelomeric rearrangements detected by MLPA technique [44]. Frequency of Xp22.3 duplication in Thai patients with unexplained ID was 1.9% (4/211). The duplication of Xp22.3 was observed in individuals with neurocognitive and behavioral abnormalities [45]. Moreover, deletion of Xp22.3 was associated with X-linked mental retardation and attention deficit hyperactivity disorder [45]. Patient 4 with deletion of 20q13.3 had only moderate ID with minor DF. Deletion of 20q13.33 was previously reported on patients with epileptic seizures. For Patient 5, significance of 3p26.3 deletion was unclear. Deletion of this region was previously described in patients with ID and atypical autism [46]; however, it was also reported in four generations of a family that were apparently healthy [47].

5. Conclusions

We detected subtelomeric rearrangements in 6.1% of the patients with ID. The sensitivity of subtelomeric FISH increases when preselection criteria were applied. We reported clinical entities observed in a patient with der(20)t(X;20)(p22.3;q13.3), a patient with der(3)t(X;3)(p22.3;p26.3), and a patient with der(10)t(7;10)(p22.3;q26.3). These rearrangements have never been or had rarely been reported in literature. Even though

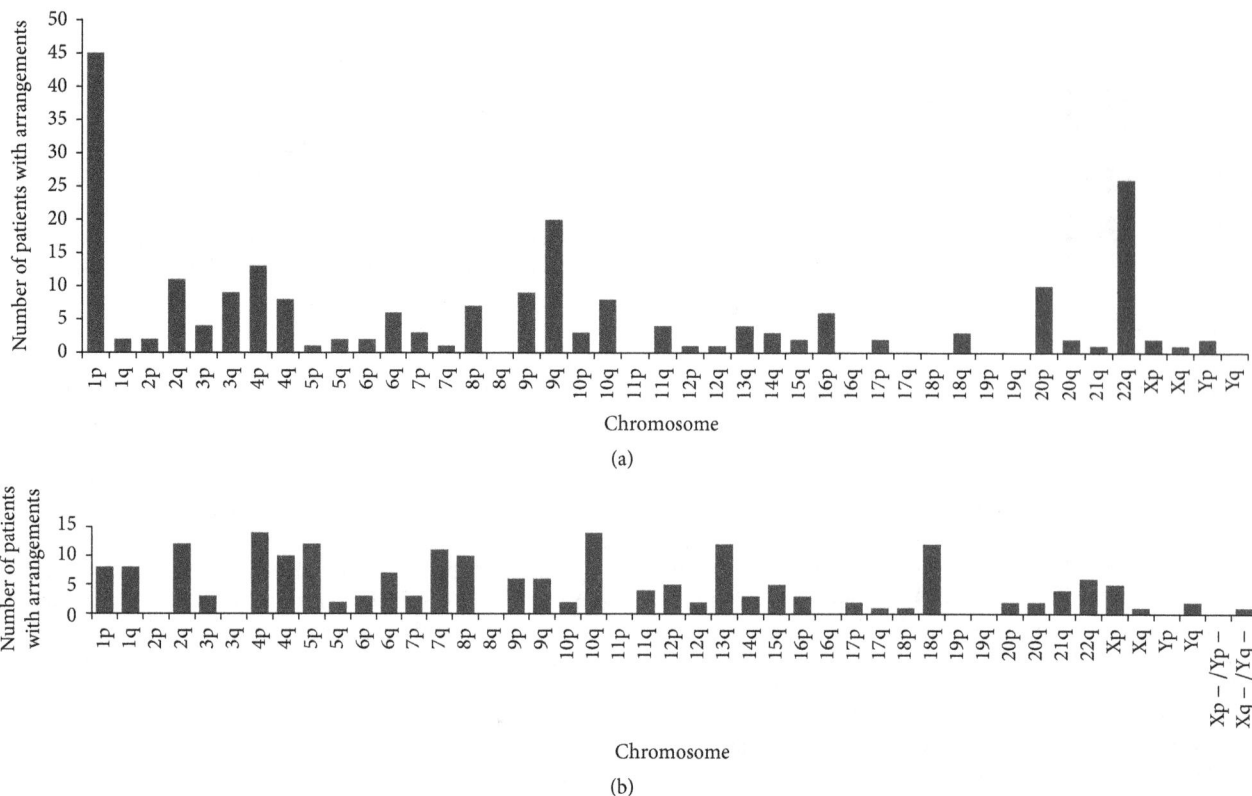

FIGURE 3: Number of patients with subtelomeric rearrangements involving each chromosome region. (a) Terminal deletion ($n = 226$). (b) Unbalanced derivative. The only chromosome in which the subtelomere region was monosomy is depicted here ($n = 204$).

subtelomeric FISH is a very robust technique, as it can detect and locate rearrangements that involve small and specific regions of the chromosomes, uncover low level mosaicism, and identify balanced chromosomal rearrangements, CMA is currently recommended as a first-tier test and replaces the standard karyotype and subtelomeric FISH analyses for patients with ID of unknown causes [48]. However, in countries where CMA is unavailable or unaffordable, FISH is a useful technique to screen for subtelomeric rearrangements. In this study, we added information regarding clinical entities of our patients with subtelomeric rearrangements. This information requires accumulation from case reports and is useful for genetic counselling.

Competing Interests

All authors declare that they have no financial or other relevant conflict of interests regarding this paper.

Acknowledgments

The authors thank all participating families for their generous contribution of time and biological materials. They appreciate all the clinicians for their collaboration. This study is supported by the Thailand Research Fund (Grant no. PDF/98/2544), the National Center for Genetic Engineering and Biotechnology (Grant no. BT-B-01-MG-79-4701), and the Graduate School Research Dissertation Funding, Prince of Songkla University.

References

[1] WHO, *ICD-10 Guide for Mental Retardation*, World Health Organization, Geneva, Switzerland, 1996.

[2] J. Xu and Z. Chen, "Advances in molecular cytogenetics for the evaluation of mental retardation," *American Journal of Medical Genetics Part C*, vol. 117, no. 1, pp. 15–24, 2003.

[3] H. C. Mefford and B. J. Trask, "The complex structure and dynamic evolution of human subtelomeres," *Nature Reviews Genetics*, vol. 3, no. 2, pp. 91–102, 2002.

[4] S. J. L. Knight, R. Regan, A. Nicod et al., "Subtle chromosomal rearrangements in children with unexplained mental retardation," *The Lancet*, vol. 354, no. 9191, pp. 1676–1681, 1999.

[5] C. A. Joyce, N. R. Dennis, S. Cooper, and C. E. Browne, "Subtelomeric rearrangements: results from a study of selected and unselected probands with idiopathic mental retardation and control individuals by using high-resolution G-banding and FISH," *Human Genetics*, vol. 109, no. 4, pp. 440–451, 2001.

[6] Y.-S. Fan, Y. Zhang, M. Speevak, S. Farrell, J. H. Jung, and V. M. Siu, "Detection of submicroscopic aberrations in patients with unexplained mental retardation by fluorescence in situ hybridization using multiple subtelomeric probes," *Genetics in Medicine*, vol. 3, no. 6, pp. 416–421, 2001.

[7] C. Sismani, J. A. L. Armour, J. Flint, C. Girgalli, R. Regan, and P. C. Patsalis, "Screening for subtelomeric chromosome

abnormalities in children with idiopathic mental retardation using multiprobe telomeric FISH and the new MAPH telomeric assay," *European Journal of Human Genetics*, vol. 9, no. 7, pp. 527–532, 2001.

[8] B.-M. Anderlid, J. Schoumans, G. Annerén et al., "Subtelomeric rearrangements detected in patients with idiopathic mental retardation," *American Journal of Medical Genetics*, vol. 107, no. 4, pp. 275–284, 2002.

[9] E. Baker, L. Hinton, D. F. Callen et al., "Study of 250 children with idiopathic mental retardation reveals nine cryptic and diverse subtelomeric chromosome anomalies," *American Journal of Medical Genetics Part A*, vol. 107, no. 4, pp. 285–293, 2002.

[10] S. Popp, B. Schulze, M. Granzow et al., "Study of 30 patients with unexplained developmental delay and dysmorphic features or congenital abnormalities using conventional cytogenetics and multiplex FISH telomere (M-TEL) integrity assay," *Human Genetics*, vol. 111, no. 1, pp. 31–39, 2002.

[11] A. J. Dawson, S. Putnam, J. Schultz et al., "Cryptic chromosome rearrangements detected by subtelomere assay in patients with mental retardation and dysmorphic features," *Clinical Genetics*, vol. 62, no. 6, pp. 488–494, 2002.

[12] B. Clarkson, K. Pavenski, L. Dupuis et al., "Detecting rearrangements in children using subtelomeric FISH and SKY," *American Journal of Medical Genetics*, vol. 107, no. 4, pp. 267–274, 2002.

[13] C. D. M. van Karnebeek, C. Koevoets, S. Sluijter et al., "Prospective screening for subtelomeric rearrangements in children with mental retardation of unknown aetiology: the Amsterdam experience," *Journal of Medical Genetics*, vol. 39, no. 8, pp. 546–553, 2002.

[14] M. Kirchhoff, S. Pedersen, E. Kjeldsen et al., "Prospective study comparing HR-CGH and subtelomeric FISH for investigation of individuals with mental retardation and dysmorphic features and an update of a study using only HR-CGH," *American Journal of Medical Genetics*, vol. 127, no. 2, pp. 111–117, 2004.

[15] E. Font-Montgomery, D. D. Weaver, L. Walsh, C. Christensen, and V. C. Thurston, "Clinical and cytogenetic manifestations of subtelomeric aberrations: report of six cases," *Birth Defects Research Part A—Clinical and Molecular Teratology*, vol. 70, no. 6, pp. 408–415, 2004.

[16] S. Walter, C. Sandig, G. K. Hinkel et al., "Subtelomere FISH in 50 children with mental retardation and minor anomalies, identified by a checklist, detects 10 rearrangements including a de novo balanced translocation of chromosomes 17p13.3 and 20q13.33," *American Journal of Medical Genetics*, vol. 128, no. 4, pp. 364–373, 2004.

[17] L. Rodriguez-Revenga, C. Badenas, A. Sánchez et al., "Cryptic chromosomal rearrangement screening in 30 patients with mental retardation and dysmorphic features," *Clinical Genetics*, vol. 65, no. 1, pp. 17–23, 2004.

[18] E. Bocian, Z. Hélias-Rodzewicz, K. Suchenek et al., "Subtelomeric rearrangements: results from FISH studies in 84 families with idiopathic mental retardation," *Medical Science Monitor*, vol. 10, no. 4, pp. CR143–CR151, 2004.

[19] G. V. N. Velagaleti, S. S. Robinson, B. M. Rouse, V. S. Tonk, and L. H. Lockhart, "Subtelomeric rearrangements in idiopathic mental retardation," *Indian Journal of Pediatrics*, vol. 72, no. 8, pp. 679–685, 2005.

[20] A. Baroncini, F. Rivieri, A. Capucci et al., "FISH screening for subtelomeric rearrangements in 219 patients with idiopathic mental retardation and normal karyotype," *European Journal of Medical Genetics*, vol. 48, no. 4, pp. 388–396, 2005.

[21] M. Sogaard, Z. Tümer, H. Hjalgrim et al., "Subtelomeric study of 132 patients with mental retardation reveals 9 chromosomal anomalies and contributes to the delineation of submicroscopic deletions of 1pter, 2qter, 4pter, 5qter and 9qter," *BMC Medical Genetics*, vol. 6, article 21, 2005.

[22] S. Yu, E. Baker, L. Hinton et al., "Frequency of truly cryptic subtelomere abnormalities—a study of 534 patients and literature review," *Clinical Genetics*, vol. 68, no. 5, pp. 436–441, 2005.

[23] A. Erjavec-Škerget, Š. Stangler-Herodež, A. Zagorac, B. Zagradišnik, and N. Kokalj-Vokač, "Subtelomeric chromosome rearrangements in children with idiopathic mental retardation: applicability of three molecular-cytogenetic methods," *Croatian Medical Journal*, vol. 47, no. 6, pp. 841–850, 2006.

[24] M. Palomares, A. Delicado, P. Lapunzina et al., "MLPA vs multiprobe FISH: comparison of two methods for the screening of subtelomeric rearrangements in 50 patients with idiopathic mental retardation," *Clinical Genetics*, vol. 69, no. 3, pp. 228–233, 2006.

[25] A. Rauch, J. Hoyer, S. Guth et al., "Diagnostic yield of various genetic approaches in patients with unexplained developmental delay or mental retardation," *American Journal of Medical Genetics, Part A*, vol. 140, no. 19, pp. 2063–2074, 2006.

[26] J. B. Ravnan, J. H. Tepperberg, P. Papenhausen et al., "Subtelomere FISH analysis of 11 688 cases: an evaluation of the frequency and pattern of subtelomere rearrangements in individuals with developmental disabilities," *Journal of Medical Genetics*, vol. 43, no. 6, pp. 478–489, 2006.

[27] G. E. Ütine, T. Çelik, Y. Alanay et al., "Subtelomeric rearrangements in mental retardation: Hacettepe University experience in 130 patients," *The Turkish Journal of Pediatrics*, vol. 51, no. 3, pp. 199–206, 2009.

[28] E. Mihçi, M. Özcan, S. Berker-Karaüzüm et al., "Subtelomeric rearrangements of dysmorphic children with idiopathic mental retardation reveal 8 different chromosomal anomalies," *The Turkish Journal of Pediatrics*, vol. 51, no. 5, pp. 453–459, 2009.

[29] E. F. Belligni, E. Biamino, C. Molinatto et al., "Subtelomeric FISH analysis in 76 patients with syndromic developmental delay/intellectual disability," *Italian Journal of Pediatrics*, vol. 35, article 9, 2009.

[30] J. Bogdanowicz, B. Pawłowska, A. Ilnicka et al., "Subtelomeric rearrangements in Polish subjects with intellectual disability and dysmorphic features," *Journal of Applied Genetics*, vol. 51, no. 2, pp. 215–217, 2010.

[31] S. R. dos Santos and D. V. Freire-Maia, "Absence of subtelomeric rearrangements in selected patients with mental retardation as assessed by multiprobe T FISH," *Journal of Negative Results in BioMedicine*, vol. 11, no. 1, article 16, 2012.

[32] J. W. Ahn, C. Mackie Ogilvie, A. Welch et al., "Detection of subtelomere imbalance using MLPA: validation, development of an analysis protocol, and application in a diagnostic centre," *BMC Medical Genetics*, vol. 8, article 9, 2007.

[33] A. P. A. Stegmann, L. M. H. Jonker, and J. J. M. Engelen, "Prospective screening of patients with unexplained mental retardation using subtelomeric MLPA strongly increases the detection rate of cryptic unbalanced chromosomal rearrangements," *European Journal of Medical Genetics*, vol. 51, no. 2, pp. 93–105, 2008.

[34] F. E. P. Mundhofir, W. M. Nillesen, B. W. M. Van Bon et al., "Subtelomeric chromosomal rearrangements in a large cohort of unexplained intellectually disabled individuals in Indonesia: a clinical and molecular study," *Indian Journal of Human Genetics*, vol. 19, no. 2, pp. 171–178, 2013.

[35] B. C. Ballif, S. G. Sulpizio, R. M. Lloyd et al., "The clinical utility of enhanced subtelomeric coverage in array CGH," *American Journal of Medical Genetics Part A*, vol. 143, no. 16, pp. 1850–1857, 2007.

[36] A.-C. Thuresson, M.-L. Bondeson, C. Edeby et al., "Whole-genome array-CGH for detection of submicroscopic chromosomal imbalances in children with mental retardation," *Cytogenetic and Genome Research*, vol. 118, no. 1, pp. 1–7, 2007.

[37] B. B. A. De Vries, S. M. White, S. J. L. Knight et al., "Clinical studies on submicroscopic subtelomeric rearrangements: a checklist," *Journal of Medical Genetics*, vol. 38, no. 3, pp. 145–150, 2001.

[38] A. Battaglia, J. C. Carey, and S. T. South, "Wolf-Hirschhorn syndrome," in *GeneReviews®*, R. A. Pagon, M. P. Adam, H. H. Ardinger et al., Eds., University of Washington, Seattle, Seattle, Wash, USA, 1993–2016.

[39] A. Battaglia, "1p36 deletion syndrome," in *GeneReviews®*, R. A. Pagon, M. P. Adam, H. H. Ardinger et al., Eds., University of Washington, Seattle, Seattle, Wash, USA, 1993–2016.

[40] V. K. Jordan, H. P. Zaveri, and D. A. Scott, "1p36 deletion syndrome: an update," *The Application of Clinical Genetics*, vol. 8, pp. 189–200, 2015.

[41] A. T. Vulto-van Silfhout, A. P. M. de Brouwer, N. de Leeuw, C. C. Obihara, H. G. Brunner, and B. B. A. de Vries, "A 380-kb duplication in 7p22.3 encompassing the LFNG gene in a boy with asperger syndrome," *Molecular Syndromology*, vol. 2, no. 6, pp. 245–250, 2012.

[42] K. M. Girisha, H. Abdollahpour, H. Shah et al., "A syndrome of facial dysmorphism, cubital pterygium, short distal phalanges, swan neck deformity of fingers, and scoliosis," *American Journal of Medical Genetics Part A*, vol. 164, no. 4, pp. 1035–1040, 2014.

[43] I. Y. Iourov, S. G. Vorsanova, O. S. Kurinnaia, and Y. B. Yurov, "An interstitial deletion at 10q26.2q26.3," *Case Reports in Genetics*, vol. 2014, Article ID 505832, 3 pages, 2014.

[44] S. Rujirabanjerd, O. Plong-On, T. Sripo, and P. Limprasert, "Subtelomeric aberrations in Thai patients with idiopathic mental retardation and autism detected by multiplex ligation-dependent probe amplification," *Asian Biomedicine*, vol. 9, no. 4, pp. 501–509, 2015.

[45] P. Liu, A. Erez, S. C. S. Nagamani et al., "Copy number gain at Xp22.31 includes complex duplication rearrangements and recurrent triplications," *Human Molecular Genetics*, vol. 20, no. 10, pp. 1975–1988, 2011.

[46] A. A. Kashevarova, L. P. Nazarenko, S. Schultz-Pedersen et al., "Single gene microdeletions and microduplication of 3p26.3 in three unrelated families: CNTN6 as a new candidate gene for intellectual disability," *Molecular Cytogenetics*, vol. 7, article 97, 2014.

[47] S. Moghadasi, A. van Haeringen, L. Langendonck, A. C. J. Gijsbers, and C. A. L. Ruivenkamp, "A terminal 3p26.3 deletion is not associated with dysmorphic features and intellectual disability in a four-generation family," *American Journal of Medical Genetics Part A*, vol. 164, no. 11, pp. 2863–2868, 2014.

[48] J. B. Moeschler, M. Shevell, and Committee on Genetics, "Comprehensive evaluation of the child with intellectual disability or global developmental delays," *Pediatrics*, vol. 134, no. 3, pp. e903–e918, 2014.

20

Does i-T744C P2Y12 Polymorphism Modulate Clopidogrel Response among Moroccan Acute Coronary Syndromes Patients?

Hind Hassani Idrissi,[1] **Wiam Hmimech,**[1] **Nada El Khorb,**[2] **Hafid Akoudad,**[2] **Rachida Habbal,**[3] **and Sellama Nadifi**[1]

[1]*Laboratory of Genetics and Molecular Pathology, Medical School, University Hassan II, Casablanca, Morocco*
[2]*Department of Cardiology, University Hospital Center Hassan II, Fes, Morocco*
[3]*Department of Cardiology, University Hospital Center Ibn Rochd, Casablanca, Morocco*

Correspondence should be addressed to Hind Hassani Idrissi; hassani-idrissi-hind@hotmail.fr

Academic Editor: Norman A. Doggett

Background. An interindividual variability in response to Clopidogrel has been widely described in patients with acute coronary syndromes (ACS). The contribution of genetics on modulating this response was widely discussed. The objective of our study was to investigate the potential effect of i-T744C P2Y12 polymorphism on Clopidogrel response in a sample of Moroccan ACS patients. We tried also to determine the frequency of this polymorphism among Moroccan ACS compared to healthy subjects. *Methods and Results.* 77 ACS patients versus 101 healthy controls were recruited. DNA samples were genotyped by PCR-RFLP method. The VerifyNow assay was used to evaluate platelet function among ACS patients. Our results show that the mutant allele C was more frequent among ACS ST (+) than ST (−) patients (39% versus 19.8%, resp.), when the wild-type allele was more represented in the ACS ST (−) group (80.2%). The C allele frequency was higher among resistant than nonresistant patients (30% versus 20.8%, resp.). Comparison of ACS patients and healthy controls shows higher frequency of mutant C allele among cases compared to controls (22.73% versus 19.31%, resp.); there was a statistically significant association of the recessive and additive transmission models with the ACS development risk (OR [95% CI] = 1.78 [1.58–5.05], $P = 0.01$ and OR [95% CI] = 1.23 [0.74–2.03], $P < 0.001$, resp.), increasing thus the association of this polymorphism with the pathology. *Conclusion.* Our results suggest that this polymorphism may have a potential effect on Clopidogrel response among our Moroccan ACS patients and also on ACS development.

1. Background

Because the majority of cardiovascular diseases are the result of an occlusive thrombosis, numerous antithrombotic drugs are used in there therapy, including platelet inhibitors and oral anticoagulants [1].

Clopidogrel is a thienopyridine derivative, used to inhibit the formation of blood clots in coronary artery disease, peripheral vascular disease, and stroke. It irreversibly inhibits the receptor of adenosine diphosphate (ADP), called P2Y12, expressed on the surface of platelets. This prodrug requires a hepatic oxidation step by the cytochrome P450 (CYP450) enzymes, to generate the active metabolite responsible for the irreversible blocking effect of the P2Y12 receptor during the life of the platelet [2].

In Acute Coronary Syndrome (ACS), a wide range of interindividual variations in platelet response to Clopidogrel, has been described. An important proportion of patients still experience thrombotic events even after receiving the treatment, so they do not reach the same degree of benefit from the given drug [3]. Several factors were found to be in association with this heterogeneity in response to antithrombotic agents among patients [4, 5]. The role of pharmacogenomics is to study the genetic factors that determine the response of a given individual to a given drug. This variability of response to treatment may explain both its efficacy and its adverse side effects [6].

Several genes are involved in the modulation of this response. These genes may act in absorption of the molecule:

transportation (ABCB), metabolism (cytochromes), excretion (for side effects), or targets of direct or indirect action of the molecule (receptors: P2Y12, Gp IIb/IIIa, Gp Ia/IIa, etc.) [7–15].

P2Y12. The P2Y12 is the platelet receptor for adenosine diphosphate (ADP) targeted by the active form of Clopidogrel [16]. The protein encoded by this gene belongs to the big family of G-protein coupled receptors, which contains several receptor groups with different pharmacological selectivity. It is involved in platelet aggregation and is a potential target for the treatment of thromboembolic pathologies and clotting disorders. The P2Y12 gene is localized on human chromosome 3 (3q25.1); it covers 47.97 pb of length. Mutations in this gene are implicated in bleeding disorder, platelet type 8 (BDPLT8). Many studies have assessed the functional role of the P2Y12 gene variants in modulating the response to antiplatelet drugs [11, 17]. Recently, the T744C polymorphism of the P2Y12 receptor gene has been associated with enhanced platelet aggregation in healthy volunteers, suggesting a possible mechanism for modulation of Clopidogrel response [18, 19].

The main objectives of our study is to determine the frequency of i-T744C P2Y12 polymorphism among Moroccan ACS and healthy subjects and to assess whether or not the Clopidogrel response may be influenced by this genetic polymorphism in a sample of Moroccan ASC patients.

2. Materials and Methods

2.1. Study Population. Patients were eligible for inclusion if they had documented antiplatelet therapy (Clopidogrel), a VerifyNow P2Y12 platelet function test, and no more heparin in their blood. Patients were excluded if they did not have a VerifyNow P2Y12 platelet function test or having incomplete clinical data. All patients received a baseline P2Y12 platelet function test to identify Clopidogrel resistance and determine whether they would need another loading dose to achieve P2Y12 response (PRU PRU < 208 and inhibition% ≥ 20%).

Blood samples were collected from 77 unrelated ACS Moroccan patients and 101 apparently healthy subjects showing no symptoms of coronary artery diseases. Clinical data concerning risk factors, biological parameters, and the VerifyNow test results were collected; an informed consent that was approved by the Ethical Committee of the University of Hassan II, School of Medecine, Casablanca, was signed by each patient and control before entering the study.

2.2. Study Protocol. Recruited patients received 300 mg loading dose of the generic molecule; it was replaced by 75 mg maintenance dose of Plavix for 7 days (washout period) if Clopidogrel resistance was noted on the initial platelet test (PRU > 208); a 300 mg loading dose of Plavix was prescribed if the PRU remains and the resistance persists; inhibition was considered adequate for good response to treatment if the value reached ≥20%.

The VerifyNow test (Accumetrics Inc., San Diego, California) was used to evaluate platelet function, as it is a point-of-care assay, easy to perform, and rapid and uses small volumes

of whole blood samples [20]. Two results are reported: the PRU (P2Y12 Reactive Units) and the percent inhibition. The ideal percent of platelet inhibition is ≥30% for Clopidogrel; however, 20–30% inhibition is considered as intermediate response [21, 22]. In our study, resistance to Clopidogrel was defined by <20% inhibition + PRU > 208 after many platelet function tests and nonresistance by ≥20% inhibition + PRU < 208.

2.3. DNA Extraction. Venous blood from all participants in this study was collected in EDTA tubes. Samples were treated immediately or stored at −20°C until extraction of DNA. Genomic DNA was extracted from blood leukocytes using the standard method of salting out [23].

2.4. Genotype Determination. We used PCR-RFLP to genotype samples for i-T744C P2Y12 polymorphism, as previously described by Malek et al. [16]. Genotyping of this variant was performed by amplification from 50 to 100 ng of genomic DNA, followed by digestion using *RsaI* restriction enzyme. The digestion gave rise to three profiles: wild TT homozygous (one fragment of 220 pb), TC heterozygote (two fragments of 220 and 196 bp), and mutated CC homozygous (one fragment of 196 bp). The digested product was separated on 3% agarose gel electrophoresis stained with Ethidium Bromide (BET) and visualized with UV rays.

2.5. Statistical Analysis. Statistical analysis was performed using SPSS 21.0 software. Chi square test (χ^2) was used to determine statistical significance of association/nonassociation between genotypes and classical risk factors. Hardy–Weinberg equilibrium (HWE) test was performed for cases and controls groups. Odds ratio (OR) were calculated to estimate the association between genotypes and ACS risk, using a Confidence Interval (CI) of 95%. Significance was approved at P value less than 0.05.

3. Results

3.1. Characteristics of the Study Population. The distribution of i-T744C P2Y12 polymorphism was in Hardy–Weinberg equilibrium (HWE) for controls and cases groups (Table 1). The average age was 57.33±9.7 for patients versus 32±9.87 for healthy controls. There was a predominance of male in both groups (cases: 54.54%; controls: 53.52%) (Table 2). Table 3 describes the routine pathology data for our 77 SCA patients; 79.6% of these patients were under IPP, when 20.4% were not.

3.2. VerifyNow Results versus Risk Factors. Patients were placed into resistant and nonresistant groups, based on their platelet function test results, and the baseline characteristics of these patients correlated to resistance groups are shown in Table 4: only creatinine level, fibrinogen, Pq numerisation, and IPP use show statistically significant association (P = 0.01; 0.04; 0.04; and 0.03, resp.).

3.3. Allelic Frequencies. When correlating i-T744C P2Y12 genotypes to the classical risk factors of the pathology, a statistically significant association was found with familial

TABLE 1: HWE among cases and controls.

Genotypes	EHW cases		EHW controls	
	χ^2 square	P value (P > 0.05)	χ^2 square	P value (P > 0.05)
i-T744C P2Y12 (rs2046934)	1.35	**0.51***	4.26	**0.13***

*Statistically significant.

TABLE 2: Description of ACS study population.

	ACS patients	Controls
Age (years)	57.33 ± 9.7	32 ± 9.87
Age of disease occurrence (years)	54.81 ± 10.2	
Sex		
Male	54.54%	53.52%
Female	45.46%	46.48%
Ethnicity		
Arab	88%	86%
Berber	2%	4%

TABLE 3: Routine pathology data of our ACS patients.

Parameters	SCA patients
Total cholesterol (g/L)	1.88 ± 0.73
HDL (g/L)	1.24 ± 0.9
LDL (g/L)	1.24 ± 0.64
Triglycerides TG (g/L)	1.53 ± 1.00
Glucose (g/L)	1.43 ± 0.8
Creatinine (mg/L)	9.51 ± 2.46
Fibrinogen	3.65 ± 1.06
HB (g/dL)	13.99 ± 2.63
GB (elts/mm^3)	12958.6 ± 24041.3
Pq (elts/mm^3)	237151.7 ± 102412
BMI (Kg/m^2)	26.72 ± 4.17
IPP	
(+)	79.6%
(−)	20.4%

TABLE 4: Baseline characteristics of our SCA patients versus VerifyNow test results.

	Nonresistant%	Resistant%	P value
Age	56.9 ± 8.48	62 ± 8.02	0.9
Sex			0.41
Male	58.3%	20%	
Female	41.7%	80%	
SCA type			0.2
ST (+)	36%	60%	
ST (−)	64%	40%	
Familial ACD			0.8
Presence	2.8%	0%	
Absence	97.2%	100%	
Personal ACD			0.3
Presence	41.7%	20%	
Absence	58.3%	80%	
Diabetes			0.6
Presence	40.3%	40%	
Absence	59.7%	60%	
HTA			0.46
Presence	47.2%	60%	
Absence	52.8%	40%	
Dyslipidemia			0.22
Presence	16.7%	40%	
Absence	83.3%	60%	
Smoking			0.6
Presence	40.3%	40%	
Absence	59.7%	60%	
Creatinine (mg/L)	11.6 ± 10.87	15 ± 4.34	**0.01***
Fibrinogen	3.57 ± 1.06	5.03 ± 1.14	**0.04***
Pq	242333 ± 102412	159427 ± 79609	**0.04***
IPP			**0.03***
Used	50.38%	60%	
Nonused	49.62%	40%	

*Statistically significant.

antecedent among SCA patients; no association was detected with the other risk factors (Table 5).

Table 6 shows the distribution of patients ACS type (ST (+) and ST (−)) according to i-T744C P2Y12 genotypic profiles: 62.75% of the wild-type and 76.48% of the heterozygous profiles were SCA ST (−), when the majority of the mutated profile was SCA ST (+) (66.7%). Mutant allele was more frequent among SCA ST (+) than SCA ST (−) patients (39% versus 19.8%, resp.); the wild-type allele was more frequent in SCA ST (−) group than SCA ST (+) one (80.2% versus 61%, resp.).

Distribution of resistant and nonresistant patients according to i-T744C P2Y12 genotypes is reported in Table 7: 69.45% of nonresistant patients are wild-type for this polymorphism; 19.45% are heterozygous; and 11.1% are homozygous for the mutation; in the resistant group, 60% are wild-type, 20% are heterozygous, and 20% are homozygous

mutant. The mutant allele is more frequent among resistant than nonresistant patients (30% and 20.8%, resp.).

Allelic and genotypic frequencies among cases and controls are reported in Table 8: genotypic frequencies were 66.23% TT, 20.07% TC, and 11.7% CC among cases versus 68.32% TT, 24.75% TC, and 6.93% CC among healthy controls. Allelic frequencies were 77.27% T and 22.73% C among cases versus 80.69% T and 19.31% C among controls. A statistically significant association was found with both TC and CC genotypes (OR [95% CI] = 0.92 [0.45–1.87], P = 0.0048

TABLE 5: i-T744C P2Y12 rs2046934 genotypes distribution versus risk factors.

Risk factor	SCA patients			P value (<0.05)
	Wild-type%	Heterozygous%	Mutant%	
Familial ACD				**0.014***
Presence	0	100	0	
Absence	71.4	17.2	11.4	
Personal ACD				0.6
Presence	75	17.9	7.1	
Absence	65.9	20.5	13.6	
HTA				0.8
Presence	71.4	17.2	11.4	
Absence	67.6	21.6	10.8	
Smoking				0.51
Presence	70	23.3	6.7	
Absence	69	16.7	14.3	
Diabetes				0.71
Presence	65.5	24.2	10.3	
Absence	72.1	16.3	11.6	
Dyslipidemia				0.8
Presence	71.4	21.4	7.2	
Absence	69	19	12	

*Statistically significant.

TABLE 6: i-T744C P2Y12 rs2046934 genotypes distribution VS ACS subgroups.

	SCA patients			Wild-type allele	Mutant allele	P value (<0.05)
	TT%	TC%	CC%			
ST (+)	37.25	23.52	66.7	61%	39%	
ST (−)	62.75	76.48	33.3	80.2	19.8%	0.26

TABLE 7: i-T744C P2Y12 rs2046934 genotypes distribution versus VerifyNow test results.

	SCA patients			Wild-type allele%	Mutant allele%	P value
	TT%	TC%	CC%			
Nonresistant	69.45	19.45	11.1	79.2	20.8	
Resistant	60	20	20	70	30	0.45

TABLE 8: Allelic and genotypic frequencies of i-T744C P2Y12 rs2046934 polymorphism among cases and controls.

	Genotypes/alleles	Cases (%)	Controls (%)	OR (95% CI)	P value
	TT	66.23	68.32	1	
	TC	22.07	24.75	0.92 [0.45–1.87]	**0.0048***
	CC	11.7	6.93	1.74 [1.66–5.00]	**0.03***
	TT + TC[b]	88.3	93.07	1	
i-T744C P2Y12 (rs2046934)	CC	11.7	6.93	1.78 [1.58–5.05]	**0.01***
	TT[c]	66.23	68.32	1	
	TC + CC	33.77	31.68	1.1 [1.7–2.05]	0.15
	T[d]	77.27	80.69	1	
	C	22.73	19.31	1.23 [0.74–2.03]	**<0.001***

*Statistically significant.
[b]Recessive model.
[c]Dominant model.
[d]Additive model.

and OR [95% CI] = 1.74 [1.66–5.00], P = 0.03, resp.). There was a positive correlation with the recessive and additive transmission models, but not the dominant one (OR [95% CI] = 1.78 [1.58–5.05], P = 0.01 and OR [95% CI] = 1.23 [0.74–2.03], P < 0.001, resp.), increasing thus the association of this polymorphism with the pathology.

4. Discussion

Clopidogrel is a second-generation thienopyridine, having better efficacy of ticlopidine that represents the first-generation; it has better tolerability profiles and is currently the antiplatelet treatment of choice for prevention of thrombosis events [24, 25]. An interindividual variation in platelet response to Clopidogrel has been widely reported; it may be explained by several factors, including genetics [6]. Many Single Nucleotide Polymorphisms (SNPs) of the P2Y12 receptor gene were described to be in association with this interindividual variability in ADP-induced platelet aggregation [26–28]. i-T744C polymorphism has been associated with enhanced platelet aggregation, suggesting its potential effect on modulating Clopidogrel response [29, 30].

Our study is the first to determine the frequency of i-T744C P2Y12 polymorphism among Moroccan ACS patients and healthy subjects and to evaluate the correlation between Clopidogrel resistance and genetic testing represented by i-T744C P2Y12 polymorphism, among a sample of Moroccan ACS patients.

In our study sample, there was a predominance of male in both cases and controls groups (54.54% and 53.52%, resp.); the average age was 57.33 and 32, respectively, among cases and controls (Table 2). This was in agreement with what Zoheir et al. found in their study about P2Y12 receptor gene polymorphism and antiplatelet effect of Clopidogrel in patients with coronary artery disease after coronary stenting [26].

79.6% of our patients were under PPI, when 20.4% were not (Table 3). PPI use showed a statistically significant association when being correlated to patients groups (resistant and nonresistant patients), divided based on their platelet function test results (P = 0.03) (Table 4). Proton-pump inhibitors (PPI) are known to potentially affect the Clopidogrel platelet inhibition relationship [30]; also creatinine level, fibrinogen, and Pq numerisation showed statistically significant association (P = 0.01; 0.04; and 0.04, resp.) (Table 4).

Concerning resistant and nonresistant groups of patients, our results were in agreement with Nordeen et al.'s study: they found the same range of age, with the majority of resistant group being female, compared to nonresistant one [30]. Several studies have suggested that women do not accrue equal therapeutic benefit of antithrombotic therapy [31, 32]. Although multiple contributing factors have been described (differences in vessel wall biology between men and women; the direct influence of sex hormones (oestrogens, progesterone, or androgens) on platelets and their indirect effect on the vasculature), the physiological mechanism behind this gender disparity remains unclear [33].

A statistical comparison was held between distribution of i-T744C polymorphism among ACS patients and traditional risk factors; we found significant association only with familial antecedent factor (P = 0.014; Table 5). Correlation between this polymorphism and ACS subgroups (ST+ and ST−) showed that the majority of wild-type and heterozygous profiles were SCA ST (−), when the majority of the mutated profiles were SCA ST (+) (Table 6). Mutant allele was more frequent among SCA ST (+) patients, when wild-type allele was more present in SCA ST (−) group. Distribution of resistant and nonresistant patients according to i-T744C P2Y12 genotypes showed that 69.45% of nonresistant patients had the wild-type profile; 19.45% were heterozygous; and 11.1% were homozygous mutant; in the resistant group, 60% were wild-type, 20% heterozygous, and 20% homozygous mutant. The mutant allele was more frequent among resistant than nonresistant patients (30% and 20.8%, resp.). Zoheir et al. [26] found a higher expression of C allele (heterozygous CT and homozygous CC) among nonresponder ACS patients (P < 0.001). Fontana et al. [29] also found similar results suggesting that the H2 haplotype of the P2Y12 gene is associated with increased platelet function in nonmedicated healthy volunteers. On the contrary, the results of Cuisset et al. [18] show no influence of i-T744C P2Y12 polymorphism on Clopidogrel response. Platelet function studies performed by Lev et al. [12] in the same context did not show any modulating effect of this genetic polymorphism on individual responsiveness to Clopidogrel. Furthermore, Hetherington et al. [34] reported no significant effect of i-T744C P2Y12 SNP on platelet response to ADP among subjects without cardiovascular disease history.

In our study, we tried also to investigate whether the mutant allele C of i-T744C P2Y12 polymorphism has an effect on ACS occurrence. Our results revealed that the mutant allele C was more frequent among cases than controls (22.73% versus 19.31%, resp.). A statistically significant association was found with both TC and CC genotypes (OR [95% CI] = 0.92 [0.45–1.87], P = 0.0048 and OR [95% CI] = 1.74 [1.66–5.00], P = 0.03, resp.). There was a positive correlation between the recessive and additive transmission models and ACS risk, but not the dominant one (OR [95% CI] = 1.78 [1.58–5.05], P = 0.01 and OR [95% CI] = 1.23 [0.74–2.03], P < 0.001, resp.), increasing thus the association of this polymorphism with the risk of pathology development. Our findings matches those of Zoheir et al. [26]; they reported that this polymorphism was positively correlated to increased risk of disease development (OR = 14.8, 95%, CI = 1.8–121.1, and P = 0.002). Similar findings were published by Cavallari et al. [35], who found that nonsmokers carrying the minor haplotype H2 of the gene were highly associated with significant CAD (OR = 1.83, 95% CI = 1.17–2.87, and P = 0.007). On the other side, findings of Schettert et al. [36] did not provide evidence for a strong association between H1/H1 and H1/H2 haplotypes and any increased risk of cardiovascular events in a population with CAD.

5. Conclusion

To the best of our knowledge, our study is the first in Morocco to assess whether or not Clopidogrel response may be modulated by i-T744C P2Y12 polymorphism in a sample of Moroccan ACS patients. We tried also to determine the frequency of this polymorphism among Moroccan ACS and healthy subjects. Our findings suggest that the wild-type and heterozygous genotypic profiles were more frequent among ACS ST (−) patients, when the homozygous mutant genotype was more frequent among ACS ST (+). The mutant allele C was more frequent among ACS ST (+) than ACS ST (−) patients, when the wild-type allele was more represented in the ACS ST (−) group. The C allele frequency was higher among resistant than nonresistant patients. Comparison of ACS patients and healthy controls shows higher frequency of C allele among cases than controls; there was a statistically significant association of the recessive and additive transmission models with the ACS development risk, increasing thus the association of this polymorphism with the pathology. Further studies including larger sample sizes and exploring interactions between this polymorphism and others are still be needed and may provide useful information to better understand the mechanism of Clopidogrel interindividual resistance and also the risk of ACS occurrence in the option to improve the biomedical context.

Competing Interests

The authors declare that they have no financial or nonfinancial competing interests.

References

[1] G. H. Guyatt, D. J. Cook, R. Jaeschke, S. G. Pauker, and H. J. Schünemann, "Grades of recommendation for antithrombotic agents: American College of Chest Physicians evidence-based clinical practice guidelines (8th edition)," *Chest*, vol. 133, no. 6, 2008.

[2] S. B. King III, S. C. Smith Jr., J. W. Hirshfeld Jr. et al., "2007 focused update of the ACC/AHA/SCAI 2005 Guideline Update for Percutaneous Coronary Intervention: a report of the American College of Cardiology/American Heart Association Task Force on Practice Guidelines: 2007 writing group to review new evidence and update the ACC/AHA/SCAI 2005 Guideline Update for Percutaneous Coronary Intervention, Writing on behalf of the 2005 Writing Committee," *Circulation*, vol. 117, no. 2, pp. 261–295, 2007.

[3] F. Marín, R. González-Conejero, P. Capranzano, T. A. Bass, V. Roldán, and D. J. Angiolillo, "Pharmacogenetics in cardiovascular antithrombotic therapy," *Journal of the American College of Cardiology*, vol. 54, no. 12, pp. 1041–1057, 2009.

[4] D. J. Angiolillo, A. Fernandez-Ortiz, E. Bernardo et al., "Variability in individual responsiveness to clopidogrel: clinical implications, management, and future perspectives," *Journal of the American College of Cardiology*, vol. 49, no. 14, pp. 1505–1516, 2007.

[5] T. H. Wang, D. L. Bhatt, and E. J. Topol, "Aspirin and clopidogrel resistance: an emerging clinical entity," *European Heart Journal*, vol. 27, no. 6, pp. 647–654, 2006.

[6] W. E. Evans and H. L. McLeod, "Pharmacogenomics—drug disposition, drug targets, and side effects," *New England Journal of Medicine*, vol. 348, no. 6, pp. 538–549, 2003.

[7] D. Taubert, N. von Beckerath, G. Grimberg et al., "Impact of P-glycoprotein on clopidogrel absorption," *Clinical Pharmacology and Therapeutics*, vol. 80, no. 5, pp. 486–501, 2006.

[8] D. J. Angiolillo, A. Fernandez-Ortiz, E. Bernardo et al., "Contribution of gene sequence variations of the hepatic cytochrome P450 3A4 enzyme to variability in individual responsiveness to clopidogrel," *Arteriosclerosis, Thrombosis, and Vascular Biology*, vol. 26, no. 8, pp. 1895–1900, 2006.

[9] J.-W. Suh, B.-K. Koo, S.-Y. Zhang et al., "Increased risk of atherothrombotic events associated with cytochrome P450 3A5 polymorphism in patients taking clopidogrel," *CMAJ*, vol. 174, no. 12, pp. 1715–1722, 2006.

[10] J.-S. Hulot, A. Bura, E. Villard et al., "Cytochrome P450 2C19 loss-of-function polymorphism is a major determinant of clopidogrel responsiveness in healthy subjects," *Blood*, vol. 108, no. 7, pp. 2244–2247, 2006.

[11] N. Von Beckerath, O. Von Beckerath, W. Koch, M. Eichinger, A. Schömig, and A. Kastrati, "P2Y12 gene H2 haplotype is not associated with increased adenosine diphosphate-induced platelet aggregation after initiation of clopidogrel therapy with a high loading dose," *Blood Coagulation and Fibrinolysis*, vol. 16, no. 3, pp. 199–204, 2005.

[12] E. I. Lev, R. T. Patel, S. Guthikonda, D. Lopez, P. F. Bray, and N. S. Kleiman, "Genetic polymorphisms of the platelet receptors $P2Y_{12}$, $P2Y_1$ and GP IIIa and response to aspirin and clopidogrel," *Thrombosis Research*, vol. 119, no. 3, pp. 355–360, 2007.

[13] J. Dropinski, J. Musial, B. Jakiela, W. Wegrzyn, M. Sanak, and A. Szczeklik, "Anti-thrombotic action of clopidogrel and $PI^{A1/A2}$ polymorphism of β_3 integrin in patients with coronary artery disease not being treated with aspirin," *Thrombosis and Haemostasis*, vol. 94, no. 6, pp. 1300–1305, 2005.

[14] H. H. Hassani Idrissi, W. Hmimech, N. El Khorb, H. Akoudad, R. Habbal, and S. Nadifi, "Association of the C3435T Multi-Drug Resistance gene-1 (MDR-1) polymorphism with Clopidogrel resistance among Moroccan Acute Coronary Syndromes (ACS) patients," *Journal of Thrombosis and Circulation*, vol. 2, article 115, 2016.

[15] M. V. Holmes, P. Perel, T. Shah, A. D. Hingorani, and J. P. Casas, "CYP2C19 genotype, clopidogrel metabolism, platelet function, and cardiovascular events a systematic review and meta-analysis," *The Journal of the American Medical Association*, vol. 306, no. 24, 2011.

[16] L. A. Malek, B. Kisiel, M. Spiewak et al., "Coexisting polymorphisms of P2Y12 and CYP2C19 genes as a risk factor for persistent platelet activation with clopidogrel," *Circulation Journal*, vol. 72, no. 7, pp. 1165–1169, 2008.

[17] D. J. Angiolillo, A. Fernandez-Ortiz, E. Bernardo et al., "Lack of association between the P2Y12 receptor gene polymorphism and platelet response to clopidogrel in patients with coronary artery disease," *Thrombosis Research*, vol. 116, no. 6, pp. 491–497, 2005.

[18] T. Cuisset, C. Frere, J. Quilici et al., "Role of the T744C polymorphism of the P2Y12 gene on platelet response to a 600-mg loading dose of clopidogrel in 597 patients with non-ST-segment elevation acute coronary syndrome," *Thrombosis Research*, vol. 120, no. 6, pp. 893–899, 2007.

[19] G. Hollopeter, H.-M. Jantzen, D. Vincent et al., "Identification of the platelet ADP receptor targeted by antithrombotic drugs," *Nature*, vol. 409, no. 6817, pp. 202–207, 2001.

[20] A. Malinin, A. Pokov, M. Spergling et al., "Monitoring platelet inhibition after clopidogrel with the VerifyNow-P2Y12® rapid

Does i-T744C P2Y12 Polymorphism Modulate Clopidogrel Response among Moroccan Acute Coronary...

177

analyzer: the VERIfy Thrombosis risk ASsessment (VERITAS) study," *Thrombosis Research*, vol. 119, no. 3, pp. 277–284, 2007.

[21] S. Prabhakaran, K. R. Wells, V. H. Lee, C. A. Flaherty, and D. K. Lopes, "Prevalence and risk factors for aspirin and clopidogrel resistance in cerebrovascular stenting," *American Journal of Neuroradiology*, vol. 29, no. 2, pp. 281–285, 2008.

[22] A. Y. Gasparyan, "Aspirin and clopidogrel resistance: methodological challenges and opportunities," *Vascular Health and Risk Management*, vol. 6, no. 1, pp. 109–112, 2010.

[23] S. A. Miller, D. D. Dykes, and H. F. Polesky, "A simple salting out procedure for extracting DNA from human nucleated cells," *Nucleic Acids Research*, vol. 16, no. 3, article 1215, 1988.

[24] S. C. Smith Jr., T. E. Feldman, J. W. Hirshfeld Jr. et al., "ACC/AHA/SCAI 2005 guideline update for Percutaneous Coronary Intervention: a report of the American College of Cardiology/American Heart Association Task Force on practice guidelines (ACC/AHA/SCAI Writing Committee to update the 2001 guidelines for Percutaneous Coronary Intervention)," *Journal of the American College of Cardiology*, vol. 47, no. 1, pp. e1–e121, 2006.

[25] H. J. Bouman, J. W. van Werkum, G. Rudež et al., "The relevance of P2Y12-receptor gene variation for the outcome of clopidogrel-treated patients undergoing elective coronary stent implantation: a clinical follow-up," *Thrombosis and Haemostasis*, vol. 107, no. 1, pp. 189–191, 2012.

[26] N. Zoheir, S. A. Elhamid, N. Abulata, M. E. Sobky, D. Khafagy, and A. Mostafa, "P2Y12 receptor gene polymorphism and antiplatelet effect of clopidogrel in patients with coronary artery disease after coronary stenting," *Blood Coagulation and Fibrinolysis*, vol. 24, no. 5, pp. 525–531, 2013.

[27] T. Simon, C. Verstuyft, M. Mary-Krause et al., "Genetic determinants of response to clopidogrel and cardiovascular events," *New England Journal of Medicine*, vol. 360, no. 4, pp. 363–375, 2009.

[28] O. V. Sirotkina, A. M. Zabotina, O. A. Berkovich, E. A. Bazhenova, T. V. Vavilova, and A. L. Shvartsman, "Genetic variants of platelet ADP receptor P2Y12 associated with changed platelet functional activity and development of cardiovascular diseases," *Genetika*, vol. 45, no. 2, pp. 247–253, 2009.

[29] P. Fontana, A. Dupont, S. Gandrille et al., "Adenosine diphosphate-induced platelet aggregation is associated with P2Y12 gene sequence variations in healthy subjects," *Circulation*, vol. 108, no. 8, pp. 989–995, 2003.

[30] J. D. Nordeen, A. V. Patel, and R. M. Darracott, "Clopidogrel resistance by P2Y12 platelet function testing in patients undergoing neuroendovascular procedures: incidence of ischemic and hemorrhagic complications," *Journal of Vascular and Interventional Neurology*, vol. 6, no. 1, pp. 26–34, 2013.

[31] J. S. Berger, M. C. Roncaglioni, F. Avanzini, I. Pangrazzi, G. Tognoni, and D. L. Brown, "Aspirin for the primary prevention of cardiovascular events in women and men: a sex-specific meta-analysis of randomized controlled trials," *Journal of the American Medical Association*, vol. 295, no. 3, pp. 306–316, 2006.

[32] E. Boersma, R. A. Harrington, D. J. Moliterno et al., "Platelet glycoprotein IIb/IIIa inhibitors in acute coronary syndromes: a meta-analysis of all major randomised clinical trials," *The Lancet*, vol. 359, no. 9302, pp. 189–198, 2002.

[33] N. Jochmann, K. Stangl, E. Garbe, G. Baumann, and V. Stangl, "Female-specific aspects in the pharmacotherapy of chronic cardiovascular diseases," *European Heart Journal*, vol. 26, no. 16, pp. 1585–1595, 2005.

[34] S. L. Hetherington, R. K. Singh, D. Lodwick, J. R. Thompson, A. H. Goodall, and N. J. Samani, "Dimorphism in the P2Y1 ADP receptor gene is associated with increased platelet activation response to ADP," *Arteriosclerosis, Thrombosis, and Vascular Biology*, vol. 25, no. 1, pp. 252–257, 2005.

[35] U. Cavallari, E. Trabetti, G. Malerba et al., "Gene sequence variations of the platelet P2Y12 receptor are associated with coronary artery disease," *BMC Medical Genetics*, vol. 8, article no. 59, 2007.

[36] I. T. Schettert, A. C. Pereira, N. H. Lopes, W. A. Hueb, and J. E. Krieger, "Association between platelet P2Y12 haplotype and risk of cardiovascular events in chronic coronary disease," *Thrombosis Research*, vol. 118, no. 6, pp. 679–683, 2006.

Permissions

The contributors of this book come from diverse backgrounds, making this book a truly international effort. This book will bring forth new frontiers with its revolutionizing research information and detailed analysis of the nascent developments around the world.

We would like to thank all the contributing authors for lending their expertise to make the book truly unique. They have played a crucial role in the development of this book. Without their invaluable contributions this book wouldn't have been possible. They have made vital efforts to compile up to date information on the varied aspects of this subject to make this book a valuable addition to the collection of many professionals and students.

This book was conceptualized with the vision of imparting up-to-date information and advanced data in this field. To ensure the same, a matchless editorial board was set up. Every individual on the board went through rigorous rounds of assessment to prove their worth. After which they invested a large part of their time researching and compiling the most relevant data for our readers.

The editorial board has been involved in producing this book since its inception. They have spent rigorous hours researching and exploring the diverse topics which have resulted in the successful publishing of this book. They have passed on their knowledge of decades through this book. To expedite this challenging task, the publisher supported the team at every step. A small team of assistant editors was also appointed to further simplify the editing procedure and attain best results for the readers.

Apart from the editorial board, the designing team has also invested a significant amount of their time in understanding the subject and creating the most relevant covers. They scrutinized every image to scout for the most suitable representation of the subject and create an appropriate cover for the book.

The publishing team has been an ardent support to the editorial, designing and production team. Their endless efforts to recruit the best for this project, has resulted in the accomplishment of this book. They are a veteran in the field of academics and their pool of knowledge is as vast as their experience in printing. Their expertise and guidance has proved useful at every step. Their uncompromising quality standards have made this book an exceptional effort. Their encouragement from time to time has been an inspiration for everyone.

The publisher and the editorial board hope that this book will prove to be a valuable piece of knowledge for researchers, students, practitioners and scholars across the globe.

List of Contributors

M. Z. Rahman, L. Nishat, Z. A. Yesmin and L. A. Banu
Department of Anatomy, Bangabandhu Sheikh Mujib Medical University (BSMMU), Dhaka, Bangladesh

Kristopher J. L. Irizarry
The Applied Genomics Center, Graduate College of Biomedical Sciences, College of Veterinary Medicine, Western University of Health Sciences, 309 East Second Street, Pomona, CA 91766, USA

Josep Rutllant
Molecular Reproduction Laboratory, College of Veterinary Medicine, Western University of Health Sciences, 309 East Second Street, Pomona, CA 91766, USA

K. Nithya, T. Angeline and W. Isabel
PG & Research Department of Zoology & Biotechnology, Lady Doak College, Madurai, Tamil Nadu 625 002, India

A. J. Asirvatham
Arthur Asirvatham Hospital, Madurai, Tamil Nadu 625 020, India

Vadim Stepanov and Kseniya Vagaitseva
Institute of Medical Genetics, Tomsk National Medical Research Center, Tomsk, Russia Tomsk State University, Tomsk, Russia

Andrey Marusin and Anna Bocharova
Institute of Medical Genetics, Tomsk National Medical Research Center, Tomsk, Russia

Oksana Makeeva
Institute of Medical Genetics, Tomsk National Medical Research Center, Tomsk, Russia Nebbiolo Center for Clinical Trials, Tomsk, Russia

Anthony Percival-Smith
Department of Biology, e University of Western Ontario, London, ON, Canada N6A 1B7

Altaf A. Kondkar, Taif A. Azad, Faisal A. Almobarak, Ibrahim M. Hatem Kalantan, Khaled K. Abu-Amero and Saleh A. Al-Obeidan
Glaucoma Research Chair, Department of Ophthalmology, College of Medicine, King Saud University, Riyadh, Saudi Arabia

Bahabri
King Khaled University Hospital, King Saud University, Riyadh, Saudi Arabia

Roshan Dadachanji, Nuzhat Shaikh and Srabani Mukherjee
Department of Molecular Endocrinology, National Institute for Research in Reproductive Health, J.M. Street, Parel, Mumbai 400012, India

Shamshad Ul Haq
Biotechnology Division, UP Council of Sugarcane Research, Shahjahanpur 242001, India
Interdisciplinary Programme of Life Science for Advance Research and Education, University of Rajasthan, Jaipur 302004, India
Department of Botany, University of Rajasthan, Jaipur 302015, India

Pradeep Kumar
Biotechnology Division, UP Council of Sugarcane Research, Shahjahanpur 242001, India
School of Biotechnology, Yeungnam University, Gyeongsan 712-749, Republic of Korea

R. K. Singh
Biotechnology Division, UP Council of Sugarcane Research, Shahjahanpur 242001, India

Kumar Sambhav Verma
Amity Institute of Biotechnology, Amity University Rajasthan, Jaipur 302006, India

Ritika Bhatt
Interdisciplinary Programme of Life Science for Advance Research and Education, University of Rajasthan, Jaipur 302004, India
Department of Botany, University of Rajasthan, Jaipur 302015, India

Meenakshi Sharma and Sumita Kachhwaha
Department of Botany, University of Rajasthan, Jaipur 302015, India

S. L. Kothari
Interdisciplinary Programme of Life Science for Advance Research and Education, University of Rajasthan, Jaipur 302004, India
Department of Botany, University of Rajasthan, Jaipur 302015, India
Amity Institute of Biotechnology, Amity University Rajasthan, Jaipur 302006, India

Fidelis Charles Bugoye and Elias Mulima
Department of Forensic Science and DNA Services, Government Chemist Laboratory Authority, Dar es Salaam, Tanzania

Gerald Misinzo
Department of Veterinary Microbiology, Parasitology and Biotechnology, Sokoine University of Agriculture, Morogoro, Tanzania

A. Marantidis, G. P. Laliotis and M. Avdi
Laboratory of Physiology of Reproduction of Farm Animals, Department of Animal Production, School of Agriculture, Aristotle University of Thessaloniki, 54124Thessaloniki, Greece

Bhaskar Ganguly, Tanuj Kumar Ambwani and Sunil Kumar Rastogi
Animal Biotechnology Center, Department of Veterinary Physiology and Biochemistry, College of Veterinary and Animal Sciences, G. B. Pant University of Agriculture and Technology, Pantnagar 263145, India

Pornprot Limprasert, Kanoot Jaruthamsophon and Thanya Sripo
Department of Pathology, Faculty of Medicine, Prince of Songkla University, Songkhla 90110, Thailand

Janpen Thanakitgosate
Department of Pathology, Faculty of Medicine, Ramathibodi Hospital, Mahidol University, Bangkok 10400, Thailand

Lili Zhou, Keith C. Summa, Christopher Olker, Martha H. Vitaterna and Fred W. Turek
Center for Sleep and Circadian Biology, Northwestern University, Evanston, IL 60208, USA
Department of Neurobiology, Northwestern University, Evanston, IL 60208, USA

Anthony Percival-Smith, Gabriel Ponce and Jacob J. H. Pelling
Department of Biology, University of Western Ontario, London, ON, Canada N6A 5B7

Karina Villalba, Jessy G. Dévieux and Rhonda Rosenberg
Department of Health Promotion and Disease Prevention, Robert Stempel College of Public Health and Social Work, Florida International University, Miami, FL 33181, USA

Jean Lud Cadet
Molecular Neuropsychiatry Research Branch, DHHS/NIH/NIDA Intramural Research Program, Baltimore, MD, USA

Farzaneh Rami and Mansoor Salehi
Department of Genetics and Molecular Biology, School of Medicine, Isfahan University of Medical Sciences, Isfahan 81746-73461, Iran

Halimeh Mollainezhad
Department of Immunology, School of Medicine, Isfahan University of Medical Sciences, Isfahan 81746-73461, Iran

Hussein Sabit, Mariam B. Samy and Osama A. M. Said
College of Biotechnology, Misr University for Science and Technology, Giza, Egypt

Mokhtar M. El-Zawahri
College of Biotechnology, Misr University for Science and Technology, Giza, Egypt
Center of Research and Development, Misr University for Science and Technology, Giza, Egypt

Gautham Arunachal, Divya Pachat and Sumita Danda
Department of Clinical Genetics, Christian Medical College, Vellore, Tamil Nadu 632004, India

C. George Priya Doss
School of Biosciences and Technology, VIT University, Vellore, Tamil Nadu 632014, India

Rekha Pai and Andrew Ebenazer
Department of Pathology, Christian Medical College, Vellore, Tamil Nadu 632004, India

Chariyawan Charalsawadi, Jariya Khayman and Pornprot Limprasert
Department of Pathology, Faculty of Medicine, Prince of Songkla University, Songkhla 90110, Thailand

Verayuth Praphanphoj
Medical Genetics Center, Bangkok 10220, Thailand

Hind Hassani Idrissi, Wiam Hmimech and Sellama Nadifi
Laboratory of Genetics and Molecular Pathology, Medical School, University Hassan II, Casablanca, Morocco

Nada El Khorb and Hafid Akoudad
Department of Cardiology, University Hospital Center Hassan II, Fes, Morocco

Rachida Habbal
Department of Cardiology, University Hospital Center Ibn Rochd, Casablanca, Morocco

Index